# The
# Technical
# Reader

# The Technical Reader

Readings in Technical, Business, and Scientific Communication

**W. Steve Anderson**
Northern Montana College

**Don Richard Cox**
University of Tennessee

Holt, Rinehart and Winston
New York · Chicago · San Francisco · Atlanta
Dallas · Montreal · Toronto · London · Sydney

Library of Congress Cataloging in Publication Data

Main entry under title:

The Technical Reader.

Includes index.
1. Business report writing — Addresses, essays,
lectures. 2. Technical writing — Addresses, essays,
lectures. 3. Report writing — Addresses, essays,
lectures. I. Anderson, Walter Steve. II. Cox, Don
Richard.
HF5719.T4      808'.0666021      79–29677

ISBN 0-03-048771-4

## Acknowledgments

PART ONE
Section 1
*The Modern Researcher,* Revised Edition by Jacques Barzun and Henry F. Graff, © 1970 by
   Harcourt Brace Jovanovich, Inc. Reprinted by permission of the publisher.
"Reports, Inferences, Judgments": From *Language in Thought and Action,* Third Edition by
   S. I. Hayakawa, copyright © 1972 by Harcourt Brace Jovanovich, Inc. Reprinted by permis-
   sion of the publisher.
"Commuter" from *The Lady Is Cold* by E. B. White, copyright © 1925 by E. B. White. Used by
   permission of Harper & Row, Publishers, Inc.
"What Is Technical Writing" by W. Earl Britton: From *College Composition and Communica-
   tion,* Vol. 16, May 1965. Copyright © 1965 by the National Council of Teachers of English.
   Reprinted by permission of the publisher and the author.
"Credo of a Tech Writer" by John Frye: From *Electronics World* (Popular Electronics) copy-
   right © 1971 by Ziff-Davis Publishing Company. Used by permission.
"The Art of Communication": From *Communication: Parkinson's Formula for Business
   Survival* by C. Northcote Parkinson and Nigel Rowe. Copyright © 1977 by Prentice-Hall
   International, Inc., London.
*Mortal Lessons: Notes on the Art of Surgery* by Richard Selzer, copyright © 1974, 1975, 1976
   by Richard Selzer. Reprinted by permission of Simon & Schuster, a Division of Gulf &
   Western Corporation.

Section 2
"Audiences Analysis": Reprinted with permission of Glencoe Publishing Company, Inc. from
   *Audience Analysis for Technical Writing* by Thomas Pearsall. Copyright by Thomas E.
   Pearsall.
"Gobbledygook" by Stuart Chase: From *The Power of Words,* copyright 1953, 1954 by Stuart
   Chase. Reprinted by permission of Harcourt Brace Jovanovich, Inc.
"How to Write Like a Social Scientist" by Samuel T. Williamson: From *Saturday Review,* Oc-
   tober 4, 1947. Used by permission.
"The Capacity to Generate Language Viability Destruction" by Edwin Newman: From *Strictly
   Speaking,* copyright © 1974 by Edwin H. Newman. Reprinted by permission of The Bobbs-
   Merrill Company, Inc.
"Degrees of Plain Talk" by Rudof Flesch: From *The Art of Readable Writing,* revised and
   enlarged edition, copyright 1949, 1974 by Rudolf Flesch. Reprinted by permission of
   Harper & Row, Publishers, Inc.

Section 3
"The Design Process Is a Problem-Solving Journey": Reproduced by permission from *The
   Universal Traveller* by Don Koberg and Jim Bagnall. Copyright © 1976 by William Kaufman,
   Inc., Los Altos, California 94022. All rights reserved.

# Preface

The Technical Reader is a new kind of anthology for a kind of writing course that is new on many campuses — report writing for undergraduates. In colleges and universities across the country, especially in those that stress vocational programs, report writing courses occasionally even replace at least a portion of the traditional English composition requirement.

A mainstay of traditional composition courses is the anthology of readings, because instructors recognize that the student writer must read. The same rule should apply to report writing courses — the student report writer also must read. Yet instructors of report writing have virtually no choice of anthologies to complement one of the many excellent report handbooks that are available. This anthology fills that gap.

The Technical Reader contains a wealth of readings for student writers preparing for careers in industry, business, and science. We combine fields in order to make the text consistent with the practice of the many schools that place students from programs in all three fields in the same report writing classes. And, in fact, we agree with the principle that supports the practice. In an introductory or intermediate report writing course, students gain a better understanding of the reporting done in their particular field by studying the theory and elements that underlie all reporting.

Our selections fall into three categories. In Part One, we include selections *about* report writing. Because each handbook necessarily takes a singular approach to report writing, we believe that relevant readings on reports in general as well as readings on languages, audience, and style give both instructor and student additional perspectives on the topics. Most of these selections are in the first two sections.

A second category of articles make up Part Two, where we gather approaches to problem solving. Although these selections do not bear directly on actually writing a report, they do help the student in generating report content.

Part Three consists largely of specific examples of reporting. Section One is devoted to introducing the methods of development — illustration/example, comparison/contrast, analogy, and so on — with short reports and excerpts. The following three sections sub-categorize longer selections according to the purpose that they achieve. Section Two contains informative reports, Section Three contains reports that persuade, and the reports in Section Four have an implemental purpose. The examples in each section illustrates the variety of approaches that might be taken for achieving these purposes.

We urge the use of this text as a complement — *not* a supplement — to the handbook. For experienced instructors, *The Technical Reader* brings together in a convenient volume the types of examples that previously they had to duplicate for their students — the inevitable handout. New report writing instructors (and many are new to the field) will find that the text saves them the time consuming trial-and-error process of searching out suitable examples.

But finally, *The Technical Reader* is for the student, and foremost in our mind when choosing our selections was the student reader. In order to present a broad spectrum of reports, we necessarily include some functional, workaday documents that were not written to be read for pleasure. But all selections are readable, and most are even enjoyable. They demonstrate that reports need not be tedious or boring, that report writing can be artful as well as functional. Imparting this realization to studnets should be one of the instructor's prime objectives in a report writing course.

In conclusion, we wish to point out that the good anthologies for conventional composition courses do more than provide rhetorical models. The selections also reveal to students the larger world with which they must learn to cope. We believe that our text does something like this for students who will enter the technical, business, and scientific worlds. Primarily, this text should be used to teach writing. We do not ask instructors to use it for teaching a "larger world." But if a larger view seeps through to students, then so much the better.

We would like to thank the following people whose suggestions and criticisms were valuable during the development of this book: Cortland Auser, Bronx Community College, John S. Harris, Brigham Young University, Frances Blosser Maguire, Tarrant County Community College, Thomas E. Pearsall, University of Minnesota, Nell Ann Pickett, Hinds Junior College, L. Dan Richards, North Central Technical College, Arlo Stoltenberg, North Iowa Area Community College, and Thomas Warren, Oklahoma State University.

# Contents

# Part Three   Applications

For
Dawn, Jesse, and Lesley

# Part One
## Preliminaries

# Section 1

---

# Reporting Information: An Overview

Written communication is a necessary tool in industry, business, and science. Manuals, specifications, proposals, research reports, and feasibility reports are the written documents that keep our world functioning. Reference works and informative books and articles, although not always used directly in running the world of work, either store or distribute factual information that helps us keep informed on industry, business, and science. Without writing, activities in these fields might continue, but not at the level of efficiency to which we have become accustomed.

If you are to work efficiently in these fields, you must be familiar with this writing. This text will introduce you to the broad variety of documents that keep industry, business, and science moving, and to the informative documents that store and disseminate information. Also, it will introduce you to the general principles that underlie the writing done for industry, business, and science.

In fact, we take the stand that for the introductory — or even intermediate — level the principles shared in common by the three fields of writing are as important as the specific writing requirements for the field in which the student plans to work. The student should have as broad a background as possible, because industry, business, and science overlap in their functions, and interaction among them is usual. And there is a further advantage. The student who intends to enter business can learn important rules for composition from the writing done for industry and science. The student who intends to work in science can learn from the business writer and the industrial writer. Each kind of writing can contribute some information to the perceptive student. Firmly grounded in the principles common to all fields, the student will find it easier to write in his own specialized field.

What do we call the writing done for industry, business, and science? Actually, no one term adequately covers all the writing that this anthology intro-

1

duces. Nor is there *one* adequate definition. There are several terms and partial definitions, and so for purposes of an overview we will start by examining the important ones:

*Technical writing.* On many campuses students in engineering, science, and business take a course often referred to as "tech" writing. And technical writing is the term long used to identify, loosely, the writing done for these fields. We might see technical writing as "technical" in that it deals with *technology,* or the application of scientific principles to practical ends. More specifically, it is "technical" in that it deals with the *specialized techniques* of engineering and scientific research. Further, it is on-the-job writing that deals with a specific technical problem or procedure, and it assumes a reader with a specialized background on the subject covered. Defined as such, technical writing would exclude writing that deals with company policy and personnel. It would exclude most sales literature. It would also exclude the informative books and articles on technology, business, and science that are written for laymen readers — those without specialized knowledge of the subject.

*Business writing.* This term is unsatisfactory because it limits us to a specific field rather than naming an underlying principle.

*Science writing.* Same as the above.

*Practical writing.* "Practical" suggests writing based only on the very ordinary activities of the workaday world. Because the writing that we are concerned with might deal with everything from Quasars to DNA to the Brooklyn Bridge, "practical" would seem to set unnecessarily narrow limits.

*Functional writing.* The term would seem appropriate for writing dealing with the "functional" world, the same workaday world referred to above. But functional, like practical, limits the expressive range.

*Occupational writing.* Suggesting travel reports and trip tickets, occupational writing has the same limitations as practical writing or functional writing.

*Basic writing.* Unfortunately, the association is too close to remedial. Good writing for your career field is *not* on a remedial level.

*Report writing.* The report is a factual account of an existing situation. All documents written for industry, business, and science are not, strictly speaking, reports. Yet report writing is often favored as the appropriately broad descriptive term for the writing done in all three fields.

Of course, any of these terms is partially satisfactory. Combined they are the parts of the type of writing that we are pursuing. All are members of the same family. Unfortunately, we have no adequate name for the family — other than writing.

For the sake of convenience we will settle on the term that is generally acceptable — report writing. Now we will turn from worrying about appropriate terminology to discuss a major, underlying principle of this writing.

Let us consider two possible extremes of nonfiction prose. At one extreme we might consider the following statements:

. . . to reach Oak Street, turn left on Fifth . . .

. . . I propose that we accept the $100,000 bid . . .

. . . the planet Mars has two moons . . .

. . . test results indicate that vehicle A got better gas mileage than vehicle B . . .

On the other extreme, we might find these:

In judging another man's life, I always inquire how he behaved at the last; and one of the principal aims of my life is to conduct myself well when it ends — peacefully, I mean, and with a calm mind. (Montaigne, *That No Man Should Be Called Happy Until After His Death*)

Books are for the scholar's idle times. When he can read God directly, the hour is too precious to be wasted in other men's transcripts of their readings. (Emerson, *The American Scholar*)

I therefore conclude that the meaning of life is the most urgent of questions. (Camus, *An Absurd Reasoning*)

The writing in the first examples offers the world to us as a mechanism. It assumes a world that man is capable of knowing and acting upon. This could be the humdrum world that we must cope with every day, the world of goods and services. It could also be the dramatic world of scientific discovery, or fabulous construction projects. In general, it is writing that is in a direct sense *useful*. It is writing used in accomplishing the tangible. It is the writing that we have labeled report writing.

If *useful* describes the first examples, does that mean that the writing in the second set is *useless*? *By no means*! It has a use: it enriches our lives and aids us in pondering life's mysteries and near meanings. Life might be *useless* without it. But this type of writing is for working men and women at their leisure. It isn't writing that we *must* know. It might help the mind survive, but not the body.

The second type we might call essay writing. The essay is an attempt, an effort to grasp and handle those elusive ideas that are never to be entirely understood. The essayist (when writing as an essayist) attempts to deal with the ambiguous nature of life, with the "may or may not" possibilities. The essayist even has the right to decide that there are no answers — that he can only ask the questions.

But a far more rigid purpose guides report writing: it attempts to give answers. Report writers, even when aware of the ambiguity of knowledge, of the ultimate mysteriousness of life, assume that there is a job to be done and that writing is an indispensable tool for doing that job.

Is one kind of writing better than the other? Not necessarily, of course. And there is no reason why the same writer can't do both kinds of writing. But there is a time for one and a time for the other.

Of course, we've been looking only at the differences in the extremes. Let us consider the similarities. All writing comes out of the human need to communicate — whether that need leads us into intriguing mysteries or practical certainties. Even the humblest shop manual, if it is a good one, has a great deal in common with a Shakespearean play. Skill, care, and personal integrity are necessary for both.

These are the common bonds that unite all writers; these are the virtues that lead to good writing.

Our discussion has been only a starting place. The selections chosen for our overview will expand further your knowledge of the meaning and the importance of report writing. They will also increase your awareness of the responsibilities that you must accept as a writer for industry, business, and science.

# The Modern Report

## Jacques Barzun and Henry F. Graff

Barzun and Graff discuss the report as the "means by which we try to substitute intelligence for routine and knowledge for guesswork." They consider both the practical role of report writing in accomplishing the world's work and the relation of the report to a long established academic discipline.

### The Report: A New and Fundamental Form

In a once famous book on the Middle East, the English archeologist Layard printed a letter in which a Turkish official answered an Englishman's question. It begins:

> My Illustrious Friend and Joy of My Liver!
>
> The thing you ask of me is both difficult and useless. Although I have passed all my days in this place, I have neither counted the houses nor have I inquired into the number of the inhabitants; and as to what one person loads on his mules and the other stows away in the bottom of his ship, that is no business of mine. But, above all, as to the previous history of this city, God only knows the amount of dirt and confusion that the infidels may have eaten before the coming of the sword of Islam. It were unprofitable for us to inquire into it. O my soul! O my lamb! Seek not after the things which concern thee not. Thou camest unto us and we welcomed thee: go in peace.[1]

This unruffled public servant obviously made no annual report of any kind to anybody — those were the good old days. At the distance of a century it is interesting to note the three things that he so courteously declined to provide. They are: vital statistics, business reports, and history. Life as we know it today would stop if information of these three kinds were not readily to be had on every sort of subject.

From Jacques Barzun and Henry F. Graff, *The Modern Researcher,* revised edition (New York: Harcourt, Brace and World, 1970), pp. 3–8. Title selected by editors.
[1] Austen H. Layard, *Discoveries in the Ruins of Nineveh and Babylon . . . ,* London, 1853, 663.

All over the globe, every moment of the day, someone is being asked to make a search and write a report on some state of fact, or else to read and analyze one, so that action may be taken. Reports are the means by which we try to substitute intelligence for routine and knowledge for guesswork.

This characteristic behavior of modern man makes "the report" fundamental in the conduct of affairs. It has become a familiar form, like the business letter or the sonnet.

Every report implies previous research, whether by the reporter or by someone else. Thousands of men not connected with academic life are thus turned into more or less able scholars. The Turkish official of today has dropped his hookah, leaped from his cushion, and is busy counting the houses for the Ministry of the Interior. The figures he gathers are then published as government statistics, which other researchers will use for still other reports — from the university student writing a paper on modern Turkey to the foreign businessman who wants to establish a branch office in that country.

Among the many useful documents that may strictly or loosely be classed as reports there is no essential difference of outlook or method. The student writing a book report for a Freshman English course is doing on a small scale and with a single source the same thing as the president of a corporation who prepares his annual report to the stockholders, or as the President of the United States when, with the aid of all his departments, he reports to the people on the state of the Union. Scope and purpose will of course differ, and this affects the worth of the report as an historical record. But the general form and the devices employed are identical in all three. The reader will readily think of other examples of the same truth.

The common element in all these tasks is that they present similar problems of investigation and exposition, which are solved in similar ways. Moreover, the writers of reports draw upon the same vast reservoir of information. Apart from the special facts that, to pursue the examples above, the treasurer of a corporation or the Secretary of State supply to their respective presidents, the written sources for the millions of words uttered in reports are the familiar ones — newspapers, learned journals, histories, statistical abstracts, law cases, state papers, and so on through the many categories of books found in great libraries. This huge accumulation is what the researcher must learn to use in order to satisfy his particular need. And as the conditions of the search are common to all researchers, it is possible to discuss research and reporting regardless of their occasion or subject.

## The Historical Attitude Underlies Research and Report

To regard the report as a form is further justified by the fact that the attitude and technique of the report-writer are derived in a straight line from one of the great literary and academic disciplines — History. It is from historical scholarship — originating with the antiquarian — that the world has taken the apparatus of footnotes, references, bibliography, and so on, which have become commonplace devices, not to say household words. It is from the historical study of texts by philologists and historians that writers at large have learned to sift evidence, balance testimony, and demand verified assertions.

At this point someone may object that there remains a great difference between the scholar's main interest and that of the ordinary report-writer. The former seeks to know the past; the latter is concerned with the present, generally with a view to plotting the future; hence their outlooks must be quite different. This difference is more striking than significant. Whatever its purpose, a report is invariably and necessarily historical. Insofar as it reports facts it gives an account of the past. Recorded opinions obviously belong to what has gone before, to history. Suppose a study of American foreign policy designed solely to change future action. It can do so only by criticizing principles or personnel: but to do this its arguments must lean on the evidence of what has been happening — on what is past, recorded, and beyond the reach of change. What else is this but a piece of history?

The same holds true when a report is, as we say, "purely fact-finding," for example, a survey of the conditions of the public schools in a certain town. This description of "the present" is actually a description of the past — recent it may be, but nonetheless a backward glance. Only events already gone by can disclose the prevailing state of things. Even the unassuming book report is a record of the past. It records, to begin with, what the student thought and felt at 2 A.M. the day it was due. Further, it is part autobiography, part criticism, and part literary history. The book was probably read earlier and compared with still older experiences; and the words of the book, which the report may quote, refer to a yet more remote past. Whatever else it is, every report is historical and cannot avoid being so.

The way the historian goes to work and his attitude toward sources are parts of the report-writer's equipment, no matter what his subject may be — literary, economic, political, scientific, or anything else that belongs among serious recitals of fact.

## Historical Writing in Daily Life

A few examples of the generality just advanced will show how frequently in ordinary life we are asked to give our attention to fragments of historical writing in the midst of other things. These fragments of course are not necessarily *good* historical writing. They and the reports in which they occur may be inaccurate, biased, fanciful, or downright fraudulent. That is not the point. The point is that the workaday world cannot do without historical materials put in historical form. "In a certain sense," said Carlyle, "all men are historians. . . . Most men . . . speak only to narrate." The following, for example, is part of a seventy-page report issued by the Columbia Broadcasting System to its stockholders in advance of a meeting at which a merger was to be approved:

> CBS was incorporated in the State of New York in 1927. In its early years, CBS's major activity was radio broadcasting. In 1939, CBS entered the field

of recording and marketing phonograph records. Developments in television broadcasting became commercially practicable after World War II, and led to the development and growth of CBS's television network and stations operations in the 1950's. In 1966 it entered the field of educational services and products.[2]

This terse paragraph is business history, obviously, but it is also technological and cultural history. From it, an archeologist of the future could be confident of drawing solid conclusions about mid-century America.

Now turn to the description of a college class by one of its members, in which we find a fragment of social history:

> Three years and 241 days ago Fontbonne College brought forth a new class, conceived in tradition and dedicated to the proposition that traditions can be changed if you try hard enough. Now we are engaged in a great struggle to the end, testing whether these students or any students so conceived can long endure.[3] [Notice in passing the somewhat amateurish parody of a famous historical document.]

We do not usually consider the advertiser a great champion of Fact, yet he too wants to find out what's what. Here is a New York agency making known through an advertisement its philosophy of research. The heading runs: "First, Get the Facts." Then:

> More business mistakes are due to faulty facts than to faulty judgment. As used by us, research takes many forms: motivation studies, copy testing, leadership surveys, consumer opinion research, store audits, consumer panel surveys, and dealer surveys. . . . In innumerable instances, research has provided a foundation for building a more efficient advertising program and taking the guesswork out of future planning.[4]

What emerges from the jargon is that advertisers have caught the habit and put on the trappings of modern historical investigation. One can imagine the endless reports arising from all these panels and surveys, the steady repetition of "It was found that . . ." and the scholarly comparisons, with dates and percentages, between the devotees of tooth paste and the partisans of tooth powder.

Like advertising, journalism has adapted to its use the ways of historical research and popularized its externals. Magazines such as *Life, Time, Look,* and *Newsweek* employ whole corps of persons who bear the title of Researcher and whose function is to verify every statement made in the stories turned in by those whose title is Reporter.

As a final example of the varied forms that historical writing may take — this

[2]Columbia Broadcasting System, Inc., *Notice of Special Meeting,* May 19, 1967, 17.
[3]Margie McNamee in *Fontbonne,* XIX (1969), 2.
[4]*New York Times,* July 11, 1955.

time on the geological scale — consider these few sentences taken from an issue of the *Scientific Monthly*:

> The element uranium changes gradually, through a series of transition products — the most important of which is radium — into its inactive end-product, the lead isotope with an atomic weight of 206. The transformation rate is extremely small; in any uranium mineral, only 1 percent of a given amount of uranium is transformed in 65 million years. . . . Similarly, but even more slowly, thorium is transformed. . . . The older methods of geologic age determination were based on the determination of the lead and helium content of uranium and thorium minerals.[5]

## STUDY QUESTIONS

1. Why is the attitude of the Turkish official inappropriate for our world?
2. What reports are necessary for carrying on in our everyday lives?
3. The authors state that all report writers are necessarily research scholars. Discuss the need for research in writing for industry, business, or science. Does research always mean library work? What are the basic sources of information for the field in which you will write reports?
4. In what sense does the "historical attitude" underlie report writing? Can you think of a reporting situation that would not require an historical attitude?
5. The authors state that the report "has become a familiar form, like the business letter or the sonnet." What do they mean by "form"? What is the form of a business letter or a sonnet? According to the authors, what are the characteristics of the report form?
6. Are the characteristics of the report that Barzun and Graff identify appropriate for the field in which you will write?
7. Can you think of additional characteristics?

[5]Otto Hahn, "Radioactive Methods for Geologic and Biologic Age Determinations," *Scientific Monthly*, LXXXII (May 1956), 258.

# Reports, Inferences, Judgments

## S. I. Hayakawa

Although S. I. Hayakawa does not discuss specific report formats, he does present the kinds of information that a report communicates. His selection is an excellent introduction to the theory of reporting.

For the purposes of the interchange of information, the basic symbolic act is the report of what we have seen, heard, or felt: "There is a ditch on each side of the road." "You can get those at Smith's Hardware Store for $2.75." "There aren't any fish on that side of the lake, but there are on this side." Then there are reports of reports: "The longest waterfall in the world is Victoria Falls in Rhodesia." "The Battle of Hastings took place in 1066." "The papers say that there was a smash-up on Highway 41 near Evansville." Reports adhere to the following rules: first, they are *capable of verification;* second, they *exclude,* as far as possible, *inferences* and *judgments.* (These terms will be defined later.)

### Verifiability

Reports are verifiable. We may not always be able to verify them ourselves, since we cannot track down the evidence for every piece of history we know, nor can we all go to Evansville to see the remains of the smash-up before they are cleared away. But if we are roughly agreed upon the names of things, upon what constitutes a "foot," "yard," "bushel," "kilogram," "meter," and so on, and upon how to measure time, there is relatively little danger of our misunderstanding each other. Even in a world such as we have today, in which everybody seems to be quarreling with everybody else, *we still to a surprising degree trust each other's reports.* We ask directions of total strangers when we are traveling. We follow directions on road signs without being suspicious of the people who put them up. We read books of information about science, mathematics, automotive engineering, travel, geography, the history of costume, and other such factual matters, and we usually assume that the author is doing his best to tell us as truly as he can what he knows. And we are safe in so assuming most of the time. With the interest given today to the discussion of biased newspapers, propagandists, and the general untrustworthiness of many of the communications we receive, we are likely to forget that we still have an enormous amount of reliable information available and that deliberate misinformation, except in warfare, is still more the exception than the rule. The desire for self-preservation that compelled men to evolve means for the exchange of information also compels them to regard the giving of false information as profoundly reprehensible.

From S. I. Hayakawa, *Language in Thought and Action,* 3rd ed. (New York: Harcourt, Brace, Jovanovich, 1972), pp. 34–47.

At its highest development, the language of reports is the language of science. By "highest development" we mean greatest general usefulness. Presbyterian and Catholic, workingman and capitalist, East German and West German agree on the meanings of such symbols as $2 \times 2 = 4$, $100°$ C, $HNO_3$, $3:35$ A.M., $1940$ A.D., $1,000$ kilowatts, Quercus agrifolia, and so on. But how, it may be asked, can there be agreement about even this much among people who disagree about political philosophies, ethical ideas, religious beliefs, and the survival of my business versus the survival of yours? The answer is that circumstances *compel men to agree,* whether they wish to or not. If, for example, there were a dozen different religious sects in the United States, each insisting on its own way of naming the time of the day and the days of the year, the mere necessity of having a dozen different calendars, a dozen different kinds of watches, and a dozen sets of schedules for business hours, trains, and television programs, to say nothing of the effort that would be required for translating terms from one nomenclature to another, would make life as we know it impossible.[1]

The language of reports, then, including the more accurate reports of science, is "map" language, and because it gives us reasonably accurate representations of the "territory," it enables us to get work done. Such language may often be dull reading: one does not usually read logarithmic tables or telephone directories for entertainment. But we could not get along without it. There are numberless occasions in the talking and writing we do in everyday life that *require that we state things in such a way that everybody will be able to understand and agree with our formulation.*

## Inferences

The reader will find that practice in writing reports is a quick means of increasing his linguistic awareness. It is an exercise which will constantly provide him with his own examples of the principles of language and interpretation under discussion. The reports should be about first-hand experience — scenes the reader has witnessed himself, meetings and social events he has taken part in, people he knows well. They should be of such a nature that they can be verified and agreed upon. For the purpose of this exercise, inferences will be excluded.

[1] According to information supplied by the Association of American Railroads, "Before 1883 there were nearly 100 different time zones in the United States. It wasn't until November 18 of that year that . . . a system of standard time was adopted here and in Canada. Before then there was nothing but local or 'solar' time. . . . The Pennsylvania Railroad in the East used Philadelphia time, which was five minutes slower than New York time and five minutes faster than Baltimore time. The Baltimore & Ohio used Baltimore time for trains running out of Baltimore, Columbus time for Ohio, Vincennes (Indiana) time for those going out of Cincinnati. . . . When it was noon in Chicago, it was 12:31 in Pittsburgh, 12:24 in Cleveland, 12:17 in Toledo, 12:13 in Cincinnati, 12:09 in Louisville, 12:07 in Indianapolis, 11:50 in St. Louis, 11:48 in Dubuque, 11:39 in St. Paul, and 11:27 in Omaha. There were 27 local time zones in Michigan alone. . . . A person traveling from Eastport, Maine, to San Francisco, if he wanted always to have the right railroad time and get off at the right place, had to twist the hands of his watch 20 times en route." Chicago *Daily News* (September 29, 1948).

Not that inferences are not important — we rely in everyday life and in science as much on *inferences* as on reports — in some areas of thought, for example, geology, paleontology, and nuclear physics, reports are the foundations; but inferences (and inferences upon inferences) are the main body of the science. An inference, as we shall use the term, is *a statement about the unknown made on the basis of the known.* We may *infer* from the material and cut of a woman's clothes her wealth or social position; we may *infer* from the character of the ruins the origin of the fire that destroyed the building; we may *infer* from a man's calloused hands the nature of his occupation; we may *infer* from a senator's vote on an armaments bill his attitude toward Russia; we may *infer* from the structure of the land the path of a prehistoric glacier; we may *infer* from a halo on an unexposed photographic plate its past proximity to radioactive materials; we may *infer* from the sound of an engine the condition of its connecting rods. Inferences may be carefully or carelessly made. They may be made on the basis of a broad background of previous experience with the subject matter or with no experience at all. For example, the inferences a good mechanic can make about the internal condition of a motor by listening to it are often startlingly accurate, while the inferences made by an amateur (if he tries to make any) may be entirely wrong. But the common characteristic of inferences is that they are statements about matters which are not directly known, made on the basis of what has been observed.[2]

The avoidance of inferences in our suggested practice in report-writing requires that we make no guesses as to what is going on in other people's minds. When we say, "He was angry," we are not reporting; we are making an inference from such observable facts as the following: "He pounded his fist on the table; he swore; he threw the telephone directory at his stenographer." In this particular example, the inference appears to be safe; nevertheless, it is important to remember, especially for the purposes of training oneself, that it is an inference. Such expressions as "He thought a lot of himself," "He was scared of girls," "He has an inferiority complex," made on the basis of casual observation, and "What Russia really wants to do is to establish a communist world dictatorship," made on the basis of casual reading, are highly inferential. We should keep in mind their inferential character and, in our suggested exercises, should substitute for them such statements as "He rarely spoke to subordinates in the plant," "I saw him at a party, and he never danced except when one of the girls asked him to," "He wouldn't apply for the scholarship, although I believe he could have won it easily," and "The Russian delegation to the United Nations has asked for *A, B,* and *C.* Last year they voted against *M* and *N* and voted for *X* and *Y.* On the basis of facts such as these, the newspaper I read makes the inference that what Russia really wants is to establish a communist world dictatorship. I agree."

Even when we exercise every caution to avoid inferences and to report only what we see and experience, we all remain prone to error, since the making of

---

[2]The behaviorist school of psychology tries to avoid inferences about what is going on in other people's minds by describing only external behavior. A famous joke about behaviorism goes: Two behaviorists meet on the street. The first says, "You're fine. How am I?"

inferences is a quick, almost automatic process. We may watch a car weaving as it goes down the road and say, "Look at that *drunken driver,*" although what we see is only *the irregular motion of the car.* I once saw a man leave a dollar at a lunch counter and hurry out. Just as I was wondering why anyone should leave so generous a tip in so modest an establishment, the waitress came, picked up the dollar, put it in the cash register as she punched up ninety cents, and put a dime in her pocket. In other words, my description to myself of the event, "a dollar tip," turned out to be not a report but an inference.

All this is not to say that we should never make inferences. The inability to make inferences is itself a sign of mental disorder. For example, the speech therapist Laura L. Lee writes, "The aphasic [brain-damaged] adult with whom I worked had great difficulty in making inferences about a picture I showed her. She could tell me what was happening at the moment in the picture, but could not tell me what might have happened just before the picture or just afterward."[3] Hence the question is not whether or not we make inferences; the question is whether or not we are aware of the inferences we make.

| Report | Can be verified or disproved |
|---|---|
| Inference | A statement about the unknown made on the basis of the known |
| Judgment | An expression of the writer's approval or disapproval |

## Judgments

In our suggested writing exercise, judgments are also to be excluded. By judgments, we shall mean *all expressions of the writer's approval or disapproval of the occurrences, persons, or objects he is describing.* For example, a report cannot say, "It was a wonderful car," but must say something like this: "It has been driven 50,000 miles and has never required any repairs." Again, statements such as "Jack lied to us" must be suppressed in favor of the more verifiable statement, "Jack told us he didn't have the keys to his car with him. However, when he pulled a handkerchief out of his pocket a few minutes later, a bunch of car keys fell out." Also a report may not say, "The senator was stubborn, defiant, and uncooperative," or "The senator courageously stood by his principles"; it must say instead, "The senator's vote was the only one against the bill."

Many people regard statements such as the following as statements of "fact": *"Jack lied to us," "Jerry is a thief," "Tommy is clever."* As ordinarily employed, however, the word "lied" involves first an inference (that Jack knew otherwise and deliberately mis-stated the facts) and second a judgment (that the speaker disapproves of what he has inferred that Jack did). In the other two instances, we may substitute such expressions as, "Jerry was convicted of theft and served two years

---

[3] "Brain Damage and the Process of Abstracting: A Problem in Language Learning," ETC.: *A Review of General Semantics,* XVI (1959), 154–62.

at Waupun," and "Tommy plays the violin, leads his class in school, and is captain of the debating team." After all, to say of a man that he is a "thief" is to say in effect, "He has stolen and *will steal again*" — which is more of a prediction than a report. Even to say, "He has stolen," is to make an inference (and simultaneously to pass a judgment) on an act about which there may be difference of opinion among those who have examined the evidence upon which the conviction was obtained. But to say that he was "convicted of theft" is to make a statement capable of being agreed upon through verification in court and prison records.

Scientific verifiability rests upon the external observation of facts, not upon the heaping up of judgments. If one person says, "Peter is a deadbeat," and another says, "I think so too," the statement has not been verified. In court cases, considerable trouble is sometimes caused by witnesses who cannot distinguish their judgments from the facts upon which those judgments are based. Cross-examinations under these circumstances go something like this:

> WITNESS: That dirty double-crosser Jacobs ratted on me.
> DEFENSE ATTORNEY: Your honor, I object.
> JUDGE: Objection sustained. (Witness's remark is stricken from the record.) Now, try to tell the court exactly what happened.
> WITNESS: He double-crossed me, the dirty, lying rat!
> DEFENSE ATTORNEY: Your honor, I object!
> JUDGE: Objection sustained. (Witness's remark is again stricken from the record.) Will the witness try to stick to the facts.
> WITNESS: But I'm telling you the facts, your honor. He did double-cross me.

This can continue indefinitely unless the cross-examiner exercises some ingenuity in order to get at the facts behind the judgment. To the witness it is a "fact" that he was "double-crossed." Often patient questioning is required before the factual bases of the judgment are revealed.

Many words, of course, simultaneously convey a report and a judgment on the fact reported, as will be discussed more fully in a later chapter. For the purposes of a report as here defined, these should be avoided. Instead of "sneaked in," one might say "entered quietly"; instead of "politician," "congressman" or "alderman" or "candidate for office"; instead of "bureaucrat," "public official"; instead of "tramp," "homeless unemployed"; instead of "dictatorial set-up," "centralized authority"; instead of "crackpot," "holder of nonconformist views." A newspaper reporter, for example, is not permitted to write, "A crowd of suckers came to listen to Senator Smith last evening in that rickety fire-trap and ex-dive that disfigures the south edge of town." Instead he says, "Between 75 and 100 people heard an address last evening by Senator Smith at the Evergreen Gardens near the South Side city limits."

## Snarl-Words and Purr-Words

Throughout this book, it is important to remember that we are not considering language as an isolated phenomenon. Our concern, instead, is with language in action – language in the full context of the nonlinguistic events which are its setting. The making of noises with the vocal organs is a muscular activity and, like other muscular activities, is often involuntary. Our responses to powerful stimuli, such as to things that make us very angry, are a complex of muscular and physiological events: the contracting of fighting muscles, the increase of blood pressure, a change in body chemistry, clutching of our hair, and the making of noises, such as growls and snarls. We are a little too dignified, perhaps, to growl like dogs, but we do the next best thing and substitute series of words, such as "You dirty double-crosser!" "The filthy scum!" Similarly, if we are pleasurably agitated, we may, instead of purring or wagging the tail, say things like "She's the sweetest girl in all the world!"

Speeches such as these are, like direct expressions of approval or disapproval, judgments in their simplest form. They may be said to be human equivalents of snarling and purring. "She's the sweetest girl in all the world!" is not a statement about the girl; it is a purr. This seems to be a fairly obvious fact; nevertheless, it is surprising how often, when such a statement is made, both the speaker and the hearer feel that something has been said about the girl. This error is especially common in the interpretation of utterances of orators and editorialists in some of their more excited denunciations of "Reds," "pigs," "Wall Street," "radicals," "foreign ideologies," and in their more fulsome dithyrambs about "our way of life." Constantly, because of the impressive sound of the words, the elaborate structure of the sentences, and the appearance of intellectual progression, we get the feeling that something is being said about something. On closer examination, however, we discover that these utterances merely say, "What I hate ('Reds,' 'Wall Street,' or whatever) I hate very, very much," and "What I like ('our way of life') I like very, very much." We may call such utterances "snarl-words" and "purr-words." They are not reports describing conditions in the extensional world in any way.

To call these judgments "snarl-words" and "purr-words" does not mean that we should simply shrug them off. It means that we should be careful to *allocate the meaning correctly* – placing such a statement as "She's the sweetest girl in the world!" as a revelation of the speaker's state of mind, and not as a revelation of facts about the girl. If the "snarl-words" about "Reds" or "pigs" are accompanied by verifiable reports (which would also mean that we have previously agreed as to who, specifically, is meant by the terms "Reds" or "pigs"), we might find reason to be just as disturbed as the speaker. If the "purr-words" about the sweetest girl in the world are accompanied by verifiable reports about her appearance, manners, character, and so on, we might find reason to admire her too. But "snarl-words" and "purr-words" as such, unaccompanied by reports, offer nothing further to discuss, except possibly the question, "Why do you feel as you do?"

It is usually fruitless to debate such questions as "Is the President a great statesman or merely a skillful politician?" "Is the music of Wagner the greatest music of all time, or is it merely hysterical screeching?" "Which is the finer sport, tennis or baseball?" "Could Joe Louis in his prime have licked Rocky Marciano in his prime?" To take sides on such issues of conflicting judgments is to reduce oneself to the same level of stubborn imbecility as one's opponents. But to ask questions of the form, "Why do you like (or dislike) the President (or Wagner, or tennis, or Joe Louis)?" is to learn something about one's friends and neighbors. After listening to their opinions and their reasons for them, we may leave the discussion slightly wiser, slightly better informed, and perhaps slightly less one-sided than we were before the discussion began.

## How Judgments Stop Thought

A judgment ("He is a fine boy," "It was a beautiful service," "Baseball is a healthful sport," "She is an awful bore") is a conclusion, summing up a large number of previously observed facts. The reader is probably familiar with the fact that students almost always have difficulty in writing themes of the required length because their ideas give out after a paragraph or two. The reason for this is that those early paragraphs contain so many judgments that there is little left to be said. When the conclusions are carefully excluded, however, and observed facts are given instead, there is never any trouble about the length of papers; in fact, they tend to become too long, since inexperienced writers, when told to give facts, often give far more than are necessary, because they lack discrimination between the important and the trivial.

   Still another consequence of judgments early in the course of a written exercise — and this applies also to hasty judgments in everyday thought — is the temporary blindness they induce. When, for example, a description starts with the words, "He was a real Madison Avenue executive" or "She was a typical hippie," if we continue writing at all, we must make all our later statements consistent with those judgments. The result is that all the individual characteristics of this particular "executive" or this particular "hippie" are lost sight of; and the rest of the account is likely to deal not with observed facts but with stereotypes and the writer's particular notion (based on previously read stories, movies, pictures, and so forth) of what "Madison Avenue executives" or "typical hippies" are like. The premature judgment, that is, often prevents us from seeing what is directly in front of us, so that clichés take the place of fresh description. Therefore, even if the writer feels sure at the beginning of a written account that the man he is describing is a "real leatherneck" or that the scene he is describing is a "beautiful residential suburb," he will conscientiously keep such notions out of his head, lest his vision be obstructed. He is specifically warned against describing *anybody* as a "beatnik" — a term (originally applied to literary and artistic Bohemians) which was blown up by sensational journalism and movies into an almost completely fictional and misleading stereotype. If a writer applies the term to any actual living human being, he will have to expend so much energy thereafter explaining what he does *not* mean by it that he will save himself trouble by not bringing it up at all. The same warning applies to "hippies" and other social classifications that tend to submerge the individual in a category.

## Slanting

In the course of writing reports of personal experiences, it will be found that in spite of all endeavors to keep judgments out, some will creep in. An account of a man, for example, may go like this: "He had apparently not shaved for several days, and his face and hands were covered with grime. His shoes were torn, and his coat, which was several sizes too small for him, was spotted with dried clay." Now, in spite of the fact that no judgment has been stated, a very obvious one is implied. Let us contrast this with another description of the same man. "Although his face was bearded and neglected, his eyes were clear, and he looked straight ahead as he walked rapidly down the road. He seemed very tall; perhaps the fact that his coat was too small for him emphasized that impression. He was carrying a book under his left arm, and a small terrier ran at his heels." In this example, the impression about the same man is considerably changed, simply by the inclusion of new details and the subordination of unfavorable ones. Even if explicit judgments are kept out of one's writing, implied judgments based on selective perception will get in.

How, then, can we ever give an impartial report? The answer is, of course, that we cannot attain complete impartiality while we use the language of everyday life. Even with the very impersonal language of science, the task is sometimes difficult. Nevertheless, we can, by being aware of the favorable or unfavorable feelings that certain words and facts can arouse, attain enough impartiality for practical purposes. Such awareness enables us to balance the implied favorable and unfavorable judgments against each other. To learn to do this, it is a good idea to write two accounts of the same subject, both strict reports, to be read side by side: the first to contain facts and details likely to prejudice the reader in favor of the subject, the second to contain those likely to prejudice the reader against it. For example:

|                        FOR                         |                        AGAINST                         |
| -------------------------------------------------- | ------------------------------------------------------ |
| He had white teeth.                                | His teeth were uneven.                                 |
| His eyes were blue, his hair blond and abundant.   | He rarely looked people straight in the eye.           |
| He had on a clean, white shirt.                    | His shirt was frayed at the cuffs.                     |
| His speech was courteous.                          | He had a high-pitched voice.                           |
| His employer spoke highly of him.                  | His landlord said he was slow in paying his rent.      |
| He liked dogs.                                     | He disliked children.                                  |

This process of selecting details favorable or unfavorable to the subject being described may be termed *slanting*. Slanting gives no explicit judgments, but it differs from reporting in that it deliberately makes certain judgments inescapable. Let us assume for a moment the truth of the statement "When Clyde was in New York last November he was seen having dinner with a show girl. . . ." The inferences that can be drawn from this statement are changed considerably when the following words are added: ". . . and her husband and their two children." Yet, if Clyde is a married man, his enemies could conceivably do him a great deal of harm by talking about his "dinner-date with a New York show girl." One-sided or biased slanting of this kind, not uncommon in private gossip and backbiting and all too common in the "interpretative reporting" of newspapers and news magazines, can be described as a technique of lying without actually telling any lies.

## Discovering One's Bias

Here, however, caution is necessary. When, for example, a newspaper tells a story in a way that we dislike, leaving out facts we think important and playing up important facts in ways that we think unfair, we are tempted to say, "Look how unfairly they've slanted the story!" In making such a statement we are, of course, making an inference about the newspaper's editors. We are assuming that what seems important or unimportant to us seems equally important or unimportant to them, and on the basis of that assumption we infer that the editors "deliberately" gave the story a misleading emphasis. Is this necessarily the case? Can the reader, as an outsider, say whether a story assumes a given form because the editors "deliberately slanted it that way" or because that was the way the events appeared to them?

The point is that, by the process of selection and abstraction imposed on us by our own interests and background, experience comes to all of us (including newspaper editors) already "slanted." If you happen to be prolabor, pro-Catholic, and a stock-car racing fan, your ideas of what is important or unimportant will of necessity be different from those of a man who happens to be indifferent to all three of your favorite interests. If, then, newspapers often side with the big business-man on public issues, the reason is less a matter of "deliberate" slanting than the fact that publishers are often, in enterprises as large as modern urban newspapers, big businessmen themselves, accustomed both in work and in social life to asso-ciating with other big businessmen. Nevertheless, the best newspapers, whether owned by "big businessmen" or not, do try to tell us as accurately as possible what is going on in the world, because they are run by newspapermen who conceive it to be part of their professional responsibility to present fairly the conflicting points of view in controversial issues. Such newspapermen are *reporters* indeed.

The writer who is neither an advocate nor an opponent avoids slanting, except when he is seeking special literary effects. The avoidance of slanting is not only a matter of being fair and impartial; it is even more importantly a matter of making good maps of the territory of experience. The profoundly biased individual cannot make good maps because he can see an enemy *only* as an enemy and a friend *only* as a friend. The individual with genuine skill in writing — one who has imagination and insight — can look at the same subject from many points of view. The following examples may illustrate the fullness and solidity of descriptions thus written:

> Adam turned to look at him. It was, in a way, as though this were the first time he had laid eyes on him. He saw the strong, black shoulders under the

red-check calico, the long arms lying loose, forward over the knees, the strong hands, seamed and calloused, holding the reins. He looked at the face. The thrust of the jawbone was strong, but the lips were heavy and low, with a piece of chewed straw hanging out one side of the mouth. The eyelids were pendulous, slightly swollen-looking, and the eyes bloodshot. Those eyes, Adam knew, could sharpen to a quick, penetrating, assessing glance. But now, looking at that slack, somnolent face, he could scarcely believe that.

Robert Penn Warren, Wilderness

Soon after the little princess, there walked in a massively built, stout young man in spectacles, with a cropped head, light breeches in the mode of the day, with a high lace ruffle and a ginger-coloured coat. This stout young man [Pierre] was the illegitimate son of a celebrated dandy of the days of Catherine, Count Bezuhov, who was now dying in Moscow. He had not yet entered any branch of the service; he had only just returned from abroad, where he had been educated, and this was his first appearance in society. Anna Pavlovna greeted him with a nod reserved for persons of the very lowest hierarchy in her drawing-room. . . .

Pierre was clumsy, stout and uncommonly tall, with huge, red hands; he did not, as they say, know how to come into a drawing-room and still less how to get out of one, that is, how to say something particularly agreeable on going away. Moreover, he was dreamy. He stood up, and picking up a three-cornered hat with the plume of a general in it instead of his own, he kept hold of it, pulling the feathers until the general asked him to restore it. But all his dreaminess and his inability to enter a drawing-room or talk properly in it were atoned for by his expression of good-nature, simplicity and modesty.

Count Leo Tolstoy, War and Peace<br>(Translated by Constance Garnett)

## Applications

I. The following statements represent the mixture of reports, inferences, and judgments we encounter daily. Try to classify each statement. Since some cannot be neatly classified in only one category, a single-word answer might not be sufficient. If any of the statements are inferences or judgments, what kinds of evidence would you seek to support them?

   1. She goes to church only in order to show off her clothes.

      *Sample Analysis:* In usual circumstances under which a statement would be made, this would be an *inference,* since people ordinarily do not admit that they go to church for that reason. A *judgment* is also strongly implied, since it is assumed that one ought to have better reasons.

   2. Stock prices were up slightly today in light trading.
   3. Overweight people should not wear stripes, plaids, or excessively bright colors.
   4. An apple a day keeps the doctor away.

5. Compared with the persecution of heresy in Europe from 1227 to 1492, the persecution of Christians by the Romans in the first three centuries after Christ was a mild and humane procedure. Making every allowance required of an historian and permitted to a Christian, we must rank the Inquisition, along with the wars and persecutions of our time, as among the darkest blots on the record of mankind, revealing a ferocity unknown in any beast.

Will Durant, The Story of Civilization, IV.

6. Commuter — one who spends his life
In riding to and from his wife;
A man who shaves and takes a train
And then rides back to shave again.

E. B. White

7. The Chinese people do not want war.
8. Many so-called religious people are hypocrites.
9. If you travel via Highway 60, the trip only takes two hours.
10. Beauty is only skin deep.
11. For rent: charming two-bedroom house in woods, easy walk to bus stop, $225 a month plus deposit.
12. And Adam lived an hundred and thirty years, and begat a son in his likeness, after his image; and called his name Seth: And the days of Adam after he had begotten Seth were eight hundred years: and he begat sons and daughters: And all the days that Adam lived were nine hundred and thirty years: and he died.

Genesis 5:3–5

13. Crisp as Jack Frost, crunchy and crackle-happy . . . redder than a fire sale of long-handle flannels. Your big delicious beauties arrive so fresh we don't guarantee they won't talk back to folks . . . in a flavor-full language all of their own. Shipping weight about 9 pounds.

Advertising material accompanying<br>a Fruit-of-the-Month Club delivery

14. William Jameson is a skinny, crippled, tuberculosis-ridden little man, weighing only 95 pounds and standing only 5 feet tall. And every ounce and inch of him is criminal — incorrigible, remorseless and vicious.

New York World-Telegram & Sun

15. If you will study the history of almost any criminal, you will find he is an inveterate cigarette smoker.

Henry Ford

16. An intelligent man makes his own opportunities.

II. Once the verifiability of a statement has been ascertained, the task of verifying still remains. If any of the following are reports, how would you verify them?

1. $A^2 + B^2 = C^2$.
2. Families of four with incomes of less than $4,000 per year live in poverty.

3. Hops can grow as much as eight inches in one day.
4. Cigarette smoking can be dangerous to your health.
5. Mixing your drinks leads to terrible hangovers.
6. America has never lost a war.
7. My car gets twenty-two miles to the gallon.

## STUDY QUESTIONS

1. What does Hayakawa mean that reports are verifiable?
2. Is it practical, or even possible, to verify all reports?
3. Discuss the author's point that report language is "useful" language.
4. What are his definitions of inference and judgment? What are the dangers in making inferences and judgments?
5. Does Hayakawa say that we must never resort to either inference or judgment? Are there instances in which the writer can use inferences and judgments?
6. How might verifiable information distort the truth?

# What Is Technical Writing?

## W. Earl Britton

Professor Britton classifies the major definitions of technical writing, then elaborates on the definition that he finds most acceptable. Although the selection is written for teachers, it should pose little difficulty for the student reader.

Although the dean of an engineering college once denied the very existence of technical writing, many of us are confident of its reality. But we are not sure that we can convince others of its uniqueness. This uncertainty deepens when we observe the variety of activities incorporated under this label, as well as those that barely elude its scope. Our schools do little to clarify the situation with such course titles as technical writing, engineering writing, engineering English, scientific English, scientific communications, and report writing. Nor do the national societies help with their emphasis upon medical writing, biological writing, science writing, and business English. Applying the general term *technical writing* to a field of such diversified activities is convenient but misleading, yet this is the current practice. In view of the confusion, there is little wonder that a teacher in this field should

W. Earl Britton, "What Is Technical Writing," *College Composition and Communication,* May 1965, pp. 113–116.

often be asked, even by colleagues, "What is this technical writing you teach, and how does it differ from any other?"

In addition to satisfying this query, a truly helpful definition should go much further and illuminate the tasks of both the teachers and authors of technical writing. This requirement has been fulfilled in varying degrees by a number of definitions already advanced, the most significant of which form four categories.

Technical writing is most commonly defined by its subject matter. Blickle and Passe say:

> Any attempt . . . to define technical writing is complicated by the recognition that exposition is often creative. Because technical writing often employs some of the devices of imaginative writing, a broad definition is necessary. Defined broadly, technical writing is that writing which deals with subject matter in science, engineering, and business.[1]

Mills and Walter likewise note that technical writing is "concerned with technical subject matter," but admit the difficulty of saying precisely what a technical subject is. For their own purpose in writing a textbook, they call a technical subject one that falls within science and engineering. They elaborate their view by adding four large characteristics of the form: namely, its concern with scientific and technical matters, its use of a scientific vocabulary and conventional report forms, its commitment to objectivity and accuracy, and the complexity of its task, involving descriptions, classifications, and even more intricate problems.[2]

The second approach is linguistic, as illustrated in an article by Robert Hays, who, admitting the existence of technical writing without actually defining it, remarks upon its conservativeness, its Teutonic subject-verb-object word order, and the fact that it shares with other forms of writing the "common" English vocabulary. However, he cites two fundamental differences between technical and other prose. The psychological difference is the writer's "attitude of utter seriousness" toward his subject, and his dedication to facts and strict objectivity. But the greater difference, at least for "teachers and students of technical writing, is linguistic," in that technical style demands a "specialized vocabulary, especially in its adjectives and nouns."[3]

The third definition concentrates on the type of thought process involved. This approach underlies some of the research being directed by A. J. Kirkman, of the Welsh College of Advanced Technology in Cardiff, who has been investigating the causes of unsatisfactory scientific and technical writing. His group is examining in particular the suggestion that a distinction exists between ways of thinking and writing about literary and scientific subjects. The theory postulates two types of thinking, each with its own mode of expression. Associative thought belongs to

[1] *Readings for Technical Writers*, ed. Margaret D. Blickle and Martha E. Passe (New York, 1963), p. 3.
[2] Gordon H. Mills and John A. Walter, *Technical Writing* (New York, 1954), pp. 3–5.
[3] Robert Hays, "What Is Technical Writing?" *Word Study*, April 1961, p. 2.

history, literature, and the arts. Statements are linked together by connectives like *then* and *rather*, indicating chronological, spatial, or emotional relationships. Sequential thought belongs to mathematics and science. Statements are connected by words like *because* and *therefore*, revealing a tightly logical sequence. Professor Kirkman suggests that the weakness of much scientific writing results from forcing upon scientific material the mode of expression appropriate to the arts. He adds:

> The important distinction is that sequential contexts call for comparatively inflexible lines of thought and rigid, impersonal forms of expression, whereas associative contexts permit random and diverse patterns of thought which can be variously expressed.[4]

Finally, technical writing is sometimes defined by its purpose. This approach rests upon the familiar differentiation between imaginative and expository prose, between DeQuincey's literature of power and literature of knowledge. Brooks and Warren find the primary advantage of the scientific statement to be that of "absolute precision." They contend that literature in general also represents a "specialization of language for the purpose of precision" but add that it "aims at treating kinds of material different from those of science," particularly attitudes, feelings, and interpretations.[5]

Reginald Kapp pursues a similar line by dividing writing into imaginative and functional literature. Imaginative literature involves personal response and is evocative; functional literature concerns the outer world that all can see. Functional English, he says, presents "all kinds of facts, of inferences, arguments, ideas, lines of reasoning. Their essential feature is that they are new to the person addressed." If imaginative literature attempts to control men's souls, functional English should control their minds. As he writes, the technical and scientific author "confers on the words the power to make those who read think as he wills it."[6]

All of these approaches are significant and useful. Kirkman's suggestions are certainly intriguing, but I find Kapp's classification particularly helpful and want to extend it slightly.

I should like to propose that the primary, though certainly not the sole, characteristic of technical and scientific writing lies in the effort of the author to convey one meaning and only one meaning in what he says. That one meaning must be sharp, clear, precise. And the reader must be given no choice of meanings; he must not be allowed to interpret a passage in any way but that intended by the writer. Insofar as the reader may derive more than one meaning from a passage, technical writing is bad; insofar as he can derive only one meaning from the writing, it is good.

---

[4] A. J. Kirkman, "The Communication of Technical Thought," *The Chartered Mechanical Engineer,* December, 1963, p. [2].

[5] Cleanth Brooks and Robert Penn Warren, *Understanding Poetry,* 3rd ed. (New York, 1960), pp. 4–5.

[6] Reginald O. Kapp, *The Presentation of Technical Information* (New York, 1957), chaps. I–II.

Imaginative writing — and I choose it because it offers the sharpest contrast — can be just the opposite. There is no necessity that a poem or play convey identical meanings to all readers, although it may. Nor need a poem or play have multiple meanings. The fact remains, nevertheless, that a work of literature may mean different things to different readers, even at different times. Flaubert's *Madame Bovary* has been interpreted by some as an attack on romanticism, and in just the opposite way by others who read it as an attack on realism. Yet no one seems to think the less of the novel. Makers of the recent film of *Tom Jones* saw in the novel a bedroom farce, whereas serious students of Fielding have always regarded it as an effort to render goodness attractive. Varied interpretations of a work of literature may add to its universality, whereas more than one interpretation of a piece of scientific and technical writing would render it useless.

When we enter the world of pure symbol, the difference between the two kinds of communication — scientific and aesthetic — becomes more pronounced. Technical and scientific writing can be likened to a bugle call, imaginative literature to a symphony. The bugle call conveys a precise message: get up, come to mess, retire. And all for whom it is blown derive identical meanings. It can mean only what was intended. But a symphony, whatever the intention of the composer, will mean different things to different listeners, at different times, and especially as directed by different conductors. A precise meaning is essential and indispensable in a bugle call; it is not necessarily even desirable in a symphony.

The analogy can be extended. Even though the bugle call is a precise communication, it can be sounded in a variety of styles. An able musician can play taps with such feeling as to induce tears, and it is conceivable that a magician might blow reveille so as to awaken us in a spirit of gladness. The fact that scientific writing is designed to convey precisely and economically a single meaning does not require that its style be flat and drab. Even objectivity and detachment can be made attractive.

Because technical writing endeavors to convey just one meaning, its success, unlike that of imaginative literature, is measurable. As far as I am aware, there is no means of determining precisely the effects of a poem or a symphony; but scientific analyses and descriptions, instructions, and accounts of investigations quickly reveal any communication faults by the inability of the reader to comprehend and carry on.

Objection may be raised to this distinction between the two kinds of writing because it makes for such large and broad divisions. This I readily admit, at the same time that I hold this feature to be a decided advantage, in that it removes the difficulty that usually arises when technical writing is defined by its subject matter.

Emphasis upon engineering subject matter in technical writing, for example, has implied that engineering has a monopoly on the form, and that a Ph.D. dissertation in linguistics or even certain kinds of literary criticism and a study of federal economic policy are other kinds of writing. When all such endeavors that convey single meanings are grouped under the label technical and scientific writing, or some other term, for that matter, then division of these into subject areas, instead of creating confusion, becomes meaningful. Some subjects will be far more technological than others, ranging from dietetics to nuclear fission, and in some instances being related to science only by method of approach; some subjects will offer more linguistic difficulty than others; some will require a tighter, more sequential mode of thinking; but all will have in common the essential effort to limit the reader to one interpretation.

It seems to me that this view not only illuminates the nature of technical writing but also emphasizes the kind of training required of our schools. Unfortunately, few educational institutions are meeting the needs in this field. Professor Kirkman mentions the failure of the traditional teachers to provide enough practice in writing on practical subjects. Professor Kapp has insisted that the conventional instruction in formal English courses does not equip a man to teach or practice scientific and technical writing. The Shakespearian scholar, G. B. Harrison, commenting upon his formal writing courses in England, says:

> The most effective elementary training I ever received was not from masters at school but in composing daily orders and instructions as staff captain in charge of the administration of seventy-two miscellaneous military units. It is far easier to discuss Hamlet's complexes than to write orders which ensure that five working parties from five different units arrive at the right place at the right time equipped with proper tools for the job. *One learns that the most seemingly simple statement can bear two meanings* and that when instructions are misunderstood the fault usually lies with the wording of the original order.[7] [My italics]

But our strictures should not be confined to the English teachers. All of us in education must share the responsibility for this condition. In fact, I believe that in all too many instances, at least in college, the student writes the wrong thing, for the wrong reason, to the wrong person, who evaluates it on the wrong basis. That is, he writes about a subject he is not thoroughly informed upon, in order to exhibit his knowledge rather than explain something the reader does not understand, and he writes to a professor who already knows more than he does about the matter and who evaluates the paper, not in terms of what he has derived, but in terms of what he thinks the writer knows. In every respect, this is the converse of what happens in professional life, where the writer is the authority; he writes to transmit new or unfamiliar information to someone who does not know but needs to, and who evaluates the paper in terms of what he derives and understands.

B. C. Brookes takes a similar position when he suggests that English teachers concerned with science students should ask them occasionally to explain aspects of their work which they know well so that the teacher who is acquainted with the material will understand it. Such an assignment not only is a real exercise in composition but also taxes the imagination of the student in devising illuminating analogies

---

[7]G. B. Harrison, *Profession of English* (New York, 1962), p. 149.

for effective communication. The teacher's theme should be: "If your paper is not plain and logical to me, then it is not good *science.*"[8]

Both Harrison and Brookes recommend the kinds of exercises that are often viewed skeptically at the college level. This is a regrettable attitude, especially since it usually derives from unfamiliarity with the nature and need of such work and from unawareness of its difficulty and challenge. Teachers oriented primarily toward literature see little of interest in this field, but those who enjoy composition — especially its communicative aspect — can find considerable satisfaction here. Of one thing they can always be sure: deep gratitude from those they help.

## STUDY QUESTIONS

1. What four basic definitions of technical writing does Britton discuss? Are the definitions easy to locate in the article?
2. Which definition does the author prefer? Why?
3. Why does the author discuss the definition of imaginative writing? How does this definition help him in precisely defining technical writing?
4. Britton states that an essential characteristic of technical writing is that it "endeavors to convey just one meaning." Is this characteristic contradictory to either Barzun and Graff's or Hayakawa's definition of a report?
5. What is the appropriate subject matter for technical writing as Britton defines it?

# Credo of a Tech Writer

## John Frye

Unlike law or medicine, technical writing is not a licensed or formally regulated profession. Still, the technical writer has professional responsibilities, some of which Frye lists in his Credo — or statement of belief. Although he focuses on technical writing, his observations are generally valid for all report writing fields. The student writer will perhaps find in the selection some worthwhile goals.

"First off, I'll assume you're not interested in the ABC business of how you query an editor, prepare a manuscript, do illustrations, etc. That primer stuff you can get from any book on writing for publication. An experienced writer takes it

[8]B. C. Brookes, "The Teaching of English to Scientists and Engineers," *The Teaching of English,* Studies in Communication 3 (London, 1959), pp. 146-7.

From John Frye, "Credo of a Tech Writer," *Electronics World,* June 1971, pp. 50–51.

for granted his manuscript should be pleasing to the eye, require a minimum of editing, and be easy to set in type. He does not expect the editor to have to do his work for him. I think you want to know something about how I feel about my profession and try to put into my writing."

. . . .

"I think a good tech writer is a true professional and acts like one at all times. He doesn't expect to write only when he feels like it. A deadline is a serious obligation to him and he makes every effort to meet it despite personal problems he may encounter. He is a responsible person. He realizes an editor publishing his writing is laying his editorial neck and the reputation of his magazine on the line in doing so. An editor cannot — and should not need to — double-check every statement the writer makes. He must have writers he can trust to be painstaking and accurate. The true professional is such a writer because he is just as concerned about what appears under his byline as the editor is about what appears in his magazine.

"My second belief is that a good tech writer is constantly 'stocking the pantry shelves.' By that I mean he must be constantly taking in and storing his mind with a wide range of knowledge from many sources: experience, magazines, books, newspapers, radio, TV, and talking with experts like yourselves. Note this knowledge doesn't necessarily pertain to a topic he plans to write about in the immediate future. A good writer is interested in *everything,* but it's amazing how much of what he learns he eventually converts into grist for his mill.

"Keeping informed is probably the most important activity of a writer, but he must not fall into the trap of becoming just a fact-collector in his own field. It's essential he perceive the relationship between what is going on there and what is happening in the rest of the world. He must develop a kind of side-looking radar that keeps him constantly aware of major developments in not only adjacent but in seemingly unrelated fields. Right now the biologist is calling on the computer and the electron microscope to help decode the secrets of the chromosomes, and store management is starting to rely on CCTV and the video recorder to help combat shoplifting.

"The tech writer must do a great deal of reading in his own special line, but he should also study how technical articles have been handled down through the ages. For example, reading a couple of 'construction' articles from the Bible, the story of Creation and the building of the Ark, should convince anyone you don't need five thousand words to tell how to put together a transistorized audio oscillator. Perusing Leonardo da Vinci's notebooks or Benvenuto Cellini's gripping account of the casting of the bronze Perseus will show the writer how a master handles a technical story.. If he needs more to take the conceit out of him, let him read Ben Franklin's account of his early experiments with electricity or see how the Sage of Philadelphia got the jump on modern ecologists with his treatise on *The Cause and Cure of Smoky Chimneys.*

"That brings me to another belief of mine. I think good technical writing should have warmth and humor and reflect the basic philosophy of the writer. I want a story to read as though *someone*  — not a committee — wrote it! Some technical stories I read are about as colorless as computer-composed music. By saying a story should have warmth and humor, I don't mean the writer should deliberately try to be funny or precious. I simply mean the writing should show the author does not take himself nor his subject too seriously. Just because the subject is technical, the writing style does not have to be dry, stilted, and formal. The scientific writings of H. G. Wells and Julian Huxley prove the contrary. As a matter of fact, the more difficult and recondite the subject, the more the reader needs those little recesses for the mind provided by touches of humor that reassure him he is still in touch with another human being — the writer. Encouraged and rested, he can then press on to deeper understanding of the abstruse subject.

"At the risk of belaboring the point, I think technical writers have a deep responsibility, especially at this moment in time, to demonstrate technology is not necessarily cold and impersonal and anti-human. Right now we are on the brink of a revolt against technology. Just last week I heard a college president on TV expressing concern over the larger number of students turning away from engineering and going into the humanities and social studies. He said people are distrustful of technology because it has failed to bring universal happiness and because many feel it has led us down the primrose path into a polluted world.

"There's reason for this feeling. Technology, as presented by many writers, seems to be trying to get rid of humanity. They speak of 'eliminating human error' as though man himself were a mistake. It's high time technical writers display technology as a servant of mankind, not its master; for as people see it, so will it become. Erich Fromm in *The Revolution of Hope* quotes Marx as saying '. . . I can only relate myself in a human way to a thing when the thing is related in a human way to man.' If the rising tide of antipathy to technology is to ebb, we tech writers must relate things in a human way to men. Man must be plugged into any system planned. Mankind needs technology, but technology also needs man."

## STUDY QUESTIONS

1. Why should the technical writer keep informed on many topics?
2. What is the value of reading into the technical literature of the past?
3. Why should the writer "not take himself nor his subject too seriously"?
4. How would you adapt Frye's Credo for the science writer or the business writer?
5. Frye says that the technical writer must show that "technology . . . needs man." How is this done when writing routine, workaday reports? What would be an equivalent responsibility for the business writer?

# The Art of Communication

## C. Northcote Parkinson and Nigel Rowe

Parkinson and Rowe focus on business communication in this selection, but their observations will also apply to writing for industry and science. The authors remind us that because communication is between people, the writer cannot neglect the personal element. For this reason they stress motivation and trust (among others) as significant elements in communication. Parkinson and Rowe are English, which is reflected in some of the terminology (for example, telephone *kiosk* instead of telephone *booth*) and spelling (*neighbour* instead of *neighbor*). But the principles discussed are fully applicable to American practices.

When we begin to communicate, whether as a society or as individuals, our efforts are initially confused by our desire to express ourselves. We have an emotion and we want to give it an outlet. A lion roars (one would guess) because it feels like roaring, not because it has any profound thought to express. A dog barks without attempting to explain its philosophy or political outlook. We ourselves often express an emotion without any real purpose, save to unburden our minds of what we feel. When we drop a hammer on our toe we say 'Damn!' or 'Blast!' merely to relieve our feelings. Our exclamation does no good and may not even be heard by anyone else. We curse when we are alone. We complain about the cussedness of things — the door that jams, the tool that breaks, the screw that is lost, the light that fuses. The first thing to realize, therefore, is that most people want to talk but few of them have anything to say. Anyone who doubts this should watch someone in a glass-sided telephone kiosk. Note the gestures and expressions, the pointing finger, the clutched fist, the smile, the frown, the tapping foot. All these signs are wasted on the telephone. We realize, in fact, that they are not seriously meant to convey anything to anybody. They are means of self-expression and that is all. They are older than speech, and when the gesture contradicts the word, the gesture is more likely to be right. What is true of the individual is often as true of the organization. Those who sit in head offices may relieve their feelings in letters, telegrams, notices and memoranda, but these are often as meaningless as the gesture made when we are on the telephone. What people say or write or print often means nothing at all and is not even meant to mean anything. It is merely the bureaucratic equivalent of breaking wind.

Once we stop relieving our emotions and show the wish to communicate, there is a different situation altogether. Our feelings are set aside and we have a thought or wish or instruction to convey. We are now called upon to use our imagination. We have to visualize the person or persons we are seeking to influence. He may be — he often is — the man on the factory floor. We have to ask ourselves

From C. Northcote Parkinson and Nigel Rowe, *Communicate: Parkinson's Formula for Business Survival* (New Jersey: Prentice-Hall, 1977), pp. 9–15.

what he already knows, what he already believes (perhaps incorrectly), what his chief interests are, where he is likely to live and what newspaper (if any) he is likely to read. Does all this sound obvious? But it is far from obvious and is all too easily forgotten. Our basic mistake, eternally repeated, is to assume that what is known to us is known to everyone else.

If your directions are to be followed correctly, they must be explained clearly. There must be careful choice of words. Behind what is said there must be the wish to communicate. Behind the bad direction there is often a muddled motive: a desire to humiliate, a desire to reveal somebody's stupidity, a desire to show that the coming disaster was not our fault. In other words, the *will* to communicate is absent. Its place is taken by the will to mystify. Does that sound improbable? It happens every day. It happens all the time. There is a technique in communicating but no sort of technique will help us if the will is absent. All too often the will to explain is simply not there.

So much for the role of imagination, for the will to put the message over. Come now to the creation of trust. In all human relationships trust is a vital factor. For success in negotiation, it is vitally important that people will believe what you say and assume that any promise you make will be kept. But it is no good saying 'Trust me. Rely on my word.' Only politicians say that. Among businessmen trust is not given. It has to be earned. After years in the trade you may gain a reputation for being reliable and honest. People will have found that you mean what you say. It takes times to establish such a reputation. Remember, however, that you can establish yourself as a friend and neighbour before you have done any business at all. Take, for example, the ritual that surrounds our visit to the village shop. There are people who rush in and say 'A kilo of tomatoes, please.' But that is mere bad manners. The right method is to begin, 'Good morning, Mr. Whatever [or George, or Sally, as the case may be]. Nice weather we are having. How are you keeping? [etc.]' You start a conversation and end up with an inquiry about tomatoes. Where the matter is of greater importance — the hiring of a truck or a senior executive, say — the approach may be more leisurely still. Why? Because you aim to be friends before the serious discussion starts. Where the matter is one of some delicacy you will spread your visits over a number of weeks. When the bargaining begins you will not be a complete stranger but someone who is known by name, a neighbour, a man who may be thought trustworthy. Now, in creating a good relationship between one firm and another, between the company and the government, between management and personnel, we have to realize that it is going to take time. The contract that we are anxious to obtain must be the reward for the good work we have done in the past.

The industrial dispute we aim to prevent is one that would take place in five or ten years' time. To prevent it, we begin talking now — not about wage differentials but about other things, horses, fishing, music or football. We have first to establish that we (the managers) are men with names, personalities, interests and, possibly, weaknesses. We are not 'they', the nameless representatives of capital, but Tom, Dick and Harry. We have next to show that the men on the shop floor are not mere personnel but are characters known to us by name and personality. They are Bill, Sam and Bob, one of them a charge hand, another a shop steward and a third (possibly) a communist. It should be possible to create a man-to-man relationship, but it will take time and effort. When you are known as a man among men you can set out to prove that you are also a man to be trusted. When you promise to be there at eight you are there at eight. When you promise to keep quiet about something, you keep quiet about it. You prove over the years that what you say is the truth, that what you promise you will do. Then, at long last, a crisis comes and you realize that an industrial dispute is possible. Knowing the men as you do, you hear of the trouble long before it starts. The probable result is that it never starts at all. You are there first and you are a man they will believe. It may seem, and it is, a laborious process, but there is no other way. The dispute that begins today is the result of what someone failed to do ten years ago. The mischief is when the company is sold, the board replaced, the managers changed and the whole process put back to the starting point. This happens too often and is the cause of much that goes wrong. There has been no time to establish a relationship of mutual trust.

We have also to consider the relationship between an industrial organization and its suppliers, its rivals, its shareholders, customers, bankers, government and public. The principles that govern policy in each case may be the same but the message in each instance is likely to be different, not because our ways are devious but because different people have different interests. Shareholders want to see the balance sheet, customers want to hear of any new product, government may be interested in export plans and the public mostly concerned about pollution and perhaps profits. So the sorts of information that are to be supplied may be varied indeed. As against that, the underlying purpose must always be the same; to create a reputation for honesty, competence, public spirit and courtesy. We know, however, that such a reputation is not based upon what we say but upon what we do. Realizing this, some businessmen resolve to say little, pointing silently at their past achievement. This is, however, a serious mistake, for our past good deeds do not, of necessity, speak for themselves. If we are silent, other people will be voluble and perhaps at our expense. To talk and do nothing will give us, admittedly, a bad reputation, but to do all we should and say nothing about it may end at best with our having no reputation at all. So our task is to explain as well as to act, and we will do well to remember that the explanations should be as good as the policy.

Suppose we are now ready to convey a message, describe a situation, explain a new development. We must decide at the outset to keep it brief. Too much information is as bad as too little and has, in fact, the same effect — that is, probably, none. But how is the message to be phrased? In other words, what precisely are we trying to do? We have ceased (let us hope) to relieve our minds of what we feel or think. We shall speak with a purpose to a public we have clearly in mind. What shall we say in order to persuade these particular people to do something or perhaps to refrain from doing something? It does not matter how the message sounds to us. How will it sound to them? Begin with the content of the message. What, briefly, do you want to say? Remember that the commonest cause of disaster, in all human affairs, is an initial failure to decide what you are trying to do. Suppose there are two projects, A and B. You will often find that one of them, project B, has attracted a thick file of untidy correspondence, with minutes, reports, committee recommendations and specialists' advice. Project A, achieving more, has attracted less paper, the file being thin, clean and neat. Go back to enclosure 1A on File A and you will find that the policy was well defined from the outset. Go back to enclosure 1A on File B and you will find that the object in view has never been properly defined. On that shaky foundation there has been built up a mountain of correspondence and little real achievement. Go wrong at the beginning and nothing will afterwards go right.

The basic question is always 'What are we trying to do — and why?' It is told of Peter Drucker that he was called in as consultant by a firm that manufactured glass bottles. At his first meeting with the board he asked them, 'In what business are you engaged?' Slightly annoyed (for he should, after all, have read the literature they had given him), the President replied, 'We are in the business of manufacturing glass bottles.' To this Drucker is said to have replied, 'No you are not. You are in the packaging business.' The directors were suddenly made to see a great truth. Beer must be packaged but not, of necessity, in glass. In two short sentences Drucker had made the directors think, perhaps (who knows?) for the first time. Apply the same question to a school. If we ask the teachers, 'What are you doing and why?' we shall be surprised at the variety of answers we receive. We may not know what the right answer should be but we shall discover, in minutes, why so little is being achieved. The object in view has never been defined. At the present time Britain is fighting a minor civil war in Northern Ireland, a confused conflict in which politicians, soldiers, police and informers are all more or less involved. Will they succeed? How can they? Succeed in what? Nobody has ever made a clear definition of a feasible task. Success is therefore impossible and will remain so until those responsible go back to the beginning, ask the right questions and give the right answers. All current operations are a waste of time. Nor are commercial or industrial projects any different. We must do our thinking first. What is our aim, our object; what, exactly, are we trying to do?

Let us assume that we have a message to put over and have decided exactly what it is; there are still some rules to observe. First of all, we must refrain from putting it over too often. Internally we should avoid issuing new notices every day, new rules every week, a plan for reorganization every month and a change of policy every year. Too much communication is as harmful as too little. Tell people what they need to know but do not swamp them with information and exhortation. Second, remember that communication is a two-way process. You must test the reaction and you must sense when there is a feeling of resentment on the factory floor. It is a sign of failure when a deputation comes to complain about something. You should have known sooner about the complaint and taken action to remedy or explain it. The good manager can smell unrest just as the good sea captain can sense panic or mutiny. Third, we should never forget that it is often possible to distract attention from a sore subject, giving people something else to think and talk about.

When sufficiently excited about something like a football match or a band performance, we can forget about a lot of discomfort. The player, coming off the field, finds that he has a bruise on the shin — he did not notice the kick that made the bruise in the first place. A part of the art of management lies in the ability to distract attention. That is how to reconcile people to a necessary inconvenience, by finding a partial remedy before people have even expressed their grievance.

Externally, the task is more complex. There will be a reduced dividend this year and a larger transfer of funds to reserve. The shareholders will want to know why. Some employees have become redundant, the result of automation. The explanation must be candid and convincing, and the treatment of those displaced must be fair and must be seen to be fair. There is to be a new factory at X despite the rumours that have been current of an even bigger development at Y. Assurances must be given about the employment situation at Y and the environmental disaster impending at X. There is a public outcry about the danger of fire said to be inherent in plastic product Z. The danger is non-existent but we need to conciliate the well-meaning folk who made the complaint. With a little ingenuity we can turn the adverse publicity into an actual advertisement for the product. The ordeal by fire might even be televised. . . . Our goods are blacked by trade union W because we insist on exporting to our old business partners in Afrasia. How can we bring pressure to bear upon the officials of union W, who are thus about to cause unemployment among the members of union V? The situations multiply in which we have to explain ourselves with lucidity, honesty, conviction, eloquence, brevity and speed.

Come now to the question of style in communication. Style is, first and foremost, the imprint of character upon action. A warship comes into port at speed, swings alongside the wharf with a flourish and all with a minimum number of orders. In a matter of minutes she is secure, a gangway is down and a sentry is at the end of it. Watching the process, a knowledgeable onlooker will say, 'That will be Captain Dashing — I can recognize his style.' An announcement, a message, can and should convey a sense of character. It should not be a colourless or civil service-type pronouncement. While precise and terse, it should go beyond precision and brevity. It should be coloured by the personality of the chief executive. This personal touch can be achieved in several ways or by a combination of several methods. Here are three, to begin with. First, it should come from the man himself, not from a nameless and faceless authority, commonly described as 'they'. It should be signed by the chief executive, a man who should be known to all by sight and to some at least by daily contact. Second, it should never include long words and involved constructions. Words of one syllable are best, each word a hammer blow rather than a handful of cotton wool. Third, it should sometimes include a touch of humour, and this for two reasons. In the first place, humour makes it human rather than impersonal. In the second place, people remember a joke when they forget everything else.

The challenge to businessmen is, in essence, the future survival of business enterprise. This requires a vigorous defence of its legitimate freedoms and a mature acceptance of change in industrial policy and practice to be in step and not in conflict with the social and political environment of which it is a key element. Effective communication, in its many forms, is the one essential catalyst to meeting this challenge. Thus far it has proved to be the illusive missing link between business enterprise and both its public acceptance and understanding, between the legitimate objectives of industry and both the aspirations and expectations of those who work in it.

### STUDY QUESTIONS

1. Why is trust important in business? How is it obtained? What is its role in communication?
2. Why is "too much communication . . . as harmful as too little"? Explain.
3. The authors assert that business communication should have a style "coloured by the personality of the chief executive." Is this requirement contradictory to the principles of objectivity in writing? Explain.
4. The authors also assert that business communication "should sometimes include a touch of humour." Would humor be appropriate for writing in technical or scientific fields? Why or why not? Does the current selection include any humor? If so, does it contribute to effective communication?

# Why a Surgeon Would Write

## Richard Selzer

We make no apologies for presenting writing as a job related tool. But we wouldn't restrict writing to this function alone. Within the professions, we find people who write to clarify their ideals and their aspirations. So for the sake of perspective, we conclude this section with an article by a surgeon who reveals his personal feelings towards writing about his medical practice.

Someone asked me why a surgeon would write. Why, when the shelves are already too full? They sag under the deadweight of books. To add a single adverb is to risk exceeding the strength of the boards. A surgeon should abstain. A surgeon, whose fingers are more at home in the steamy gullies of the body than they are tapping the dry keys of a typewriter. A surgeon, who feels the slow slide of intestines against the back of his hand and is no more alarmed than were a family of snakes

From Richard Selzer, *Mortal Lessons: Notes on the Art of Surgery* (New York: Simon and Schuster, 1976), pp. 15–23. Title selected by editors.

taking their comfort from such an indolent rubbing. A surgeon, who palms the human heart as though it were some captured bird.

Why should he write? Is it vanity that urges him? There is glory enough in the knife. Is it for money? One can make too much money. No. It is to search for some meaning in the ritual of surgery, which is at once murderous, painful, healing, and full of love. It is a devilish hard thing to transmit — to find, even. Perhaps if one were to cut out a heart, a lobe of the liver, a single convolution of the brain, and paste it to a page, it would speak with more eloquence than all the words of Balzac. Such a piece would need no literary style, no mass of erudition or history, but in its very shape and feel would tell all the frailty and strength, the despair and nobility of man. What? Publish a heart? A little piece of bone? Preposterous. Still I fear that is what it may require to reveal the truth that lies hidden in the body. Not all the undressings of Rabelais, Chekhov, or even William Carlos Williams have wrested it free, although God knows each one of those doctors made a heroic assault upon it.

I have come to believe that it is the flesh alone that counts. The rest is that with which we distract ourselves when we are not hungry or cold, in pain or ecstasy. In the recesses of the body I search for the philosophers' stone. I know it is there, hidden in the deepest, dampest cul-de-sac. It awaits discovery. To find it would be like the harnessing of fire. It would illuminate the world. Such a quest is not without pain. Who can gaze on so much misery and feel no hurt? Emerson has written that the poet is the only true doctor. I believe him, for the poet, lacking the impediment of speech with which the rest of us are afflicted, gazes, records, diagnoses, and prophesies.

I invited a young diabetic woman to the operating room to amputate her leg. She could not see the great shaggy black ulcer upon her foot and ankle that threatened to encroach upon the rest of her body, for she was blind as well. There upon her foot was a Mississippi Delta brimming with corruption, sending its raw tributaries down between her toes. Gone were all the little web spaces that when fresh and whole are such a delight to loving men. She could not see her wound, but she could feel it. There is no pain like that of the bloodless limb turned rotten and festering. There is neither unguent or anodyne to kill such a pain yet leave intact the body.

For over a year I trimmed away the putrid flesh, cleansed, anointed, and dressed the foot, staving off, delaying. Three times each week, in her darkness, she sat upon my table, rocking back and forth, holding her extended leg by the thigh, gripping it as though it were a rocket that must be steadied lest it explode and scatter her toes about the room. And I would cut away a bit here, a bit there, of the swollen blue leather that was her tissue.

At last we gave up, she and I. We could not longer run ahead of the gangrene. We had not the legs for it. There must be an amputation in order that she might live — and I as well. It was to heal us both that I must take up knife and saw, and cut the leg off. And when I could feel it drop from her body to the table, see the blessed *space* appear between her and that leg, I too would be well.

Now it is the day of the operation. I stand by while the anesthetist administers the drugs, watch as the tense familiar body relaxes into narcosis. I turn then to uncover the leg. There, upon her kneecap, she has drawn, blindly, upside down for me to see, a face; just a circle with two ears, two eyes, a nose, and a smiling upturned mouth. Under it she has printed SMILE, DOCTOR. Minutes later I listen to the sound of the saw, until a little crack at the end tells me it is done.

So, I have learned that man is not ugly, but that he is Beauty itself. There is no other his equal. Are we not all dying, none faster or more slowly than any other? I have become receptive to the possibilities of love (for it is love, this thing that happens in the operating room), and each day I wait, trembling in the busy air. Perhaps today it will come. Perhaps today I will find it, take part in it, this love that blooms in the stoniest desert.

All through literature the doctor is portrayed as a figure of fun. Shaw was splenetic about him; Molière delighted in pricking his pompous medicine men, and well they deserved it. The doctor is ripe for caricature. But I believe that the truly great writing about doctors has not yet been done. I think it must be done *by* a doctor, one who is through with the love affair with his technique, who recognizes that he has played Narcissus, raining kisses on a mirror, and who now, out of the impacted masses of his guilt, has expanded into self-doubt, and finally into the high state of wonderment. Perhaps he will be a nonbeliever who, after a lifetime of grand gestures and mighty deeds, comes upon the knowledge that he has done no more than meddle in the lives of his fellows, and that he has done at least as much harm as good. Yet he may continue to pretend, at least, that there is nothing to fear, that death will not come, so long as people depend on his authority. Later, after his patients have left, he may closet himself in his darkened office, sweating and afraid.

There is a story by Unamuno in which a priest, living in a small Spanish village, is adored by all the people for his piety, kindness, and the majesty with which he celebrates the Mass each Sunday. To them he is already a saint. It is a foregone conclusion, and they speak of him as Saint Immanuel. He helps them with their plowing and planting, tends them when they are sick, confesses them, comforts them in death, and every Sunday, in his rich, thrilling voice, transports them to paradise with his chanting. The fact is that Don Immanuel is not so much a saint as a martyr. Long ago his own faith left him. He is an atheist, a good man doomed to suffer the life of a hypocrite, pretending to a faith he does not have. As he raises the chalice of wine, his hands tremble, and a cold sweat pours from him. He cannot stop for he knows that the people need this of him, that their need is greater than his sacrifice. Still . . . still . . . could it be that Don Immanuel's whole life is a kind of prayer, a paean to God?

A writing doctor would treat men and women with equal reverence, for what is the "liberation" of either sex to him who knows the diagrams, the inner geographies of each? I love the solid heft of men as much as I adore the heated capaciousness of women — women in whose penetralia is found the repository of existence. I would have them glory in that. Women are physics and chemistry. They are matter. It is their bodies that tell of the frailty of men. Men have not their cellular, enzymatic wisdom. Man is albuminoid, proteinaceous, laked pearl; woman is yolky, ovoid, rich. Both are exuberant bloody growths. I would use the defects and deformities of each for my sacred purpose of writing, for I know that it is the marred and scarred and faulty that are subject to grace. I would seek the soul in the facts of animal economy and profligacy. Yes, it is the exact location of the soul that I am after. The smell of it is in my nostrils. I have caught glimpses of it in the body diseased. If only I could tell it. Is there no mathematical equation that can guide me? So much pain and pus equals so much truth? It is elusive as the whippoorwill that one hears calling incessantly from out the night window, but which, nesting as it does low in the brush, no one sees. No one but the poet, for he sees what no one else can. He was born with the eye for it.

Once I thought I had it: Ten o'clock one night, the end room off a long corridor in a college infirmary, my last patient of the day, degree of exhaustion suitable for the appearance of a vision, some manifestation. The patient is a young man recently returned from Guatemala, from the excavation of Mayan ruins. His left upper arm wears a gauze dressing which, when removed, reveals a clean punched-out hole the size of a dime. The tissues about the opening are swollen and tense. A thin brownish fluid lips the edge, and now and then a lazy drop of the overflow spills down the arm. An abscess, inadequately drained. I will enlarge the opening to allow better egress of the pus. Nurse, will you get me a scalpel and some . . . ?

What happens next is enough to lay Francis Drake avomit in his cabin. No explorer ever stared in wilder surmise than I into that crater from which there now emerges a narrow gray head whose sole distinguishing feature is a pair of black pincers. The head sits atop a longish flexible neck arching now this way, now that, testing the air. Alternately it folds back upon itself, then advances in new boldness. And all the while, with dreadful rhythmicity, the unspeakable pincers open and close. Abscess? Pus? Never. Here is the lair of a beast at whose malignant purpose I could but guess. A Mayan devil, I think, that would soon burst free to fly about the room, with horrid blanket-wings and iridescent scales, raking, pinching, injecting God knows what acid juice. And even now the irony does not escape me, the irony of my patient as excavator excavated.

With all the ritual deliberation of a high priest I advance a surgical clamp toward the hole. The surgeon's heart is become a bat hanging upside down from his rib cage. The rim achieved — now thrust — and the ratchets of the clamp close upon the empty air. The devil has retracted. Evil mocking laughter bangs back and forth in the brain. More stealth. Lying in wait. One must skulk. Minutes pass, perhaps an hour. . . . A faint disturbance in the lake, and once again the thing upraises, farther and farther, hovering. Acrouch, strung, the surgeon is one with his instrument; there is no longer any boundary between its metal and his flesh. They are joined in a single perfect tool of extirpation. It is just for this that he was born. Now — thrust — and clamp — and *yes*. Got him!

Transmitted to the fingers comes the wild thrashing of the creature. Pinned and wriggling, he is mine. I hear the dry brittle scream of the dragon, and a hatred seizes me, but such a detestation as would make of Iago a drooling sucktit. It is the demented hatred of the victor for the vanquished, the warden for his prisoner.

It is the hatred of fear. Within the jaws of my hemostat is the whole of the evil of the world, the dark concentrate itself, and I shall kill it. For mankind. And, in so doing, will open the way into a thousand years of perfect peace. Here is Surgeon as Savior indeed.

Tight grip now . . . steady, relentless pull. How it scrabbles to keep its tentacle-hold. With an abrupt moist plop the extraction is complete. There, writhing in the teeth of the clamp, is a dirty gray body, the size and shape of an English walnut. He is hung everywhere with tiny black hooklets. Quickly . . . into the specimen jar of saline . . . the lid screwed tight. Crazily he swims round and round, wiping his slimy head against the glass, then slowly sinks to the bottom, the mass of hooks in frantic agonal wave.

"You are going to be all right," I say to my patient. "We are *all* going to be all right from now on."

The next day I take the jar to the medical school. "That's the larva of the botfly," says a pathologist. "The fly usually bites a cow and deposits its eggs beneath the skin. There, the egg develops into the larval form which, when ready, burrows its way to the outside through the hide and falls to the ground. In time it matures into a fullgrown botfly. This one happened to bite a man. It was about to come out on its own, and, of course, it would have died."

The words *imposter, sorehead, servant of Satan* sprang to my lips. But now he has been joined by other scientists. They nod in agreement. I gaze from one gray eminence to another, and know the mallet-blow of glory pulverized. I tried to save the world, but it didn't work out.

No, it is not the surgeon who is God's darling. He is the victim of vanity. It is the poet who heals with his words, stanches the flow of blood, stills the rattling breath, applies poultice to the scalded flesh.

Did you ask me why a surgeon writes? I think it is because I wish to be a doctor.

## STUDY QUESTIONS

1. What sort of job-related reporting would you expect a surgeon to do?
2. Is Selzer's article a report (you should apply some of the principles introduced in the earlier selections)? Discuss.
3. What does he state as his purpose for writing? Would this stated purpose contradict the objectives of report writing?
4. Discuss the selection as an essay.
5. Would it be possible to rewrite the article as a report? What would we gain, and what would we lose?
6. By the conclusion, we are fully aware that when Selzer talks about writing, he does not mean work-related reporting. Yet he says that he must write in order to be a doctor. What does he mean?
7. Can you understand the engineer, technician, businessman, or scientist having a similar need to establish professional identity through writing? Is it necessary to have such motivation in order to write effective reports for your job?

# Section 2

---

# Language, Style, and Audience

Noted scientists such as Albert Einstein and P. W. Bridgman have written on how they were often forced to examine language and its communicative power in order to resolve problems that seemed at first to be purely scientific. Naturally, most students do not need to worry over the relationship between language and physics, or concern themselves with the philosophical foundations of science. Language *is* important to student writers, however, because on a practical level the ideas you wish to communicate, the plans, instructions, and analyses you wish to transmit, will be limited by the language you use in the transmission of these ideas.

Consider for a moment trying to explain to a five-year-old how an automobile works. Obviously, the limited vocabulary of the child would be the chief limitation you would have to overcome. You might be able to describe a carburetor as a "mixing bowl" of sorts, but how would you deal with the ignition system? Does the word *burn* really describe accurately the process of *combustion* that takes place within the engine? Is *torque* completely explained when we call it a "twisting force"? Assuming for the moment that you would be able to create an explanation of how an automobile works, using a vocabulary the child could understand, would the explanation have altered or distorted the facts in any way? That is, would the child understand "what really happens," or would he have obtained a "child's version" that misrepresented the facts slightly? The important issue here, of course, is not whether we shall ever to able to make children understand a highly technical world: what is important is that language itself limits our ability to frame and transmit our ideas. It may be, for example, that gorillas understand nuclear physics better than we, but since they cannot speak (at least not to us), their inability to use language is preventing this information from benefiting anyone.

Simply put, language *is* important in technical communication for it is the medium through which our thoughts flow. If we are not adept in our use of language our communication will be hampered. Skill in writing, however, does not depend solely upon possessing an adequate vocabulary or the ability to create interesting sentences. A writer must know his audience and adjust his writing to that audience's need and comprehension, an adjustment that requires an understanding of both audience analysis and style. This is an intricate assignment, but it is perhaps not as complex as it sounds, for all of us adjust our communicative style to fit the needs of our audience many times every day. Consider your own interactions with your close friends, classmates, instructors, or fellow workers. Chances are you do not act the same way around all these people, do not talk the same way around all of them. Students, for example, very frequently become "different people" when they enter a classroom; sometimes, in fact, these "classroom personalities" vary widely in each class a student takes. Actions, mannerisms, and language shift in each situation as we adjust to our surroundings and the people with whom we are interacting. There is nothing necessarily "phony" or "false" about this. The behavior and language we find in the student cafeteria is not the behavior and language of the job interview, and we would not expect it to be. The style and tone you adopt when you write a letter to a close friend does not project the same image as the style and tone of a letter you write to the president of a large corporation — unless that president is also your very close friend.

These adjustments that we tend to make naturally everday in our language and behavior also need to be made consciously in our writing. To dismiss such adjustments because they are "only linguistic" and therefore unimportant is a mistake, for language does have the power to alter and shape our perceptions and emotions. Mixing sugar and water and lemon juice might produce *sugar hydrolysis* in the laboratory, but it is only *lemonade* in our glass. An insect called the *harlequin bug* would seem to be charmingly colorful, and it is. But it is also known as a *stink bug,* and when we call it that this insect does not seem quite so charming.

Analyzing an audience incorrectly or adopting the wrong style and language for a particular occasion can cause a perfectly good report to fail in its mission. The essays in this section should help you grasp the nature of language as a communicative medium, and understand the relationships between language, style, and audience.

# Audience Analysis

## Thomas Pearsall

Thomas Pearsall, who teaches technical writing at the University of Minnesota, has
written several highly respected technical writing textbooks. The selection on
audience analysis that is reprinted here is typical of Pearsall's contributions to the
study of technical writing. It is one of the most important essays in this text.

### The Audiences in Technical Writing

Probably no other kind of writing matches technical writing in the importance —
and, in some ways, the ease — of mating a particular piece of writing to a particular
audience. Anyone who has ever picked up a piece of technical writing knows that a
report written by an expert for an expert might as well be written in a foreign
language, as far as the layman is concerned. Or, conversely, that an expert will become bored by a "popular" article in his field. He may well even be a little condescending toward it. The need for mating audience to report has long been
recognized. . . .

But perhaps you have never been told to think about the *purpose of your
audience* in reading your paper. Perhaps, to this point, your only audience has been
a teacher and his purpose, as far as you could see, was to put red marks on your
baby and give it a grade. When you move beyond the classroom, you will have to
consider why a reader wants to read your paper and the fund of knowledge he
brings to the reading.

You must consider what the reader is going to *do* with the information you
give him.

> *Is he a layman?* Then perhaps all he wants to do is read and enjoy your paper
> and file away a few interesting facts to add to his awareness of the world
> around him.
>
> *Is he an executive?* He has a profit motive. He may use your paper to make
> stock-buying decisions or to explore new markets for his company.
>
> *Is he an expert in the subject?* Then he wants new information to add to the
> large store he already has. Your information may stimulate him to further
> research or to design a new piece of equipment, or it may help him do a
> familiar job better.
>
> *Is he a technician?* He wants information that will help him understand and
> maintain the equipment the engineer or scientist has given him to work with.
>
> *Is he an operator of equipment?* Then he wants clear, unequivocal instruc-
> tions, step by step, in how to get the most out of the equipment he operates.

From Thomas Pearsall, "Audience Analysis," *Audience Analysis for Technical Writing*
(Beverly Hills: Glencoe, 1969), pp. ix–xxi.

The true expert on a subject is in an enviable yet difficult position when he sits down to write. He has a mass of information on a specific subject available to him. The difficulty? What to draw from that mass to interest, to inform, to satisfy a particular audience. . . .

As well as understanding his reader's purpose, the writer must understand his reader's knowledge. The writer must know who his reader is, what he already knows, and what he doesn't know. He must know what the reader will understand without background and without definitions. He must know what information he must elaborate, perhaps with simple analogies. He must know when he can use a specialized word or term and when he cannot. He must know when and how to define specialized words that he can't avoid using. All this is asking a great deal of the writer. But the good writer knows that each particular reader brings his experience *and his experience only* to his reading. Not to understand this is to miss the whole point of writing.

To be a good writer you must know your audience — its purpose and its knowledge.

## A Caution

Before I elaborate about the five audiences, let me caution you. No audience is uniform and classified simply into a neat category. I speak of a lay audience, or an executive audience, but these audiences are by no means totally homogeneous units. An audience is much more analogous to an aggregate of hard, sharp, different-sized rocks than it is to a mass of smooth marble.

There are laymen and there are laymen. A physicist reading a paper on biology is something of a layman because he is out of his specialty. Yet because he understands the scientific process, he is much less of a layman than a musician reading the same paper. There are executives who deal only in market research. There are executives who are personnel experts. The first is interested in a new piece of equipment because it may open up a new market. The second is interested in the same piece of equipment because it may change the way he staffs a plant.

Even at the operator level you will not find a comfortable, single audience. Here the educational level may vary tremendously. The operator may be a high school dropout trained to run a lathe and little else. Or he may be an astronaut with an M.S. in mechanical engineering who must be shown how to operate a hand-held maneuvering unit while deep in space.

Take my generalizations and narrow them as much as possible to fit the individuals you are dealing with. Fortunately, in technical writing this can often be done fairly precisely.

I speak of writing for *the executive,* thinking of the term as representing *most* executives. Someday you may be writing for *one* executive. Get to know him. Talk with him. Find out his education, his job experience. If you work at it you can know his background as well as you know your own. . . .

## The Layman

Who is the layman? He is the fourth-grade boy reading a simplified explanation of atomic fission in terms of mouse traps and ping pong balls. He is the bank clerk reading a Sunday newspaper supplement story about desalination. He is the biologist with a Ph.D. reading an article in *Scientific American* entitled "The Nature of Metals." In short, the layman is everyman (and every woman) once he is outside his own particular field of specialization.

We can make only a few generalizations about him. He is reading for interest. He is reading to tune in more accurately on the universe. He is not very expert in the field (or else he would not be reading a layman's article). Probably, his major interest is practical. He is much more concerned with what things do than in how they work. He is more interested in how computers will affect his daily life than in the fact that they work on a binary number system. He is probably more attuned to fiction and television than scientific exposition. He likes drama. For this reason the use of narration, when possible, is often an effective device when writing for the layman. Relate anecdotes and incidents to illustrate what something is and what it does.

Beyond these few simple statements the layman presents a bewildering complexity of interests, skills, educational levels, and prejudices. How then can we define exactly how to write for him? The truth is that we cannot – completely. But we can make some general statements about his needs, interests, likes, and dislikes that – to paraphrase Lincoln – apply to all of the laymen some of the time, some of the laymen all of the time, but not to all of the laymen all of the time.

To simplify matters a little, let's get a picture in our minds of what might be an average layman: He is an individual fairly bright and interested in science and technology. He has at least a high school education. He reads fairly well, and he has a smattering of mathematics and science, but he is a little vague about both subjects. How do we treat him? What approaches are best when we write for him?

### Background

To begin with, the layman needs to be given background material in the subject. We must assume he knows little or nothing about the specialty. An AEC booklet entitled *Atomic Energy in Use* is a simplified layman's explanation of how a nuclear reactor works. Chapter I gives a history of uranium, and Chapter II explains nuclear radiation. The beginning of Chapter II is instructive in how to give a layman background information. It begins:

> Light is radiation that we can see. Heat is radiation that we can feel. Radio and television waves and X-rays are electromagnetic waves of radiation that we can neither see nor feel, but with whose usefulness we are well acquainted.
>
> Now we are hearing more and more about another kind of radiation as a result of man's continuing scientific and engineering achievements.
>
> This is nuclear radiation.

Nuclear radiation consists of a stream of fast-flying particles or waves originating in and coming from the nucleus, or heart, of an atom. It is a form of energy we have come to call atomic, or nuclear, energy.[1]

The author begins with the familiar — light radiation and heat radiation. There is no pretense that the reader knows all about light and heat radiation; that he could, for example, construct mathematical models of such phenomena in the way a physicist could. Nor does the writer attempt to give the reader such complicated theory. The writer knows that the average layman dimly understands that light, heat, and electromagnetic waves are forms of energy that somehow travel from point A to point B and can therefore be *used* in various practical ways. Nuclear radiation is a similar form of radiation and that's the only point the writer wants immediately to make.

To make his point the writer has relied upon analogy — comparing the unfamiliar to the familiar. There is no better device to help the lay reader. In the everyday world around us — in light bulbs, radios, garden hoses, faucets, windows, mirrors, trees, tennis rackets, baseballs, hot air registers, clay, loam, granite, ocean waves — are countless things known and somewhat understood by the layman that the writer can use to explain about every law of science. It's a question of imagination and of being willing to talk in lay terms without being condescending.

Give the layman, then, a grounding in your subject using familiar things as points of comparison.

## Definitions

Laymen need specialized words and terms defined. There are at least two reasons why you shouldn't force a reader to go to the dictionary while he is reading your work.

First, you are the host. You have invited your reader to come to you. You owe him every courtesy and defining difficult terms is a courtesy. Remember that the layman reads primarily for interest. If you force him to the dictionary every fourth line, his interest will soon flag.

Second, if you give the definition, you can limit or expand the term in the way that is most useful to you. In Chapter II of *Atomic Energy in Use* the author's emphasis in the second half of the chapter is upon radiation control and radiation safety. His definition of an alpha particle reflects that emphasis: "Alpha particles are comparatively heavy particles given off by the nuclei of heavy radioactive materials such as uranium, thorium, and radium. They can travel about an inch in air and can readily be stopped by the skin or by a thin sheet of paper."[2]

If the writer had left the reader to look up the word himself, the emphasis wanted would have been lost. Compare this definition from *Webster's Seventh New Collegiate Dictionary* and you'll see the difference: "Alpha particle: a positively

[1] *Atomic Energy in Use.* Washington, D.C.: United States Atomic Energy Commission, 1967, p. 11.
[2] *Ibid.*, pp. 11–12.

charged nuclear particle identical with the nucleus of a helium atom that consists of 2 protons and 2 neutrons and is ejected at high speed in certain radioactive transformations." Nothing is said about the relative ease with which alpha particles can be blocked, a fact important to the writer's later emphasis on radiation control.

Define terms, then, both for courtesy and understanding and to aid your own exposition.

## Simplicity

There are several ways to keep an article simple for the layman. Two of them I have already discussed: give him needed background in a way he can understand and define those specialized terms that you must use. Most often, avoid specialized words for which you can find simple substitutes. The expert can read certain meanings into the word *homeostasis* and you should use it for him, but for the layman *stable state* or *equilibrium* will serve as well. But, another caution here. Most laymen like to enrich their vocabulary. So, don't avoid technical terms altogether. Just don't put them together in incomprehensible strings with a "reader-be-damned" attitude.

Some scientific specialties are loaded with mathematics. Others, such as biochemistry, are full of formulas and complicated charts and diagrams incomprehensible to the layman. Mathematics, formulas, and diagrams are useful shorthand expressions for the expert. He knows he can attain a precision with them that he can attain in no other way. But what the expert sometimes forgets is that he has spent years learning how to handle such precise tools. The layman, however, most likely cannot handle them. He either has had no training in them beyond high school, or he learned them so long ago that he has forgotten them.

## Conclusion

The layman is the hardest audience to write for because he is so difficult to pin down and define. We can generalize that for him we should provide ample background — without mathematics or complicated formulas — definitions, photographs and simple charts. Remember, he is reading mainly for interest and his interest is mainly practical: What effect will this scientific development have on me? Your cardinal rules in writing for the layman are to keep things uncomplicated, interesting, practical, and, if possible, a touch dramatic.

## The Executive

Much of what I said about the layman applies directly to the executive. You cannot assume that he possesses very much knowledge in the field you are writing about. While most executives have college degrees and many have technical experience, they represent many disciplines, and not necessarily the one you are writing about. Some may have little technical background but have been trained in management, psychology, social science, or the humanities.

Like the layman, the executive's chief concern is with practical matters. He is more concerned with what things do rather than in how they work. He wants to know what effects a technological development will bring. He needs simple background. However, he can probably handle and wants a bit more technical background than does the layman. James W. Souther, who has made an extensive study of the executive and his needs, suggests that writing for him be approximately on the level found in *Scientific American.*[3] The executive will want most purely technical terms defined for him. You should avoid shop jargon when addressing him. He is a busy man; don't force him to a dictionary any more than you would the layman.

When writing for the executive, write in plain language, using sentences averaging about 21 words. Avoid mathematics. Use simple graphs of the layman type: bar graphs, pie charts, and pictographs.

But while the executive resembles the lay reader in many ways, there is a significant difference. The layman reads primarily for interest. The executive is interested, too, but he has a much more vital concern. As an executive *he must make decisions based upon what he reads.* And his decisions revolve around two poles: *profits and people.* How much money will a technological development cost? What new markets will it open? How much money will a technological development make? Does the development call for restaffing a plant? Will new people have to be hired or old people retrained? There are many other ramifications in the executive's decision-making process, but people and profits are the two poles about which all decisions revolve.

### The Executive's Needs

What is the executive interested in? What questions does he want you to answer in a report written for him?

He wants to know how a new process or piece of equipment can be used. What new markets will they open up? What will they cost, and why is the cost justified? What are the alternatives?

Why was your final choice chosen over the other alternatives? Give some information about the also-rans. Convince the executive that you have explored the problem thoroughly. For all the alternatives include comments on cost, size of the project, time to completion, future costs in upkeep and replacement, and the effects on productivity, efficiency, and profits. Consider such things as new staffing, competition, experimental results, troubles to be expected. What are the risks involved?

Be honest. Remember that if your ideas are bought, they are *your* ideas. Your reputation will stand or fall on their success. Therefore, don't overstate your case. Qualify your statements where necessary.

Give your conclusions and recommendations clearly. In writing any report for

[3]James W. Souther, "What Management Wants in the Technical Report," *Journal of Engineering Education,* 52 (8), 498–503 (1962).

the executive, remember that you must interpret your material and present its implications, not merely give the facts. Souther points out that "the manager seldom uses the detail, though he often wants it available. It is the *professional judgment* of the writer that the manager wants to tap."[4] . . .

## The Expert

To many laymen, writing meant for the expert seems incredibly dull. But to the expert nothing could be more exciting. In the well-written report in his specialty, he has facts to chew — the expert is a man in love with facts. He also has inferences drawn from those facts to hail as new and true or to dispute as examples of faulty reasoning. He is akin to the symphony conductor who can read a score and *hear it* and judge its potential. The layman, like the musical audience, must wait for the interpretation by the conductor and his full orchestra before he can appreciate the same score.

Who is the expert? For our purposes I will define him as a scientist or engineer with either an M.S. or a Ph.D. in his field or a B.S. and years of experience. He may be a college professor, an industrial researcher, or an engineer who designs and builds. Whichever of these he is, he knows his field intimately. When he reads he seldom looks for background information. Rather, he is looking for a new body of information, new conclusions, or perhaps new and better techniques to help him in his work. . . .

## The Technician

The technician is the man at the heart of any operation. He is the man who finally brings the scientist's imaginative research and the engineer's calculations and drawings to life. He builds equipment and after it is built he maintains it. He is an intensely practical man, perhaps with years of experience behind him. He is the one who can say, "you know, sir, if we used a wing nut here, instead of a hexagonal, the operator would have a lot easier job getting that plate on and off," and he'll be right. He's a man well worth listening to and certainly he is a man for whom you should write well.

The technician's educational level varies. Most typically he will range anywhere from a high school graduate to a junior engineer with a B.S. The high school graduate will probably have a great deal of on-the-job training and experience. The junior engineer may be better trained in theory but have less practical experience. The technician has limitations. He most likely will not be able to follow complicated mathematics and he'll grow restive with too much theory. Also you will want to take care with your sentence length when writing for the technician. Research cited by Rudolf Flesch indicates that sentences over 17 words long can be classified as difficult.[5] Other research indicates that high school students write

[4] *Ibid.*

sentences 17 to 19 words long.[6] Unless you are sure that your technicians are college trained, give them sentences that they are used to and can handle — about 17 words or less on the average.

## The Operator

In many ways, the operator is a cross between the layman and the technician. Like the technician, he works with technical equipment, operating it and sometimes performing simple maintenance on it. Like the layman, he may have little real technical and scientific knowledge. Also, like laymen, operators may come from many educational levels. For example, some of the divers in undersea exploration are Naval petty officers with a high school education. Others are oceanographers with Ph.D.'s. Both groups need equally simple instructions in how to operate Naval diving equipment.

Because of the similarities between operators, technicians, and laymen, much of what I have said already applies in this section as well. I will here merely attempt to point up some of the things that apply primarily to writing for operators.

### Background

Normally you must assume that the operator brings no necessary background to reading an operator's manual. You must furnish the background. Considering that the operator is essentially a layman, the background must be kept simple. Use analogy and define all or most technical terms. Avoid mathematics altogether. Normally, avoid sending the operator to other manuals for additional information. The operator's manual should be self-contained.

Some operator's manuals are mere lists of instructions containing no background at all. In these cases the writer considers background unessential to the performance of the job. In other cases, perhaps where the typical operator represents a higher educational level or the task is more than usually complex, a good deal of background is provided.

## STUDY QUESTIONS

1. Pearsall identifies five different technical writing audiences. Choose any essay from this book and try to distinguish which audience it was written for. What aspects of the article help you make this decision?
2. As a student writer, which audience do you most frequently find yourself addressing? How might you, as a student, gain experience in writing for all the audiences Pearsall describes?
3. *Writing Assignment:* Choose an article that is written for an expert. Rewrite

---

[5] Rudolf Flesch, *The Art of Plain Talk.* New York: Harper and Brothers, 1946, p. 38.
[6] Porter G. Perrin and George H. Smith, *Handbook of Current English.* New York: Scott, Foresman and Company, 1955, p. 211.

part of this article, addressing the new version to a layman. What changes have you made? Next, choose an article that is written for a layman. Rewrite part of this article, addressing the new version to an expert. Now what kinds of changes have you made? Is it possible to "translate" any essay in this way? What limitations (if any) do you forsee?

# Gobbledygook

## Stuart Chase

Some writers seem to believe that writing about complex topics requires the adoption of a complex style. *Gobbledygook,* the stilted, elaborate style that often masquerades as serious prose, is the subject of this discussion by Stuart Chase.

Said Franklin Roosevelt, in one of his early presidential speeches: "I see one-third of a nation ill-housed, ill-clad, ill-nourished." Translated into standard bureaucratic prose his statement would read:

> It is evident that a substantial number of persons within the Continental boundaries of the United States have inadequate financial resources with which to purchase the products of agricultural communities and industrial establishments. It would appear that for a considerable segment of the population, possibly as much as 33.3333* of the total, there are inadequate housing facilities, and an equally significant proportion is deprived of the proper types of clothing and nutriment.
> *Not carried beyond four places.

This rousing satire on gobbledygook — or talk among the bureaucrats — is adapted from a report[1] prepared by the Federal Security Agency in an attempt to break out of the verbal squirrel cage. "Gobbledygook" was coined by an exasperated Congressman, Maury Maverick of Texas, and means using two, or three, or ten words in the place of one, or using a five-syllable word where a single syllable would suffice. Maverick was censuring the forbidding prose of executive departments in Washington, but the term has now spread to windy and pretentious language in general.

"Gobbledygook" itself is a good example of the way a language grows. There

[1]This and succeeding quotations from F.S.A. report by special permission of the author, Milton Hall.
    Stuart Chase, "Gobbledygook," *Power of Words* (New York: Harcourt Brace Jovanovich, 1954), pp. 249–259.

was no word for the event before Maverick's invention; one had to say: "You know, that terrible, involved, polysyllabic language those government people use down in Washington." Now one word takes the place of a dozen.

A British member of Parliament, A. P. Herbert, also exasperated with bureaucratic jargon, translated Nelson's immortal phrase, "England expects every man to do his duty":

> England anticipates that, as regards the current emergency, personnel will face up to the issues, and exercise appropriately the functions allocated to their respective occupational groups.

A New Zealand official made the following report after surveying a plot of ground for an athletic field:[2]

> It is obvious from the difference in elevation with relation to the short depth of the property that the contour is such as to preclude any reasonable developmental potential for active recreation.

Seems the plot was too steep.

An office manager sent this memo to his chief:

> Verbal contact with Mr. Blank regarding the attached notification of promotion has elicited the attached representation intimating that he prefers to decline the assignment.

Seems Mr. Blank didn't want the job.

> A doctor testified at an English trial that one of the parties was suffering from "circumorbital haematoma."

Seems the party had a black eye.

> In August 1952 the U.S. Department of Agriculture put out a pamphlet entitled: "Cultural and Pathogenic Variability in Single-Condial and Hyphal-tip Isolates of Hemlin-Thosporium Turcicum Pass."

Seems it was about corn leaf disease.

On reading the top of the Finsteraarhorn in 1845, M. Dollfus-Ausset, when he got his breath, exclaimed:

> The soul communes in the infinite with those icy peaks which seem to have their roots in the bowels of eternity.

Seems he enjoyed the view.

A government department announced:

> Voucherable expenditures necessary to provide adequate dental treatment required as adjunct to medical treatment being rendered a pay patient in in-

[2]This item and the next two are from the piece on gobbledygook by W. E. Farbstein, New York *Times,* March 29, 1953.

patient status may be incurred as required at the expense of the Public Health Service.

Seems you can charge your dentist bill to the Public Health Service. Or can you?

## Legal Talk

Gobbledygook not only flourishes in government bureaus but grows wild and lush in the law, the universities, and sometimes among the literati. Mr. Micawber was a master of gobbledygook, which he hoped would improve his fortunes. It is almost always found in offices too big for face-to-face talk. Gobbledygook can be defined as squandering words, packing a message with excess baggage and so introducing semantic "noise." Or it can be scrambling words in a message so that meaning does not come through. The directions on cans, bottles, and packages for putting the contents to use are often a good illustration. Gobbledygook must not be confused with double talk, however, for the intentions of the sender are usually honest.

I offer you a round fruit and say, "Have an orange." Not so an expert in legal phraseology, as parodied by editors of *Labor:*

> I hereby give and convey to you, all and singular, my estate and interests, right, title, claim and advantages of and in said orange, together with all rind, juice, pulp and pits, and all rights and advantages therein . . . anything hereinbefore or hereinafter or in any other deed or deeds, instrument or instruments of whatever nature or kind whatsoever, to the contrary, in any wise, notwithstanding.

The state of Ohio, after five years of work has redrafted its legal code in modern English, eliminating 4,500 sections and doubtless a blizzard of "whereases" and "hereinafters." Legal terms of necessity must be closely tied to their referents, but the early solons tried to do this the hard way, by adding synonyms. They hoped to trap the physical event in a net of words, but instead they created a mumbo-jumbo beyond the power of the layman, and even many a lawyer, to translate. Legal talk is studded with tautologies, such as "cease and desist," "give and convey," "irrelevant, incompetent, and immaterial." Furthermore, legal jargon is a dead language; it is not spoken and it is not growing. An official of one of the big insurance companies calls their branch of it "bafflegab." Here is a sample from his collection:[3]

> One-half to his mother, if living, if not to his father, and one-half to his mother-in-law, if living, if not to his mother, if living, if not to his father. Thereafter payment is to be made in a single sum to his brothers. On the one-half payable to his mother, if living, if not to his father, he does not bring in his mother-in-law as the next payee to receive, although on the one-half to his mother-in-law, he does bring in the mother or father.

[3] Interview with Clifford B. Reeves by Sylvia F. Porter, New York *Evening Post,* March 14, 1952.

You apply for an insurance policy, pass the tests, and instead of a straight-forward "here is your policy," you receive something like this:

> This policy is issued in consideration of the application therefor, copy of which application is attached hereto and made part hereof, and of the payment for said insurance on the life of the above-named insured.

## Academic Talk

The pedagogues may be less repetitious than the lawyers, but many use even longer words. It is a symbol of their calling to prefer Greek and Latin derivatives to Anglo-Saxon. Thus instead of saying: "I like short clear words," many a professor would think it more seemly to say: "I prefer an abbreviated phraseology, distinguished for its lucidity." Your professor is sometimes right, the longer word may carry the meaning better — but not because it is long. Allen Upward in his book *The New Word* warmly advocates Anglo-Saxon English as against what he calls "Mediterranean" English, with its polysyllables built up like a skyscraper.

Professional pedagogy, still alternating between the Middle Ages and modern science, can produce what Henshaw Ward once called the most repellent prose known to man. It takes an iron will to read as much as a page of it. Here is a sample of what is known in some quarters as "pedageese":

> Realization has grown that the curriculum or the experiences of learners change and improve only as those who are most directly involved examine their goals, improve their understandings and increase their skill in performing the tasks necessary to reach newly defined goals. This places the focus upon teacher, lay citizen and learner as partners in curricular improvement and as the individuals who must change, if there is to be curriculum change.

I think there is an idea concealed here somewhere. I think it means: "If we are going to change the curriculum, teacher, parent, and student must all help." The reader is invited to get out his semantic decoder and check on my translation. Observe there is no technical language in this gem of pedageese, beyond possibly the word "curriculum." It is just a simple idea heavily ententeverbalized.

In another kind of academic talk the author may display his learning to conceal a lack of ideas. A bright instructor, for instance, in need of prestige may select a common sense proposition for the subject of a learned monograph — say, "Modern cities are hard to live in" and adorn it with imposing polysyllables: "Urban existence in the perpendicular declivities of megalopolis . . ." et cetera. He coins some new terms to transfix the reader — "mega-decibel" or "strato-cosmopolis" — and works them vigorously. He is careful to add a page or two of differential equations to show the "scatter." And then he publishes, with 147 footnotes and a bibliography to knock your eye out. If the authorities are dozing, it can be worth an associate professorship.

While we are on the campus, however, we must not forget that the technical language of the natural sciences and some terms in the social sciences, forbidding as they may sound to the layman, are quite necessary. Without them, specialists could not communicate what they find. Trouble arises when experts expect the un-initiated to understand the words; when they tell the jury, for instance, that the defendant is suffering from "circumorbital haematoma."

Here are two authentic quotations. Which was written by a distinguished modern author, and which by a patient in a mental hospital? You will find the answer at the end of the chapter.

1. Have just been to supper. Did not knowing what the woodchuck sent me here. How when the blue blue blue on the said anyone can do it that tries. Such is the presidential candidate.
2. No history of a family to close with those and close. Never shall he be alone to be alone to be alone to be alone to be alone to lend a hand and leave it left and wasted.

## Reducing the Gobble

As government and business offices grow larger, the need for doing something about gobbledygook increases. Fortunately the biggest office in the world is working hard to reduce it. The Federal Security Agency in Washington,[4] with nearly 100 million clients on its books, began analyzing its communication lines some years ago, with gratifying results. Surveys find trouble in three main areas: correspondence with clients about their social security problems, office memos, official reports.

Clarity and brevity, as well as common humanity, are urgently needed in this vast establishment which deals with disability, old age, and unemployment. The surveys found instead many cases of long-windedness, foggy meanings, clichés, and singsong phrases, and gross neglect of the reader's point of view. Rather than talking to a real person, the writer was talking to himself. "We often write like a man walking on stilts."

Here is a typical case of long-windedness:

*Gobbledygook as found:* "We are wondering if sufficient time has passed so that you are in a position to indicate whether favorable action may now be taken on our recommendation for the reclassification of Mrs. Blank, junior clerk-stenographer, CAF 2, to assistant clerk-stenographer CAF 3?"
*Suggested improvement:* "Have you yet been able to act on our recommendation to reclassify Mrs. Blank?"

Another case:

Although the Central Efficiency Rating Committee recognizes that there are many desirable changes that could be made in the present efficiency rating system in order to make it more realistic and more workable than it now is, this committee is of the opinion that no further change should be made in the present system during the current year. Because of conditions prevailing

[4] Now the Department of Health, Education, and Welfare.

throughout the country and the resultant turnover in personnel, and difficulty in administering the Federal programs, further mechanical improvement in the present rating system would require staff retraining and other administrative expense which would seem best withheld until the official termination of hostilities, and until restoration of regular operations.

The F.S.A. invites us to squeeze the gobbledygook out of this statement. Here is my attempt:

The Central Efficiency Rating Committee recognizes that desirable changes could be made in the present system. We believe, however, that no change should be attempted until the war is over.

This cuts the statement from 111 to 30 words, about one-quarter of the original, but perhaps the reader can do still better. What of importance have I left out?

Sometimes in a book which I am reading for information — not for literary pleasure — I run a pencil through the surplus words. Often I can cut a section to half its length with an improvement in clarity. Magazines like *The Reader's Digest* have reduced this process to an art. Are long-windedness and obscurity a cultural lag from the days when writing was reserved for priests and cloistered scholars? The more words and the deeper the mystery, the greater their prestige and the firmer the hold on their jobs. And the better the candidate's chance today to have his doctoral thesis accepted.

The F.S.A. surveys found that a great deal of writing was obscure although not necessarily prolix. Here is a letter sent to more than 100,000 inquirers, a classic example of murky prose. To clarify it, one needs to *add* words, not cut them:

In order to be fully insured, an individual must have earned $50 or more in covered employment for as many quarters of coverage as half the calendar quarters elapsing between 1936 and the quarter in which he reaches age 65 or dies, whichever first occurs.

Probably no one without the technical jargon of the office could translate this: nevertheless, it was sent out to drive clients mad for seven years. One poor fellow wrote back: "I am no longer in covered employment. I have an outside job now."

Many words and phrases in officilese seem to come out automatically, as if from lower centers of the brain. In this standardized prose people never *get jobs,* they "secure employment": *before* and *after* become "prior to" and "subsequent to"; one does not *do,* one "performs"; nobody *knows* a thing, he is "fully cognizant"; one never *says,* he "indicates." A great favorite at present is "implement."

Some charming boners occur in this talking-in-one's-sleep. For instance:

The problem of extending coverage to all employees, regardless of size, is not as simple as surface appearances indicate.
Though the proportions of all males and females in ages 16–45 are essentially the same . . .
Dairy cattle, usually and commonly embraced in dairying . . .

In its manual to employees, the F.S.A. suggests the following:

*Instead of*                                    *Use*
give consideration to . . . . . . . . . . . . . consider
make inquiry regarding. . . . . . . . . . . . inquire
is of the opinion . . . . . . . . . . . . . . . . believes
comes into conflict with . . . . . . . . . . . conflicts
information which is of a
confidential nature . . . . . . . . . . . . . . confidential information

Professional or office gobbledygook often arises from using the passive rather than the active voice. Instead of looking you in the eye, as it were, and writing "This act requires . . ." the office worker looks out of the window and writes: "It is required by this statute that . . ." When the bureau chief says, "We expect Congress to cut your budget," the message is only too clear; but usually he says, "It is expected that the departmental budget estimates will be reduced by Congress."

> *Gobbled:* "All letters prepared for the signature of the Administrator will be single spaced."
> *Ungobbled:* "Single space all letters for the Administrator." (Thus cutting 13 words to 7.)

## Only People Can Read

The F.S.A. surveys pick up the point, stressed in Chapter 15, that human communication involves a listener as well as a speaker. Only people can read, though a lot of writing seems to be addressed to beings in outer space. To whom are you talking? The sender of the officialese message often forgets the chap on the other end of the line.

A woman with two small children wrote the F.S.A. asking what she should do about payments, as her husband had lost his memory. "If he never gets able to work," she said, "and stays in an institution would I be able to draw any benefits? . . . I don't know how I am going to live and raise my children since he is disable to work. Please give me some information. . . ."

To this human appeal, she received a shattering blast of gobbledygook, beginning, "State unemployment compensation laws do not provide any benefits for sick or disabled individuals . . . in order to qualify an individual must have a certain number of quarters of coverage . . ." et cetera, et cetera. Certainly if the writer had been thinking about the poor woman he would not have dragged in unessential material about old-age insurance. If he had pictured a mother without means to care for her children, he would have told her where she might get help — from the local office which handles aid to dependent children, for instance.

Gobbledygook of this kind would largely evaporate if we thought of our messages as two way — in the above case, if we pictured ourselves talking on the doorstep of a shabby house to a woman with two children tugging at her skirts, who in her distress does not know which way to turn.

### Results of the Survey

The F.S.A. survey showed that office improvements could be cut 20 to 50 percent, with an improvement in clarity and a great saving to taxpayers in paper and payrolls.

A handbook was prepared and distributed to key officials.[5] They read it, thought about it, and presently began calling section meetings to discuss gobbledygook. More booklets were ordered, and the local output of documents began to improve. A Correspondence Review Section was established as a kind of laboratory to test murky messages. A supervisor could send up samples for analysis and suggestions. The handbook is now used for training new members; and many employees keep it on their desks along with the dictionary. Outside the Bureau some 25,000 copies have been sold (at 20 cents each) to individuals, governments, business firms, all over the world. It is now used officially in the Veterans Administration and in the Department of Agriculture.

The handbook makes clear the enormous amount of gobbledygook which automatically spreads in any large office, together with ways and means to keep it under control. I would guess that at least half of all the words circulating around the bureaus of the world are "irrelevant, incompetent, and immaterial" — to use a favorite legalism; or are just plain "unnecessary" — to ungobble it.

My favorite story of removing the gobble from gobbledygook concerns the Bureau of Standards at Washington. I have told it before but perhaps the reader will forgive the repetition. A New York plumber wrote the Bureau that he had found hydrochloric acid fine for cleaning drains, and was it harmless? Washington replied: "The efficacy of hydrochloric acid is indisputable, but the chlorine residue is incompatible with metallic permanence."

The plumber wrote back that he was mighty glad the Bureau agreed with him. The Bureau replied with a note of alarm: "We cannot assume responsibility for the production of toxic and noxious residues with hydrochloric acid, and suggest that you use an alternate procedure." The plumber was happy to learn that the Bureau still agreed with him.

Whereupon Washington exploded: "Don't use hydrochloric acid; it eats hell out of the pipes!"

### STUDY QUESTIONS
1. What chase calls *gobbledygook* has also been called *bureaucratese, governmentese,* and *academese.* Why does this kind of language seem to orignate and prosper in government and education?
2. What corrective measures can be taken to prevent the spread of gobbledygook?
3. Chase provides some examples of gobbledygook that he has clarified. Find some examples of your own and help stamp out gobbledygook by revising them.

*Note:* The second quotation on page 53 comes from Gertrude Stein's *Lucy Church Amiably.*
[5]By Milton Hall.

4. Simple contracts such as insurance policies and loan agreements have traditionally been written with a lot of gobbledygook (according to the layman who
attempt to understand them). Today, however, some companies are rewriting
such documents so that an average person can understand the agreement he is
signing. The advantages of these more easily understood contracts are clear. Are
there any disadvantages to them? Why or why not?

# How to Write Like a Social Scientist

## Samuel T. Williamson

Today we frequently hear that there are many people in our society who do not
know how to write well. We usually believe that poor writing is the end product of
poor education. The more education one has, we assume, the more literate one becomes. This article by Samuel T. Williamson, however, exposes a flaw in this
assumption, by demonstrating that poor writing, paradoxically, may frequently be
the product of an excellent education.

During my years as an editor, I have seen probably hundreds of job applicants who
were either just out of college or in their senior years. All wanted "to write." Many
brought letters from their teachers. But I do not recall one letter announcing that
its bearer could write what he wished to say with clarity and directness, with economy of words, and with pleasing variety of sentence structure.

Most of these young men and women could not write plain English. Apparently their noses had not been rubbed in the drudgery of putting one simple,
well-chosen word behind the other. If this was true of teachers' pets, what about
the rest? What about those going into business and industry? Or those going into
professions? What about those who remain at college — first for a Master of Arts
degree, then an instructorship combined with work for a Ph.D., then perhaps an
assistant professorship, next a full professorship and finally, as an academic crown
of laurel, appointment as head of a department or as dean of a faculty?

Certainly, faculty members of a front-rank university should be better able
to express themselves than those they teach. Assume that those in the English department have this ability: Can the same be said of the social scientists — economists, sociologists, and authorities on government? We need today as we never
needed so urgently before all the understanding they can give us of problems of

Samuel T. Williamson, "How to Write Like a Social Scientist," *Saturday Review*,
October 4, 1947.

earning a living, caring for our fellows, and governing ourselves. Too many of them, I find, can't write as well as their students.

I am still convalescing from over-exposure some time ago to products of the academic mind. One of the foundations engaged me to edit the manuscripts of a socio-economic research report designed for the thoughtful citizen as well as for the specialist. My expectations were not high — no deathless prose, merely a sturdy, no-nonsense report of explorers into the wilderness of statistics and half-known fact. I knew from experience that economic necessity compels many a professional writer to be a cream-skimmer and a gatherer of easily obtainable material; for unless his publisher will stand the extra cost, he cannot afford the exhaustive investigation which endowed research makes possible. Although I did not expect fine writing from a trained, professional researcher, I did assume that a careful fact-finder would write carefully.

And so, anticipating no literary treat, I plunged into the forest of words of my first manuscript. My weapons were a sturdy eraser and several batteries of sharpened pencils. My armor was a thesaurus. And if I should become lost, a near-by public library was a landmark, and the Encyclopedia of Social Sciences on its reference shelves was an ever-ready guide.

Instead of big trees, I found underbrush. Cutting through involved, lumbering sentences was bad enough, but the real chore was removal of the burdocks of excess verbiage which clung to the manuscript. Nothing was big or large; in my author's lexicon, it was "substantial." When he meant "much," he wrote "to a substantially high degree." If some event took place in the early 1920's, he put it "in the early part of the decade of the twenties." And instead of "that depends," my author wrote, "any answer to this question must bear in mind certain peculiar characteristics of the industry."

So it went for 30,000 words. The pile of verbal burdocks grew — sometimes twelve words from a twenty-word sentence. The shortened version of 20,000 words was perhaps no more thrilling than the original report; but it was terser and crisper. It took less time to read and it could be understood quicker. That was all I could do. As S. S. McClure once said to me, "An editor can improve a manuscript, but he cannot put in what isn't there."

I did not know the author I was editing; after what I did to his copy, it may be just as well that we have not met. Aside from his cat-chasing-its-own-tail verbosity, he was a competent enough workman. Apparently he is well thought of. He has his doctorate, he is a trained researcher and a pupil of an eminent professor. He has held a number of fellowships and he has performed competently several jobs of economic research. But, after this long academic preparation for what was to be a life work, it is a mystery why so little attention was given to acquiring use of simple English.

Later, when I encountered other manuscripts, I found I had been too hard on this promising Ph.D. Tone-deaf as he was to words, his report was a lighthouse of clarity among the chapters turned in by his so-called academic betters. These brethren — and sister'n — who contributed the remainder of the foundation's study were professors and assistant professors in our foremost colleges and universities. The names of one or two are occasionally in newspaper headlines. All of them had, as the professorial term has it, "published."

Anyone who edits copy, regardless of whether it is good or bad, discovers in a manuscript certain pet phrases, little quirks of style and other individual traits of its author. But in the series I edited, all twenty reports read alike. Their words would be found in any English dictionary, grammar was beyond criticism, but long passages in these reports demanded not editing but actual translation. For hours at a time, I foundered in brier patches like this: "In eliminating wage changes due to purely transitory conditions, collective bargaining has eliminated one of the important causes of industrial conflict, for changes under such conditions are almost always followed by a reaction when normal conditions appear."

I am not picking on my little group of social scientists. They are merely members of a caste; they are so used to taking in each other's literary washing that it has become a habit for them to clothe their thoughts in the same smothering verbal garments. Nor are they any worse than most of their colleagues, for example:

> In the long run, developments in transportation, housing, optimum size of plant, etc., might tend to induce an industrial and demographic pattern similar to the one that consciousness of vulnerability would dictate. Such a tendency might be advanced more effectively if the causes of urbanization had been carefully studied.

Such pedantic Choctaw may be all right as a sort of code language or shorthand of social science to circulate among initiates, but its perpetrators have no right to impose it on others. The tragedy is that its users appear to be under the impression that it is good English usage.

Father, forgive them; for they know not what they do! There once was a time when everyday folk spoke one language, and learned men wrote another. It was called the Dark Ages. The world is in such a state that we may return to the Dark Ages if we do not acquire wisdom. If social scientists have answers to our problems yet feel under no obligation to make themselves understood, then we laymen must learn their language. This may take some practice, but practice should become perfect by following six simple rules of the guild of social science writers. Examples which I give are sound and well tested, they come from manuscripts I edited.

> RULE 1 *Never use a short word when you can think of a long one.* Never say "now" but "currently." It is not "soon" but "presently." You did not have "enough" but a "sufficiency." Never do you come to the "end" but to the "termination." This rule is basic.

> RULE 2 *Never use one word when you can use two or more.* Eschew "probably." Write, "it is improbable," and raise this to "it is not improbable." Then you'll be able to parlay "probably" into "available evidence would tend to indicate that it is not unreasonable to suppose."

RULE 3 *Put one-syllable thought into polysyllabic terms.* Instead of observing that a work force might be bigger and better, write, "In addition to quantitative enlargement, it is not improbable that there is need also for qualitative improvement in the personnel of the service." If you have discovered that musicians out of practice can't hold jobs, report that "the fact of rapid deterioration of musical skill when not in use soon converts the employed into the unemployable." Resist the impulse to say that much men's clothing is machine made. Put it thus: "Nearly all operations in the industry lend themselves to performance by machine, and all grades of men's clothing sold in significant quantity involve a very substantial amount of machine work."

RULE 4 *Put the obvious in terms of the unintelligible.* When you write that "the product of the activity of janitors is expended in the identical locality in which that activity takes place," your lay reader is in for a time of it. After an hour's puzzlement, he may conclude that janitors' sweepings are thrown on the town dump. See what you can do with this: "Each article sent to the cleaner is handled separately." You become a member of the guild in good standing if you put it like this: "Within the cleaning plant proper the business of the industry involves several well-defined processes, which, from the economic point of view, may be characterized simply by saying that most of them require separate handling of each individual garment or piece of material to be cleaned."

RULE 5 *Announce what you are going to say before you say it.* This pitcher's wind-up technique before hurling towards — not at — home plate has two varieties. First is the quick wind-up: "In the following sections the policies of the administration will be considered." Then you become strong enough for the contortionist wind-up: "Perhaps more important, therefore, than the question of what standards are in a particular case, there are the questions of the extent of observance of these standards and the methods of their enforcement." Also, you can play with reversing Rule 5 and *say what you have said after you have said it.*

RULE 6 *Defend your style as "scientific."* Look down on — not up to — clear, simple English. Sneer at it as "popular." Scorn it as "journalistic." Explain your failure to put more mental sweat into your writing on the ground that "the social scientists who want to be scientific believe that we can have scientific description of human behavior and trustworthy predictions in the scientific sense only as we build adequate taxonomic systems for observable phenomena and symbolic systems for the manipulation of ideal and abstract entities."

For this explanation I am indebted to Lyman Bryson in an *SRL [Saturday Review of Literature]* article (Oct. 13, 1945) "Writers: Enemies of Social Science." Standing on ground considerably of his own choosing, Mr. Bryson argued against judging social science writing by literary standards.

Social scientists are not criticized because they are not literary artists. The trouble with social science does not lie in its special vocabulary. Those words are doubtless chosen with great care. The trouble is that too few social scientists take enough care with words outside their special vocabularies.

It is not too much to expect that teachers should be more competent in the art of explanation than those they teach. Teachers of social sciences diligently try to acquire knowledge; too few of them exert themselves enough to impart it intelligently.

Too long has this been excused as "the academic mind." It should be called by what it is: intellectual laziness and grubby-mindedness.

## STUDY QUESTIONS

1. Williamson focuses on social scientists. Look through your textbooks from the social sciences (and other fields as well). Has the linguistic situation changed any in the years since 1947 when this essay was written?
2. Williamson's rules are not to be taken seriously (unless you really do wish to write like a social scientist). Remove the satire and write rules so they state what Williamson really believes a good writer should do. Do you presently violate any of these rules in your own reports?

# The Capacity to Generate Language Viability Destruction

## Edwin Newman

At one time Edwin Newman was best known as a television newsman. In recent years, however, his books on the decline of English have also made him one of the country's more prominent authorities on the use and misuse of language. The following excerpt is taken from his first best-selling book *Strictly Speaking.*

The business instinct is by no means to be sneered at. I have had only one money-making idea in my life. It came to me like a flash (though not from Mr. Tash, the manager of a jewelry store in Washington, D.C., after the Second World War, and the inspiration of a radio commercial which began, "Now here's a flash from Mr. Tash — If you'll take a chance on romance, then I'll take a chance on you," meaning that he sold engagement and wedding rings on credit). It came to me like a flash one day when I was thinking about the growth of the population and the domination of American life by the automobile.

I fell to wondering, as any red-blooded American would, how some money might be made from that combination of factors, and I conceived the idea that be-

Edwin Newman, "The Capacity to Generate Language Viability Destruction," *Strictly Speaking* (New York: Warner Books, 1974), pp. 155–177.

cause walking as a pleasure was becoming a lost art, a great deal of money might be made by setting up a pedestrians' sanctuary, a place where people could walk. I saw in my mind's eye the name *Walkorama*, or *Strollateria*, or something of the sort, and a place that would require little in the way of outlay or upkeep — just some space, grass, trees, and quiet. Obviously it would need a parking lot so that people could drive to it and park their cars before entering the walkorama to walk, and I intended to hold on to the parking concession for myself.

Nothing came of it. It was a typically footless newsman's dream, like the little weekly with which to get back to real people, dispense serene wisdom, and go broke, in Vermont.

I do not, therefore, sneer at the men and women of business. If they were not buying time on NBC, the world might or might not be a poorer place, but I would unquestionably be a poorer inhabitant of it.

However, the contributions of business to the health of the language have not been outstanding. Spelling has been assaulted by Duz, and E-Z Off, and Fantastik, and Kool and Arrid and Kleen, and the tiny containers of milk and cream catchily called the Pour Shun, and by products that make you briter, so that you will not be left hi and dri at a parti, but made welkom. . . .

In many such monstrosities, the companies involved know what they are doing. In others they often do not, especially when it is a matter of grammar. New York remains the business capital of the United States, and on a typical day there you may pick up the *New York Times,* or that paragon of eastern sophistication, the *New Yorker* magazine, and find a well-known Fifth Avenue jeweler telling the world that "The amount of prizes Gübelin has won are too numerous to be pure chance." I happen to know that this was a straw man Gübelin was knocking down because nobody had said it were pure chance. The sentiment in the circles I travel in was that the amount of prizes were fully deserved.

In the same advertisement Gübelin also gives us the following: "Sculpture II, an 18 carat white-gold ring with 24 diamond baguettes and two smoky quartzes, fancy cut, is a unique work of art to be worn on one finger, and without doubt rightly among the Gübelin creations that have taken the Diamonds-International Award." Turning the word rightly into a verb is no small achievement, but it should have been rightlies, so that the advertisement would read, ". . . and without doubt rightlies among the Gübelin creations that have taken the Diamonds-International Award."

Another possible verb is gübelin. "I have gübelined," he said, hanging his head, "and I no longer rightly among you, winners all of the Diamonds-International Award." He turned and walked falteringly toward the door.

"For a moment it seemed that the high priest, or Tiffany, was about to forgive him, but it was not to be. 'Go,' the Tiffany said, pointing to the outer darkness, 'go and gübelin no more.'". . .

Most business language is not so evocative. It is simply wrong. Gulf Oil used to speak of "one of the most unique roadways ever built," which of course helped

Gulf to be ready for what it so long claimed to be ready for — "Whatever the work there is to be done." TWA has long had it Amarillio, not Amarillo, Texas; B. Altman in New York advertises sweaters that are "definitly for a young junior"; Bergdorf Goodman makes it known that "an outstanding selection of luxurious furs are now available at tremendous reductions"; Cartier believes that a memorandum pad, a stationery holder, and a pencil cup make a triumverate; the Great Lakes Mink Association wrote a letter to a New York store, the Tailored Woman, referring to its "clientel," and the Tailored Woman was happy to print it in an advertisement, though I do not say that this is what caused the Tailored Woman to close down; the chain of men's stores, Broadstreet's, capitalizing on the growing interest in food, tried to sell some of its wares by spreading the word that "Good taste is creme sengelese soup in a mock turtleneck shirt from Broadstreet's," but the number of people in New York familiar with, or curious about, sengelese cooking must have been small, and even the later announcement, "We shrunk the prices on our premium men's stretch hose," did not keep Broadstreet's from disappearing from the New York scene. Hunting World, a New York shop, sells Ella Phant, "pride and joy of the Phant family," and say, "She's only 9″ tall, and every little people you know will love her and you will too." Perhaps that depends on the kind of little people you know. Every little people that some of us know probably would be more interested in the Selig Imperial Oval Sofa, advertised by the Selig Manufacturing Company of Leominster, Massachusetts, which noted that "an orgy of 18 pillows, all shapes and colors, make a self-contained environment." An orgy do a lot of other things also.

Business language takes many forms. Camaraderie: "Us Tareyton smokers would rather fight than switch." Pomposity: When Morgan Guaranty Trust announced that negotiable securities worth $13,000,000 were missing from its vaults, it said, "A thorough preliminary search for the securities has been made, and a further search is now being made." All it needed to say was, "We're looking for them" — if indeed it couldn't expect its distinguished clients to take for granted that it was looking.

Pseudo science: "You are about to try the most technologically advanced shaving edge you can buy. Wilkinson Sword, with a world-wide reputation for innovation, brings you still another advance in razor blade technology, the first third-generation stainless steel blade. First, a microscopically thin layer of pure chromium is applied to the finely ground and stropped edge. Then, another layer of a specially developed chromium compound is applied. This special layer of chromium compound adds extra qualities of hardness, durability and corrosion resistance. Finally, a thin polymer film is coated onto the edge. This coating allows the blade to glide smoothly and comfortably over your face." Shaving seems an inadequate employment for so distinguished a product of razor blade technology, but even technology cannot hold back the dawn, and the razor is going the way of the reaper and the cotton gin. We are now invited to use the Trac-2 shaving system, which apparently is to the razor as the weapons system is to the bow and arrow. Much more of this might make you want to use the first third-generation stainless steel blades, or even the Trac-2, to slit your throat.

Stainless steel I may not be, but I was the first third-generation American in my family, on either trac, to hear life jackets carried on airliners referred to as articles of comfort. It was on a flight from London to New York in 1966, and the stewardess began her little lecture by saying," Because of our interest in your comfort, we will now demonstrate your life jackets." It was a wonderful notion, classifying the gadget to be used after a plane has gone down in the North Atlantic as part of the comfort of flying. Euphemistic business language can go no further. Only calling used cars preowned has, in my experience, equaled it. . . .

When business turns its attention from customers to shareholders, the change in tone is drastic. Customers must be tempted and/or bullied; shareholders must be impressed and intimidated, wherefore the annual corporate reports. Something like six or seven thousand of these are issued every year, but the language is so nearly uniform that they may all be written by a single team, as paperback pornographic novels are written wholesale in porno novel factories. (I was about to say sweatshops, but I assume that for reasons already made clear, the sweatshop either is no more or exists only where perverseness bordering on un-Americanism lingers on.)

In the pornos, what counts is the detailed description of sexual enterprise. In corporate reports it is growth, which at the very least should be significant, and with any luck at all will be substantial. The ultimate for growth is to be dynamic. Whether it is, and whether it occurs at all, depends largely on growth opportunities; if they occur often enough, a consistent growth pattern may be achieved, brought about, perhaps, by an upward impetus that makes things move not merely fast but at an accelerated rate.

No company can grow, of course, without having a growth potential. To realize that potential, the company must have capabilities: overall capabilities, systems capabilities, flexible capabilities, possibly nuclear services capabilities, generating capabilities, environmental control capabilities, predictability capabilities. If all of these are what they should be, and the company's vitality, viability, and critical reliability are what *they* should be, the growth potential will be realized, and profitability should result.

There are, however, other factors that must mesh. Outlooks, solutions, and systems must be sophisticated, or, if possible, highly sophisticated or optimal. Innovative products are requisite; they, in turn, are the consequence of innovative leadership that keeps its eye firmly on target areas, on inputs and outputs, on components and segments and configurations. Innovative leadership does this because capabilities are interrelated so that requirements, unwatched, may burgeon. For example, after a corporation has identified the objective of getting a new facility into start-up, environmental-impact reporting requirements must be met so that the facility can go on-stream within the envisaged time-span.

Even this tells only the bare bones of the story. Multiple markets and multi-target areas may well be penetrated, but not without impact studies, market strategies, cost economies, product development and product packaging, and consumer acceptance. Product packaging sounds simple enough, but it may call for in-house box-making capability. Box-making in turn is a process; that calls for process equipment capability; and *that* calls for process development personnel.

If all this is to be done, management teams must be sound and prudent and characterized by vision, enterprise, and flexibility. In a surprising number of companies, the corporate reports assure us, management teams are.

Business puts enormous pressure on language as most of us have known it. Under this pressure, triple and quadruple phrases come into being — high retention characteristics, process knowledge rate development, anti-dilutive common stock equivalents. Under this pressure also, adjectives become adverbs; nouns become adjectives; prepositions disappear; compounds abound.

In its report on 1972, American Buildings Company told its shareholders that its new products included "improved long-span and architectural panel configurations which enhance appearance and improve weatherability." Despite the travail concealed behind those simple words, the achievement must have been noteworthy on the cutting edge of the construction industry.

A statement by the Allegheny Power System was, on the other hand, hardly worth making: "In the last analysis the former, or front-end process seems the more desirable because the latter, or back-end, process is likely to create its own environmental problems." This is an old story, for the front-end process often does not know what the back-end process is doing.

In its annual report for 1972, Continental Hair Products drove home two lessons. One was that "Depreciation and amortization of property, plant and equipment are provided on the straight-line and double declining balance methods at various rates calculated to extinguish the book values of the respective assets over their estimated useful lives."

Among Continental's shareholders, one suspects, sentimentalists still quixotically opposed to the extinguishing of book values may have forborne to cheer. But not the others, and they must have been roused to still greater enthusiasm by the outburst of corporate ecstasy which was the second point: "Continental has exercised a dynamic posture by first establishing a professional marketing program and utilizing that base to penetrate multi-markets."

For myself, looking at this array of horrors, I forbear to cheer. People are forever quoting Benjamin Franklin, coming out of the Constitutional Convention in 1787, being asked what kind of government the Convention was giving the country, and replying, "A republic — if you can keep it." We were also given a language, and there is a competition in throwing it away. Business is in the competition and doing nicely.

## STUDY QUESTIONS

1. Scan a newspaper or magazine for the language mistakes Newman cites. Has the language of business deteriorated as much as Newman implies? Has he selected

representative examples to make his point? What other errors does your search turn up?

2. Has Newman criticized any terms, expressions, or clichés that you normally use? What changes or substitutions can you make to avoid the use of these terms?

3. Whenever a critic such as Newman attacks the language, other critics (or those being attacked) respond by saying he is too strict or pedantic. How do you feel about this issue? Is there anything wrong with allowing people to speak or write as they please?

# Degrees of Plain Talk

## Rudolph Flesch

Rudolf Flesch is the author of numerous books on writing, reading, and what Flesch has called "readability." His analyses of reading levels and reading difficulty are often referred to by technical writers. In this selection from his book *The Art of Readable Writing,* Flesch explains what readability is all about.

Popularization is a mysterious business. In November 1941, the *Journal of the American Medical Association* printed a paper by Drs. Rovenstine and Wertheim, in which the authors reported on a new kind of anesthesia called "therapeutic nerve block." This was obviously of interest to doctors, but nobody bothered to tell the general public about it. The nerve block was not then considered news.

Six years later, the popular magazines broke out into a rash of nerve-block articles. On October 25, 1947, the *New Yorker* began a three-part profile of Dr. Rovenstine; two days later, *Life* published a four-page picture-story of his work. Other magazines followed. Suddenly, the nerve block had become something everybody ought to know about.

I came across this mystery when I was looking for a good example of what popularization does to language and style. The nerve-block articles are perfect specimens. On its way from the *A.M.A. Journal* to *Life* and the *New Yorker,* the new method of anesthesia underwent a complete change of coloring, tone, and style. A study of the three articles is a complete course in readability by itself.

On the following pages are excerpts from the three articles. Nothing has been changed; but to show clearly the differences in sentence length, I have put / between the sentences, and to show the differences in human interest, I have capitalized the "personal words" and italicized the "personal sentences" (You will notice the difference in word length without my pointing it up.)

Rudolf Flesch, "Degrees of Plain Talk," *The Art of Readable Writing,* Rev. Ed. (New York: Harper and Row, 1974), pp. 166–174.

This is the beginning of "Therapeutic Nerve Block" by E. A. Rovenstine, M.D. and H. M. Wertheim, M.D. (*Journal of the American Medical Association, vol. 117, no. 19, Nov. 8, 1941*):

> "Therapeutic nerve block" is but one of the many ramifications of regional analgesia./ The history of the introduction and development of perineural injections of analgesic and neurolytic agents for therapy coincides with that of similar types of injections to control the pain associated with surgical procedures./ The use of surgical analgesic nerve blocks has eclipsed by far similar procedures employed to cure or alleviate pain or symptoms resulting from disease or injury . . ./

The paper ends as follows:

> The most interesting and probably more promising and fruitful results from therapeutic nerve blocking are the techniques for interrupting sympathetic pathways with analgesic or neurolytic solutions./ This recent practice has already gained wide application and produced many favorable reports./ A comparison of the value of the chemical destruction of sympathetic pathways or surgical section cannot be made accurately with present knowledge and experience, but there are indications that for many conditions the former are to be preferred./
>
> Interruption of the sympathetic pathways at the stellate ganglion is used to cure hyperhidrosis of the upper extremity./ It is useful to relieve sympathalgia of the face and causalgia./ It has been employed successfully to treat posttraumatic spreading neuralgia, the pain of amputation stumps and vasomotor disturbances./ The treatment of angina pectoris after medical remedies have failed to relieve pain is now conceded to include alcohol injections of the upper thoracic sympathetic ganglions./ The same procedure has been effective in controlling or alleviating the distressing pain from an aneurysm of the arch or the descending aorta./
>
> Interruption of the lumbar sympathetic pathways is indicated for conditions in the lower extremities similar to those enumerated for the upper extremities./ This therapeutic nerve block has been employed also to treat thrombophlebitis of the lower extremity./ The results from these injections have been dramatic and largely successful./ Not only is the pain relieved immediately but the whole process subsides promptly./ This remedy represents so much of an improvement over previous therapeutic efforts that it should be used whenever the condition develops./

In *Life* (October 27, 1947) the article about the nerve block carried the heading

## Pain-Control Clinic

### New York Doctors Ease Suffering by Blocking Off Nerves with Drugs.

Eight pictures were accompanied by the following text:

> Except in the field of surgery, control of pain is still very much in the primitive stages./ Countless thousands of patients suffer the tortures of cancer,

angina pectoris and other distressing diseases while THEIR physicians are helpless to relieve THEM./ A big step toward help for these sufferers is now being made with a treatment known as nerve-blocking./ This treatment, which consists of putting a "block" between the source of pain and the brain, is not a new therapy./ But its potentialities are just now being realized./ Using better drugs and a wider knowledge of the mechanics of pain gained during and since the war, Doctors E. A. ROVENSTINE and E. M. PAPPER of the New York University College of Medicine have been able to help two-thirds of the patients accepted for treatment in THEIR "pain clinic" at Bellevue Hospital./

The nerve-block treatment is comparatively simple and does not have serious aftereffects./ It merely involves the injection of an anesthetic drug along the path of the nerve carrying pain impulses from the diseased or injured tissue to the brain./ Although its action is similar to that of spinal anesthesia used in surgery, nerve block generally lasts much longer and is only occasionally used for operations./ The N.Y.U doctors have found it effective in a wide range of diseases, including angina pectoris, sciatica, shingles, neuralgia and some forms of cancer./ Relief is not always permanent, but usually the injection can be repeated./ Some angina pectoris patients have had relief for periods ranging from six months to two years./ While recognizing that nerve block is no panacea, the doctors feel that results obtained in cases like that of MIKE OSTROICH (*next page*) will mean a much wider application in the near future./

The *New Yorker* (October 25, 1947) in its profile of Dr. Rovenstine describes the nerve block like this:[1]

. . . Recently, HE [Rovenstine] devoted a few minutes to relieving a free patient in Bellevue of a pain in an arm that had been cut off several years before./ The victim of this phantom pain said that the tendons ached and that HIS fingers were clenched so hard HE could feel HIS nails digging into HIS palm./ Dr. ROVENSTINE'S assistant, Dr. E. M. PAPPER, reminded ROVENSTINE that a hundred and fifty years ago the cure would have been to dig up the MAN'S arm, if its burial place was known, and straighten out the hand./ ROVENSTINE smiled./ *"I tell YOU,"* HE said./ *"WE'll use a two-per-cent solution of procaine, and if it works, in a couple of weeks, WE'll go on with an alcohol solution./ Procaine, YOU know, lasts a couple of weeks, alcohol six months or longer./ In most cases of this sort, I use the nerve block originated by LABAT around 1910 and improved on in New Orleans about ten years back, plus one or two improvisations of MY own."/* (Nerve blocking is a method of anesthetizing a nerve that is transmitting pain.) ROVENSTINE does little anesthetizing HIMSELF these days, except when HE is demonstrating HIS methods at HIS lectures./ HE carries on only a small practice outside Bellevue./ If HE is called in on routine cases, HE asks extremely high fees./ HE proceeds on the principle that a person who wants HIM to handle a routine operation ought to pay well for HIM./ If HE is asked

[1] From an article by Mark Murphy in the *New Yorker*. Copyright 1947 The New Yorker Magazine Inc.

to apply HIS specialized knowledge to an unusual case, HE doesn't care what the fee is./ Like a great many other doctors, HE feels that only millionaires and indigents get decent medical care./ PEOPLE of these two classes are the only ones who feel that THEY can call on the leading surgeons and ROVENSTINE./

The MAN with the pain in the non-existent hand was an indigent, and ROVENSTINE was working before a large gallery of student anesthetists and visitors when HE exorcised the ghosts that were paining HIM./ Some of the spectators, though THEY felt awed, also felt inclined to giggle./ Even trained anesthetists sometimes get into this state during nerve-block demonstrations because of the tenseness such feats of magic induce in THEM./ The patient, thin, stark-naked, and an obvious product of poverty and cheap gin mills, was nervous and rather apologetic when HE was brought into the operating theatre./ HE lay face down on the operating table./ *ROVENSTINE has an easy manner with patients, and as HIS thick, stubby hands roamed over the MAN'S back, HE gently asked, "How YOU doing?"/ "MY hand, it is all closed together, DOC," the MAN answered, startled and evidently a little proud of the attention HE was getting./ "YOU'll be O.K. soon," ROVENSTINE said, and turned to the audience./ "One of MY greatest contributions to medical science has been the use of the eyebrow pencil," HE said./* HE took one from the pocket of HIS white smock and made a series of marks on the patient's back, near the shoulder of the amputated arm, so that the spectators could see exactly where HE was going to work./ With a syringe and needle HE raised four small weals on the MAN's back and then shoved long needles into the weals./ The MAN shuddered but said HE felt no pain./ ROVENSTINE then attached a syringe to the first needle, injected the procaine solution, unfastened the syringe, attached it to the next needle, injected more of the solution, and so on./ The patient's face began to relax a little./ *"LORD, DOC," HE said. "MY hand is loosening up a bit already."/ "YOU'll be all right by tonight, I think," ROVENSTINE said./* HE was.

That the language of the three articles is different, everybody can see. But it's not so easy to tell *how* different. For that, let's look at a few figures:

|  | A.M.A. Journal | Life | New Yorker |
|---|---|---|---|
| Average sentence length in words | 20.5 | 22 | 18 |
| Average word length in syllables (per 100 words) | 194 | 165 | 145 |
| Percent "personal words" | 0 | 2 | 11 |
| Percent "personal sentences" | 0 | 0 | 41 |

What has happened is this: *Life* magazine naturally had to be more readable than the *A.M.A. Journal.* So it avoided words like *analgesic* and *thrombophlebitis* and otherwise presented the facts in more or less newspaper fashion. Effect: The words in *Life* are 15 per cent shorter than those in the *A.M.A. Journal.* (The sentences would be shorter too if Drs. Rovenstine and Wertheim hadn't written exceptionally short sentences to begin with — for two doctors, that is.) But *Life* didn't bother to dramatize the facts and make them humanly interesting. The *New Yorker* began *its* popularization of the nerve block where *Life* left off: its sentences are a good bit shorter than those in *Life* (19 percent), and aside from that, there's a story, there's drama, there's something that's interesting to read. The nerve block has become an experience to the reader.

This gives us a good clue to the baffling question of what readability means. In most dictionaries, *readable* is defined as "easy *or* interesting to read." (It also has another meaning *legible*, but we'll skip that here.) Actually, to most people, readability means ease of reading *plus* interest. They want to make as little effort as possible while they are reading, and they also want something "built in" that will automatically carry them forward like an escalator. Structure of words and sentences has to do with one side of readability, "personal words" and "personal sentences" with the other.

That's why, in this book, a piece of writing is given not one readability score but two: a "reading ease" score and a "human interest" score. Length of words and sentences are combined into one, "personal" words and sentences into the other.

Working out the scores of our three articles on nerve blocking, we get this picture:

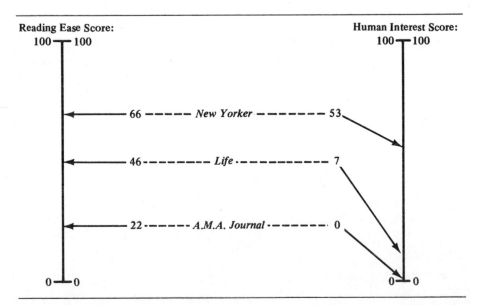

*Life* is much easier to read than the *A.M.A. Journal,* but gets hardly "off the ground" with human interest; the *New Yorker* is still easier than *Life,* and, in addition, is as interesting to read as fiction.

All this doesn't mean, of course, that the writers for *Life* and the *New Yorker* *consciously* did something about their sentences and words. Naturally not. But if we want to find out something about any art or skill, we must analyze the work of leading performers, and then laboriously imitate their seemingly effortless performance. There's no guarantee that we'll ever become champions this way, but at least we can try.

## STUDY QUESTIONS

1. Flesch is attempting to define what he means by readability. In your judgment, what role does audience analysis play in determining readability?
2. Flesch suggests that the writers he quotes did not consciously choose their styles. Do you agree with that assumption?
3. Select a news story that has appeared in national magazines such as *Time* or *Newsweek,* and your local newspaper. Analyze these stories as Flesch has done, comparing them for "reading ease" and "human interest." Is there a noticeable difference?
4. Obtain a copy of Flesch's readability scale and apply it to an essay in this book. Then use the scale to measure the readability level of an article in a national magazine. What differences do you find?

# Section 3

# Problem - Solving

Let us say that your boss has told you to report, in writing, on how the interoffice system of mail distribution could be improved. An obvious part of the assignment will be the rhetorical task of developing and arranging your findings and suggestions into a clear report. The problem at this advanced stage is one of using the appropriate rhetorical means, and there are many good handbooks that can help you with it.

But if we back up one step we realize that there is another problem involved here — one for which there may be no handy reference. That is, *how* can the distribution of mail be improved? Perhaps you are an experienced hand at such problems, and the solution is readily apparent to you. Then you have to worry only about the rhetoric of the report, which itself can be formidable enough. Or maybe the solution, if there is one at all, isn't so apparent. In this case, the problem of rhetoric isn't so immediate — or maybe seems to be no problem at all. Effective communication is, of course, crucial, but first you must have something to communicate. You must solve the problem.

A great deal of the activity in industry, business, and science is exactly this: solving problems for which there are no readily apparent solutions. In fact, if we didn't solve problems the world that we have built would come to a halt — we would be "stuck." Our technological world necessarily admires people who can get us "unstuck," and we usually reward them generously.

Effectiveness at problem-solving often comes with experience. As you become familiar with your field, the resources at your disposal, and the strategies that your colleagues try, you will find that problem-solving is a natural part of your working life — maybe you are not even aware that you are solving problems. But even the experienced businessman, technician, or scientist will find that occasional problem that isn't readily cracked. And he is still better off than the beginner in the field who doesn't even have experience to back him up, who is trying to solve problems while gaining experience. So it would seem that both the old hand and the novice could benefit from knowing something about the problem-solving processes.

But how are problems solved? To be sure, there is rarely a routine method. Each problem has its own twists and turns, and requires individual consideration. But even though there is probably no universal method that will solve all problems, there are guidelines that will at least help us approach most problems. For example, *analysis,* or the taking apart of a problem, enables us to reduce a problem into pieces that will bear closer examination. *Induction* gives us a method of forming conclusions on the basis of specific examples. *Deduction,* on the other hand, is a method of making conclusions about the specific examples on the basis of general principles. Other problem-solving approaches include *causal connections, analogy,* and *visual thinking.* These methods and others are introduced by the readings in this section.

Even though the selections that make up this section are not focused on writing (although most of the selections are very well written), we believe that student writers will find that problem-solving is one of the most pertinent topics in this text. You are now training for a position in a scientific, industrial, or business field, and you will be expected to write. But you will not necessarily be a full-time writer. Your writing will be a product of your work, not the work itself. As an everyday part of your job you will be solving problems. So it is to your advantage to have in mind some problem-solving strategies. You will be valuable to your organization to the extent that you can further its goals, which means that you must be able to solve its problems. Sometimes you will write up your solutions in the form of a report, sometimes not. But it is a certainty that a report that is supposed to solve a problem cannot be written until the problem is solved.

The following selections will not tell you everything about problem solving, but they are a worthy start. They might give you a clue as to how to improve the interoffice mail system — then you can get on with writing your report.

# The Design Process Is a
# Problem - Solving Journey

## Don Koberg and Jim Bagnall

Many of us perhaps assume that problem-solving is by inspiration, that solutions come only in sudden flashes of light (the proverbial light bulb in the mind). Of course, *some* solutions are arrived at spontaneously. But it is a haphazard method, and not generally reliable. In problem-solving, as in most job skills, a systematic approach can increase your effectiveness. Although no one system will be suitable for everybody, Koberg and Bagnall's overview on problem-solving introduces a general scheme and useful terminology that you can adapt to your own needs.

Don Koberg and Jim Bagnall, *The Universal Traveller* (Los Altos, California: Kaufmann, 1976), pp. 16–21.

# The DESIGN PROCESS is a Problem-Solving JOURNEY

Since all of us, more or less, embark on many
problem-solving journeys, each of us is, more
or less, involved in the design process.  The
more we understand about DESIGN PROCESS, the
more interesting and meaningful will be our
problem-solving journeys.

Design can be defined as the process of
creative problem-solving; a process of
creative, constructive behavior.  Designers
are people who behave creatively relative to
problem situations; people who generate
uniquely satisfying solutions to such situations.  Gym teachers and geology majors,
free-lance writers and truck farmers, movie
makers and motorcyclists, audiophiles and
elevator operators, xylophonists and ZOOM
fans are all problem-solvers.  Everyone is a
problem-solver.  Some just do it better than
others.

The PROCESS of DESIGN or creative problem-
solving describes a series or sequence of
events, stages, phases or ENERGY STATES, as
we will call them.  Before completing a total
design journey, it is usually necessary to
pass through each of those phases.  Once
each phase is known through experience, the
design process as a whole can be appreciated
as a round-trip which includes intentions,
decisions, solutions, actions and evaluations.
There are seven such ENERGY STATES given.
Each has been chosen as a synthesis from many
models of process.*

*Wallas, Dewey, Rossman, Guilford, Osborn,
Stanislawski, Parnes, Gordon, Kepner-Tregoe,
Arnold, Churchman, Zwicky, General Electric,
the Military, Pert, etc.

The logical sequence of events included in the design process is:

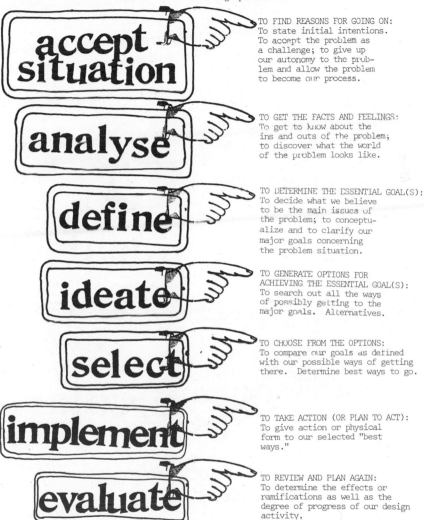

**accept situation**

TO FIND REASONS FOR GOING ON: To state initial intentions. To accept the problem as a challenge; to give up our autonomy to the problem and allow the problem to become our process.

**analyse**

TO GET THE FACTS AND FEELINGS: To get to know about the ins and outs of the problem; to discover what the world of the problem looks like.

**define**

TO DETERMINE THE ESSENTIAL GOAL(S): To decide what we believe to be the main issues of the problem; to conceptualize and to clarify our major goals concerning the problem situation.

**ideate**

TO GENERATE OPTIONS FOR ACHIEVING THE ESSENTIAL GOAL(S): To search out all the ways of possibly getting to the major goals. Alternatives.

**select**

TO CHOOSE FROM THE OPTIONS: To compare our goals as defined with our possible ways of getting there. Determine best ways to go.

**implement**

TO TAKE ACTION (OR PLAN TO ACT): To give action or physical form to our selected "best ways."

**evaluate**

TO REVIEW AND PLAN AGAIN: To determine the effects or ramifications as well as the degree of progress of our design activity.

The basic function of using a conscious format
for increased creativity within the problem-
solving activity of design is to free us from
uncertainty, anxiety, confusion and other in-
securities as we travel along our journey.  It
is an organizational guide which can help us
to better enjoy the trip by allowing us to concentrate more on our creative participation
rather than on whether or not we are moving
toward or away from planned destinations.

Knowledge and use of the DESIGN PROCESS provides a fuller, richer and productive life by
allowing us to take conscious control of our
own life process--as opposed to being the passive
victim of the decisions of others and/or the
consequences of nature.

The design process can be viewed in a variety
of ways.  Some see it as a linear thing; others
see it as a circular configuration.

# linear

Where one thing follows
another in a straight line.

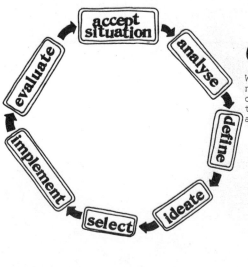

# circular

Where there is continuity, but
never a beginning and end.  As
one problem situation appears
to be resolved, another one
appears to begin.

# feedback

Others see it as a constant feedback system where you never go forward without always looping back to check on yourself; where one progresses by constant backward relationships and where the stages of the process advance somewhat concurrently until some strong determining variable terminates the process (time, money, energy, etc.)

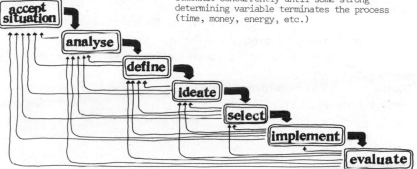

# branching

And still others see the design process as a branching system where certain events determine more than one direction and where directional progress is achieved via a many-branched excursion.

As a matter of fact and practicality, it really doesn't make too much difference how you view it. The design process is a round-trip of phases which begins at one point and goes outward from there and back again. If the excursion has been logically pursued, a systematic trek through all of the energy states has been made. If the trip has been meaningful, the return to the place of beginning shows it to be a different place than when you left. The journey has become, because of your new experiences, a new and exciting thing. From this point of view, the design process can be seen as not merely circular or helical, but rather as a SPIRAL CONTINUUM of sequential round trips which becomes a basic model, perhaps, like the DNA molecule, for all existence.

## STUDY QUESTIONS

1. What do Koberg and Bagnall mean by *design*? Why is design important in problem-solving?
2. Think back on some problems you've solved. Were you systematic in your approach, or did the solution "just come to you"?
3. If you have a system, list its essential elements. Do you have any terms in common with Koberg and Bagnall's? Discuss any similarities.
4. What advantages in problem-solving do the *feedback system* and the *branching system* offer?
5. Discuss the effectiveness of the artwork. Does it add to, or detract from, the content of the selection?

# Does the Man Go Round the Squirrel or Not ?

## William James

Although James uses the following anecdote to illustrate a philosophical issue, it is easily applied to what Koberg and Bagnall call problem definition: "to decide what we believe to be the main issues of the problem; to conceptualize and to clarify our major goals concerning the problem situation." We might conclude from James's selection that the problem-solver must avoid philosophical disputes in favor of practical consequences.

Some years ago, being with a camping party in the mountains, I returned from a solitary ramble to find every one engaged in a ferocious metaphysical dispute. The *corpus* of the dispute was a squirrel — a live squirrel supposed to be clinging to one side of a tree-trunk; while over against the tree's opposite side a human being was imagined to stand. This human witness tries to get sight of the squirrel by moving rapidly round the tree, but no matter how fast he goes, the squirrel moves as fast in the opposite direction, and always keeps the tree between himself and the man, so that never a glimpse of him is caught. The resultant metaphysical problem now is this: *Does the man go round the squirrel or not?* He goes round the tree, sure enough, and the squirrel is on the tree; but does he go round the squirrel? In the unlimited leisure of the wilderness, discussion had been worn threadbare. Every one had taken sides and was obstinate; and the numbers on both sides were even.

From William James, *Pragmatism: A New Name for Some Old Ways of Thinking*, 1907. Title selected by editors.

Each side, when I appeared therefore appealed to me to make it a majority. Mindful of the scholastic adage that whenever you meet a contradiction you must make a distinction, I immediately sought and found one, as follows: "Which party is right," I said, "depends on what you *practically mean* by 'going round' the squirrel. If you mean passing from the north of him to the east, then to the south, then to the west, and then to the north of him again, obviously the man does go round him, for he occupies these successive positions. But if on the contrary you mean being first in front of him, then on the right of him, then behind him, then on his left, and finally in front again, it is quite as obvious that the man fails to go round him, for by the compensating movements the squirrel makes, he keeps his belly turned towards the man all the time, and his back turned away. Make the distinction, and there is no occasion for any farther dispute. You are both right and both wrong according as you conceive the verb 'to go round' in one practical fashion or the other."

Although one or two of the hotter disputants called my speech a shuffling evasion, saying they wanted no quibbling or scholastic hair-splitting, but meant just plain honest English "round," the majority seemed to think that the distinction has assuaged the dispute.

I tell this trivial anecdote because it is a peculiarly simple example of what I wish now to speak of as *the pragmatic method.* The pragmatic method is primarily a method of settling metaphysical disputes that otherwise might be interminable. Is the world one or many? — fated or free? — material or spiritual? — here are notions either of which may or may not hold good of the world; and disputes over such notions are unending. The pragmatic method in such cases is to try to interpret each notion by tracing its respective practical consequences. What difference would it practically make to any one if this notion rather than that notion were true? If no practical difference whatever can be traced, then the alternatives mean practically the same thing, and all dispute is idle. Whenever a dispute is serious, we ought to be able to show some practical difference that must follow from one side or the other's being right.

## STUDY QUESTIONS

1. According to James, is the issue of whether or not the man goes "round the squirrel" important? Why or why not? How does the *pragmatic method* help clarify the importance of the issue?
2. Is problem-solving in industry and business always a *pragmatic* activity? Discuss, considering James's definition of pragmatic as well as the dictionary definition.

# Cultural and Environmental Blocks

## James L. Adams

Even if fully aware of what the problem is that we must solve, several obstacles
might still hamper our efforts to find a solution. James L. Adams identifies many
of the obstacles to thinking that are ingrained in our culture and environment.

Cultural Blocks are acquired by exposure to a given set of cultural patterns. Environmental blocks are imposed by our immediate social and physical environment.
Since these two types of blocks are somewhat interrelated, we will discuss both of
them in this chapter. Some examples of cultural blocks (for our culture) are:

1. Fantasy and reflection are a waste of time, lazy, even crazy
2. Playfulness is for children only
3. Problem-solving is a serious business and humor is out of place
4. Reason, logic, numbers, utility, practicality are *good*; feeling, intuition,
   qualitative judgments, pleasure are *bad*
5. Tradition is preferable to change
6. Any problem can be solved by scientific thinking and lots of money
7. Taboos

Some examples of environmental blocks are:

1. Lack of cooperation and trust among colleagues
2. Autocratic boss who values only his own ideas; does not reward others
3. Distractions – phone, easy intrusions
4. Lack of support to bring ideas into action

Let us discuss cultural blocks first. We will begin by working a problem that
will make the message clearer.

**Exercise:** Assume that a steel pipe is imbedded in the concrete floor of a bare
room as shown below. The inside diameter is .06″ larger than the diameter of
a ping-pong ball (1.50″) which is resting gently at the bottom of the pipe.
You are one of a group of six people in the room, along with the following
objects:

100′ of clothesline
A carpenter's hammer
A chisel
A box of Wheaties
A file
A wire coat hanger

James L. Adams, *Conceptual Blockbusting* (San Francisco: San Francisco Book Company, 1976), pp. 31–34.

A.money wrench
A light bulb

List as many ways you can think of (in five minutes) to get the ball out of the pipe without damaging the ball, tube, or floor.

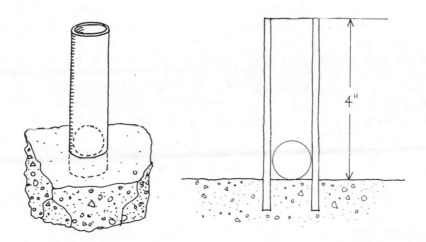

J. P. Guilford, one of the pioneers in the study of creativity, speaks a great deal about fluency and flexibility of thought. *Fluency* refers to the number of concepts one produces in a given length of time. If you are a fluent thinker, you have a long list of methods of retrieving the ball from the pipe. However, quantity is only part of the game. *Flexibility* refers to the diversity of the ideas generated. If you are a flexible thinker, you should have come up with a wide variety of methods. If you thought of filing the wire coat hanger in two, flattening the resulting ends, and making large tweezers to retrieve the ball, you came up with a solution to the problem, but a fairly common one. If you thought of smashing the handle of the hammer with the monkey wrench and using the resulting splinters to retrieve the ball, you were demonstrating a bit more flexibility of thought, since one does not usually think of using a tool as a source of splinters to do something with. If you managed to do something with the Wheaties you are an even more flexible thinker.

Did you think of having your group urinate in the pipe? If you did not think of this, why not? The answer is probably a cultural block, in this case a *taboo,* since urinating is somewhat of a closet activity in the U.S.

## Taboos

We have used this ping-pong ball exericse with many groups and the response is not only a function of our culture, but also of the particular people in the group and the particular ambiance of the meeting. A mixed group newly convened in elegant surroundings will seldom think of urinating in the pipe. Even if members in the group do come up with this as a solution, they will keep very quiet about it. A group of people who work together, especially if all-male and if it's at the end of a working session, will instantly break into delighted chortles as they think of this and equally gross solutions. The importance of this answer is not that urinating in the pipe is necessarily the best of all solutions to the problem (although it is certainly a good one), but rather that cultural taboos can remove entire families of solutions from the ready grasp of the problem-solver. Taboos therefore are conceptual blocks. This is not a tirade against taboos. Taboos usually are directed against acts which would cause displeasure to certain members of a society. They therefore play a positive cultural role. However, it is the acts themselves which would offend. If imagined, rather than carried out, the acts are not harmful. Therefore, when working on problems within the privacy of one's own mind, one does not have to be concerned with the violation of taboos.

Let us discuss a few more cultural blocks. The first two listed earlier, "Fantasy and reflection are a waste of time, lazy, even crazy" and "Playfulness is for children only," will be discussed further in the next chapter. However, a few comments should be made here. There is quite a bit of evidence to indicate that fantasy, reflection, and mental playfulness are essential to good conceptualization. These are properties which seem to exist in children, and then unfortunately are to some extent socialized out of people in our culture. A four-year-old who amuses himself with an imaginary friend, with whom he shares his experiences and communicates, is cute. A 30-year-old with a similar imaginary friend is something else again. "Daydreaming" or "woolgathering" is considered to be a symptom of an unproductive person.

As mentioned previously, environmental and cultural blocks are somewhat interrelated. People can fantasize much more easily in a supportive environment. We quite frequently ask students to fantasize as part of a design task, and when assigned the task they do quite well. However, they tend to feel quite guilty if they spend their time in fantasy if it is not an assigned part of the problem, since it often seems to be a diversion. Nevertheless, if one is attempting to solve a problem having to do with bickering children, is it not worth the time and effort to fantasize a situation in which one's children do not bicker and proceed to examine the situation closely to see how it works? If one is designing a new recreational vehicle, should he not fantasize what it would be like to use that vehicle?

Many psychologists have concluded that children are more creative than adults. One explanation for this is that the adult is so much more aware of practical constraints. Another explanation, which I believe, is that our culture trains mental playfulness, fantasy, and reflectiveness out of people by placing more stress on the value of channeled mental activities. We spend more time attempting to derive a better world directly from what we have than in imagining a better world and what it would be. Both are important.

**STUDY QUESTIONS**
1. We include the specific discussions on only a couple of the "blocks" that Adams lists. Discuss each one, considering specifically how each can interfere with creativity in problem-solving. When possible, give personal examples.
2. Why does Adams discuss the ping-pong ball exercise first?

# Analysis

## Albert Upton

Analysis is commonly defined as the process of separating into its parts an object or concept. Upton recognizes this common definition as well as two additional ones: 1) finding the larger grouping to which an object belongs and 2) identifying the steps of an operation. Each of these procedures has a role in effective problem-solving.

The Logical operations of analysis are extremely simple considering the indescribably complex organ that performs them and the astounding achievements that sometimes result from their application. For there are only three sorts of analytical questions that the rational intellect may propose and three sorts of rational answers it can get. It can ask (1) "What is this a sort of, or what are the sorts of this?" (2) "What is this a part of, or what are the parts of this?" (3) "What is this a stage of, or what are the stages of this?"

You may recall that we began Chapter 2 with an answer to the question "What is your brain a sort of?" By giving the answer "organ" to our question we took the brain into one of our most fundamental sortings — the sorting into things that are or have been alive and things that do not have the mysterious power to grow. We say that animals and plants grow. Clouds and dunes form.

Plants and animals are sorts of organisms; men and amoebas are sorts of animals. Now when we persistently and consistently ask and answer the question of sort, the results are what we call *classifications,* and we shall have more to say about them in this book. In biological science the study of classification is called *tax-*

From Albert Upton, *Design for Thinking* (Stanford: Stanford Univ. Press, 1961), pp. 22–24.

*onomy,* which in Greek means the system or law of order. Here is a simple diagram showing how the system works:

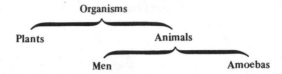

When you put the question "What is this a part of, or what are the parts of this?" you have in mind not what pigeonhole a thing belongs in, nor what sorts of things are in the same pigeonhole, but rather what the boundaries of things are and where and how they connect with the things that belong around them. You are thinking the kind of thoughts we think when we take things apart or when we put them together. We shall call this kind of analysis *structure analysis* because things that have parts connected to make wholes are called structures. An organism is a structure. A cloud is a structure, and so is a dune. In the science of biology, structure analysis goes by the name of anatomy, which comes from a Greek word meaning "to cut apart." A good anatomist, the Greeks would tell you, is like a good butcher — he knows how to cut where the joints are. When we talk about the connections or boundaries between the heart and the brain, or when we talk about the three parts of the brain that make up the organ of consciousness, we are anatomizing. Structure analysis really calls for pictures, but verbal statements of constituent parts and their arrangement may be conveniently presented thus:

$$
\text{Body}
\begin{cases}
\text{head}
\begin{cases}
\text{skull}
\begin{cases}
\text{crown} \\
\text{jaw}
\end{cases}
\begin{cases}
\text{bone} \\
\text{teeth}
\end{cases} \\
\text{organs}
\begin{cases}
\text{brain} \\
\text{sensory}
\end{cases}
\begin{cases}
\text{skin} \\
\text{eyes} \\
\text{etc.}
\end{cases} \\
\text{muscles}
\end{cases} \\
\text{trunk} \\
\text{limbs}
\end{cases}
$$

When the wholes that we take apart or put together in our minds are changing or moving wholes, the fourth dimension of time plays its part in our thinking; we then use the word *stage* or its synonyms for the constituent parts of the sort of wholes we call *operations.* Thus structures are three-dimensional and made up of parts. Events or operations are four-dimensional. Change takes time. Structures are wholes that have parts for parts. Operations are wholes that have stages for parts.

The operation or process of taking events apart we shall call *operation analysis*. Because we never have events without things to evolve — that is, structures with changing relations among their parts — analysis of an event turns out to be the tracing of the functions of the parts, just as analysis of a structure is the tracing of the boundaries and the connections of the parts.

In biological science operation analysis is called physiology, which in Greek means, as you might expect, the science of natural change. We suggest the conventional outline form for representing operation analysis. For example:

    I. Ingestion
      A. Mastication
      B. Swallowing
   II. Digestion
      A. Stomachic
      B. Intestinal
  III. Elimination

Here then are the three primary logical operations: classification, structure analysis, and operation analysis. Whenever you begin to see relationships in the world about you, one or more of these operations is going on in your head. Whenever you make a vigorous, systematic attempt to complete these operations you are being a scientist. Whenever these three operations are arbitrarily committed to some desired end and in Hamlet's phrase, "reason panders will," you are being a philosopher. When you revel in the richness of your observation of the world about you and display grace or skill in the manipulation of your environment, you are being an artist. And whenever you combine the three activities in harmony and balance you are being a statesman in the world of "psycho-politics," which is the Greek way of saying the city government of the soul. Franklin, Jefferson, and Madison were such statesmen, and two of the noblest documents of human history are partial records of their realistic solution to the problem of finding the good life.

At this stage in our discussion of language in mind and society, you are quite justified if you think that we are going to put great stress upon the importance of scientific behavior. Before taking our next step we must therefore explain clearly our theory of the status of science in the modern world. We think that there are two unfortunate, and ultimately unscientific, concepts concerning the nature of science and its relation to civilization. One is the idea of specialism, or as Professor Munro has said, "knowing more and more about less and less until you become intellectually unbuttoned"; the other is the idea that specialization is good because it makes for civilization which is, of course, good.

We suggest making a distinction between a highly developed or complex culture and a civilization. Culture may be defined in passing as the relatively rigid and unreasoned type of social behavior found in hives, lodges, and sometimes even in pentagons. It may be impressive and its achievements marvelous. We suggest, however, a meaning for "civilization" that would make it rather a special sort of complex culture in which the constituent parts have developed linguistic, that is to say, parliamentary, techniques for resolving the inevitable disputes that arise from conflicts of interest, real or apparent.

**STUDY QUESTIONS**

1. Test the usefulness of Upton's approach by applying his three questions to a specific problem. You might try them on the problem introduced earlier: the need to improve the interoffice mail system. We could consider the larger category to which an interoffice mail system belongs, the component parts of the mail system, and the steps of the mail system. How could such information help us in improving the system? Discuss.

# The Scientific Method

## Thomas H. Huxley

Convinced that scientific and technological progress were the keys to the betterment of mankind, Englishman Thomas H. Huxley was a nineteenth century advocate of scientific and technical education. He gave many public lectures for the purpose of informing the "working classes" on scientific concepts. The following excerpt from one of his lectures is notable for its common sense approach to the logic of science. His emphasis is on induction, deduction, and hypothesis — important terms for logical problem-solving.

The method of scientific investigation is nothing but the expression of the necessary mode of working of the human mind. It is simply the mode at which all phenomena are reasoned about, rendered precise and exact. There is no more difference, but there is just the same kind of difference, between the mental operations of a man of science and those of an ordinary person, as there is between the operations and methods of a baker or of a butcher weighing out his goods in common scales, and the operations of a chemist in performing a difficult and complex analysis by means of his balance and finely-graduated weights. It is not that the action of the scales in the one case, and the balance in the other, differ in the principles of their construction or manner of working; but the beam of one is set on an infinitely finer axis than the other, and of course turns by the addition of a much smaller weight.

You will understand this better, perhaps, if I give you some familiar example.

From Thomas H. Huxley, "The Method by Which the Causes of the Present and Past Conditions of Organic Nature Are to be Discovered" (1863). Title selected by editors.

You have all heard it repeated, I dare say, that men of science work by means of induction and deduction, and that by the help of these operations, they, in a sort of sense, wring from Nature certain other things, which are called natural laws, and causes, and that out of these, by some cunning skill of their own, they build up hypotheses and theories. And it is imagined by many, that the operations of the common mind can be by no means compared with these processes, and that they have to be acquired by a sort of special apprenticeship to the craft. To hear all these large words, you would think that the mind of a man of science must be consti- tuted differently from that of his fellow men; but if you will not be frightened by terms, you will discover that you are quite wrong, and that all these terrible appa- ratus are being used by yourselves every day and every hour of your lives.

There is a well-known incident in one of Molière's plays, where the author makes the hero express unbounded delight on being told that he had been talking prose during the whole of his life. In the same way, I trust, that you will take com- fort, and be delighted with yourselves, on the discovery that you have been acting on the principles of inductive and deductive philosophy during the same period. Probably there is not one here who has not in the course of the day had occasion to set in motion a complex train of reasoning, of the very same kind, though differing of course in degree, as that which a scientific man goes through in tracing the causes of natural phenomena.

A very trivial circumstance will serve to exemplify this. Suppose you go into a fruiterer's shop, wanting an apple, — you take up one, and, on biting it, you find it is sour; you look at it, and see that it is hard and green. You take up another one, and that too is hard, green, and sour. The shopman offers you a third; but, before biting it, you examine it, and find that it is hard and green, and you immediately say that you will not have it, as it must be sour, like those that you have already tried.

Nothing can be more simple than that, you think; but if you will take the trouble to analyse and trace out into its logical elements what has been done by the mind, you will be greatly surprised. In the first place, you have performed the operation of induction. You found that, in two experiences, hardness and green- ness in apples went together with sourness. It was so in the first case, and it was confirmed by the second. True, it is a very small basis, but still it is enough to make an induction from; you generalise the facts, and you expect to find sourness in apples where you get hardness and greenness. You found upon that a general law, that all hard and green apples are sour; and that, so far as it goes, is a perfect induction. Well, having got your natural law in this way, when you are offered another apple which you find is hard and green, you say, "All hard and green apples are sour; this apple is hard and green, therefore this apple is sour." That train of reasoning is what logicians call a syllogism, and has all its various parts and terms, — its major premiss, its minor premiss, and its conclusion. And, by the help of further reasoning, which, if drawn out, would have to be exhibited in two or three other syllogisms, you arrive at your final determination, "I will not have that apple." So that, you see, you have, in the first place, established a law by induction, and upon that you have founded a deduction, and reasoned out the special conclusion of the particular case. Well now, suppose, having got your law, that at some time afterwards, you are discussing the qualities of apples with a friend: you will say to him, "It is a very curious thing, — but I find that all hard and green apples are sour!" Your friend says to you, "But how do you know that?" You at once reply, "Oh, because I have tried them over and over again, and have always found them to be so." Well, if we were talking science instead of common sense, we should call that an experimental verification. And, if still opposed, you go further, and say, "I have heard from the people in Somersetshire and Devonshire, where a large number of apples are grown, that they have observed the same thing. It is also found to be the case in Normandy and in North America. In short, I find it to be the universal experience of mankind wherever attention has been directed to the subject." Whereupon, your friend, unless he is a very unreasonable man, agrees with you, and is convinced that you are quite right in the conclusion you have drawn. He believes, although perhaps he does not know he believes it, that the more extensive verifications are, — that the more frequently experiments have been made, and results of the same kind arrived at, — that the more varied the conditions under which the same results are attained, the more certain is the ultimate conclusion, and he disputes the question no further. He sees that the experiment has been tried under all sorts of conditions, as to time, place, and people, with the same result; and he says with you, therefore, that the law you have laid down must be a good one, and he must believe it.

In science we do the same thing; — the philosopher exercises precisely the same faculties, though in a much more delicate manner. In scientific inquiry it becomes a matter of duty to expose a supposed law to every possible kind of verification, and to take care, moreover, that this is done intentionally, and not left to a mere accident, as in the case of the apples. And in science, as in common life, our confidence in a law is in exact proportion to the absence of variation in the result of our experimental verifications. For instance, if you let go your grasp of an article you may have in your hand, it will immediately fall to the ground. That is a very common verification of one of the best established laws of nature — that of gravitation. The method by which men of science establish the existence of that law is exactly the same as that by which we have established the trivial proposition about the sourness of hard and green apples. But we believe it in such an extensive, thorough, and unhesitating manner because the universal experience of mankind verifies it, and we can verify it ourselves at any time; and that is the strongest possible foundation on which any natural law can rest.

So much, then, by way of proof that the method of establishing laws in science is exactly the same as that pursued in common life. Let us now turn to another matter (though really it is but another phase of the same question), and that is, the method by which, from the relations of certain phenomena, we prove that some stand in the position of causes towards the others.

I want to put the case clearly before you, and I will therefore show you what I mean by another familiar example. I will suppose that one of you, on coming down in the morning to the parlour of your house, finds that a tea-pot and some spoons which had been left in the room on the previous evening are gone, — the window is open, and you observe the mark of a dirty hand on the window-frame, and perhaps, in addition to that, you notice the impress of a hob-nailed shoe on the gravel outside. All these phenomena have struck your attention instantly, and before two seconds have passed you say, "Oh, somebody has broken open the window, entered the room, and run off with the spoons and the tea-pot!" That speech is out of your mouth in a moment. And you will probably add, "I know there has; I am quite sure of it!" You mean to say exactly what you know; but in reality you are giving expression to what is, in all essential particulars, an hypothesis. You do not *know* it at all; it is nothing but an hypothesis rapidly framed in your own mind. And it is an hypothesis founded on a long train of inductions and deductions.

What are those inductions and deductions, and how have you got at this hypothesis? You have observed, in the first place, that the window is open; but by a train of reasoning involving many inductions and deductions, you have probably arrived long before at the general law — and a very good one it is — that windows do not open of themselves; and you therefore conclude that something has opened the window. A second general law that you have arrived at in the same way is, that tea-pots and spoons do not go out of a window spontaneously, and you are satisfied that, as they are not now where you left them, they have been removed. In the third place, you look at the marks on the window-sill, and the shoe-marks outside, and you say that in all previous experience the former kind of mark has never been produced by anything else but the hand of a human being; and the same experience shows that no other animal but man at present wears shoes with hob-nails in them such as would produce the marks in the gravel. I do not know, even if we could discover any of those "mising links" that are talked about, that they would help us to any other conclusion! At any rate the law which states our present experience is strong enough for my present purpose. You next reach the conclusion, that as these kinds of marks have not been left by any other animals than men, or are liable to be formed in any other way than by a man's hand and shoe, the marks in question have been formed by a man in that way. You have, further, a general law, founded on observation and experience, and that, too, is, I am sorry to say, a very universal and unimpeachable one, — that some men are thieves; and you assume at once from all these premisses — and that is what constitutes your hypothesis — that the man who made the marks outside and on the window-sill, opened the window, got into the room, and stole your tea-pot and spoons. You have now arrived at a *vera causa;* — you have assumed a cause which, it is plain, is competent to produce all the phenomena you have observed. You can explain all these phenomena only by the hypothesis of a thief. But that is a hypothetical conclusion, of the justice of which you have no absolute proof at all; it is only rendered highly probable by a series of inductive and deductive reasonings.

I suppose your first action, assuming that you are a man of ordinary common sense, and that you have established this hypothesis to your own satisfaction, will very likely be to go off for the police, and set them on the track of the burglar, with the view to the recovery of your property. But just as you are starting with this object, some person comes in, and on learning what you are about, says, "My good friend, you are going on a great deal too fast. How do you know that the man who really made the marks took the spoons? It might have been a monkey that took them, and the man may have merely looked in afterwards." You would probably reply, "Well, that is all very well, but you see it is contrary to all experience of the way tea-pots and spoons are abstracted; so that, at any rate, your hypothesis is less probable than mine." While you are talking the thing over in this way, another friend arrives, one of that good kind of people that I was talking of a little while ago. And he might say, "Oh, my dear sir, you are certainly going on a great deal too fast. You are most presumptuous. You admit that all these occurrences took place when you were fast asleep, at a time when you could not possibly have known anything about what was taking place. How do you know that the laws of Nature are not suspended during the night? It may be there there has been some kind of supernatural interference in this case." In point of fact, he declares that your hypothesis is one of which you cannot at all demonstrate the truth, and that you are by no means sure that the laws of Nature are the same when you are asleep as when you are awake.

Well, now, you cannot at the moment answer that kind of reasoning. You feel that your worthy friend has you somewhat at a disadvantage. You will feel perfectly convinced in your own mind, however, that you are quite right, and you say to him, "My good friend, I can only be guided by the natural probabilities of the case, and if you will be kind enough to stand aside and permit me to pass, I will go and fetch the police." Well, we will suppose that your journey is successful, and that by good luck you meet with a policeman; that eventually the burglar is found with your property on his person, and the marks correspond to his hand and to his boots. Probably any jury would consider those facts a very good experimental verification of your hypothesis, touching the cause of the abnormal phenomena observed in your parlour, and would act accordingly.

Now, in this suppositious case, I have taken phenomena of a very common kind, in order that you might see what are the different steps in an ordinary process of reasoning, if you will only take the trouble to analyse it carefully. All the operations I have described, you will see, are involved in the mind of any man of sense in leading him to a conclusion as to the course he should take in order to make good a robbery and punish the offender. I say that you are led, in that case, to your conclusion by exactly the same train of reasoning as that which a man of science pursues when he is endeavouring to discover the origin and laws of the most occult phenomena. The process is, and always must be, the same; and precisely the same mode of reasoning was employed by Newton and Laplace in their endeavours to discover and define the causes of the movements of the heavenly bodies, as you, with your own common sense, would employ to detect a burglar. The only difference is, that the nature of the inquiry being more abstruse, every step has to be most carefully watched, so that there may not be a single crack or flaw in your hypothesis. A flaw or crack in many of the hypotheses of daily life may be of little or no moment as affecting the general correctness of the conclusions at which we may arrive; but, in a scientific inquiry, a fallacy, great or small, is always of importance, and is sure to be in the long run constantly productive of mischievous, if not fatal results.

Do not allow yourselves to be misled by the common notion that an hypothesis is untrustworthy simply because it is an hypothesis. It is often urged, in respect to some scientific conclusion, that, after all, it is only an hypothesis. But what more have we to guide us in nine-tenths of the most important affairs of daily life than hypotheses and often very ill-based ones? So that in science, where the evidence of an hypothesis is subjected to the most rigid examination, we may rightly pursue the same course. You may have hypotheses and hypotheses. A man may say, if he likes, that the moon is made of green cheese: that is an hypothesis. But another man, who has devoted a great deal of time and attention to the subject, and availed himself of the most powerful telescopes and the results of the observations of others, declares that in his opinion it is probably composed of materials very similar to those of which our own earth is made up: and that is also only an hypothesis. But I need not tell you that there is an enormous difference in the value of the two hypotheses. That one which is based on sound scientific knowledge is sure to have a corresponding value; and that which is a mere hasty random guess is likely to have but little value. Every great step in our progress in discovering causes has been made in exactly the same way as that which I have detailed to you. A person observing the occurrence of certain facts and phenomena asks, naturally enough, what process, what kind of operation known to occur in Nature applied to the particular case, will unravel and explain the mystery? Hence you have the scientific hypothesis; and its value will be proportionate to the care and completeness with which its basis had been tested and verified. It is in these matters as in the commonest affairs of practical life: the guess of the fool will be folly, while the guess of the wise man will contain wisdom. In all cases, you see that the value of the result depends on the patience and faithfulness with which the investigator applies to his hypothesis every possible kind of verification.

## STUDY QUESTIONS

1. What is the "method of scientific investigation"; Are you convinced that this method is really as close to everyday thinking as Huxley claims? Discuss.
2. What are Huxley's definitions for *induction* and *deduction*? What role do they play in forming a *hypothesis*?
3. Discuss how the method of scientific investigation can be of value in solving problems that aren't related to science.

# Are All Generalizations False ?

## Lionel Ruby and Robert Yarber

Generalizations can be misleading, but they are also necessary. Imagine trying to
go through a single day without relying on generalizations! Ruby and Yarber illustrate some of the pitfalls of generalizing, but they also affirm the need to generalize.

### We Begin with a Generalization: Human Beings Are Great Generalizers.

Every race has its proverbs, and proverbs are generalizations. "It never rains but it
pours." "Faint heart never won fair lady." "Familiarity breeds contempt." Sometimes, of course, these proverbs are incompatible with each other, as in "Absence
makes the heart grow fonder," and "Out of sight, out of mind."[1]

Listen attentively to those around you, and note the generalizations that float
into every conversation: Women drivers are the most careless. Professors are absent-
minded. The Irish are alcoholics. Gentlemen prefer blondes. Politicians are crooks.
The French are great lovers. People on welfare don't want to work. And so on.
After more of the same we may be tempted to agree with Justice Holmes that "the
chief end of man is to frame general propositions, and no general proposition is
worth a damn."

Our awareness of the inadequacy of "sweeping generalizations" may lead us
to say that all generalizations are false. But this is truly a sweeping generalization!
And worse: if it is true, then the witticism that "all generalizations are false, *including this one*" would appear to be justified. But this will not do either, for this
generalization asserts that it itself is false, from which it follows that it is not the
case that all generalizations are false. Or perhaps we should say that "all generalizations are half-truths — including this one"? But this is not much better. The fact of
the matter is that some generalizations are true, others are false, and still others are
uncertain or doubtful. The deadliness of this platitude may be forgiven because of
its truth.

By a "generalization" is meant a general law or principle which is inferred
from particular facts. As a sample of the way in which we arrive at such generalizations, consider the following: Some years ago I saw my first Italian movie. The
directing, the acting, the dialogue, the lighting — all were superior. Encouraged by
this initial experience, I saw another Italian movie. It, too, was enjoyable. I saw
other Italian movies, always with the same results — comedies, dramas, "Westerns,"
thrillers. I generalized: All Italian movies are enjoyable.

A generalization is a statement that *goes beyond* what is actually observed, to
a rule or law covering both the observed cases and those that have not as yet been

From Lionel Ruby and Robert Yarber, *The Art of Making Sense* (New York: Lippincott,
1974), pp. 151–156.

[1]Once translated by a foreign student as "invisible idiot."

observed. This going-beyond is called the "inductive leap." An inductive leap is a "leap in the dark," for the *generalization may not be true,* even though the *observations* on which it is based *are* true. Thus, there may be a bad Italian movie — happily I have not seen it — but if so, then I should not say that *all* are good.

A generalization involves an "inductive leap." The word *induction,* from Latin roots meaning "to lead in," means that we examine particular cases and "lead in" to a generalization. Induction is the method we use when we learn lessons from our experience; we generalize from particular cases. *Deduction,* on the other hand, refers to the process of "drawing out" the logical consequences of what we already know (or assume) to be true. By induction we learn that Italian movies are enjoyable. If a friend tells us that he saw a bad movie, then by deduction we know that he did not see an Italian movie. Both induction and deduction are essential characteristics of rational thinking.

A generalization is a statement of the form: "All A's are B's." "All" means exactly what is says: *all* without exception. A single exception overthrows a generalization of this kind. Before we proceed further we must first dispose of a popular confusion concerning the expression, "The exception proves the rule." This is a sensible statement when properly interpreted, but it is sometimes understood in a manner that makes it nonsense. If I say that "all A's are B's," a single exception will make my statement false. Now, suppose that someone says, "The fact that there is a bad Italian movie proves that *all* are good because *it* is an exception, and the exception proves the rule!" Does a wicked woman prove that all women are saints? The sensible interpretation of the expression, "The exception proves the rule," is this: When we *say* that a certain case *is* an exception," we imply that there is a rule which generally holds. When a mother tells her daughter, "Have a good time at the prom, and, for tonight, you have my permission to stay out until 3 A.M.," she implies that this is an exception to the rule which requires earlier reporting. A statement that *creates* an exception implies a rule for all nonexceptional cases, but a generalization that is stated as a rule without exceptions (all A's are B's) would be overthrown by a single exception.

Scientific laws, stated in the form "All A's are B's," or some variation thereof, are never "violated." When an exception to a law is definitely established, the law in its previous form is abandoned, but it may be possible to revise it to exclude the "exception" as a special case because of special circumstances. The revised law: "All A's, under such and such conditions, are B's." Water freezes at 32°F. *at sea level.*

All too often "general propositions are not worth a damn," as Holmes remarked. This is because we generalize too hastily on the basis of insufficient evidence. The fallacy called the "hasty generalization" simply refers to the fact that we jump too quickly to conclusions concerning "all." For example, we see a woman driving carelessly, and generalize: "All women are poor drivers." We see a car weaving in and out of traffic, and note that it has a California license: "Wouldn't you know," we say. "A California driver. That's the way they all drive out there." Anita Loos's gay heroine thought that gentlemen preferred blondes because she was a blonde and men were attracted to her.

We learn that Napoleon got along on five hours of sleep. From this we may conclude that "five hours of sleep is all that anybody really needs." Our assumption is that what Napoleon could do, anybody can do, until we learn that we are not Napoleons. (If we don't learn this eventually, we aren't permitted to circulate freely.) The next example is undoubtedly the worst example of generalizing ever committed: A man declared that all Indians walk in single file. When challenged for his evidence, he replied, "How do I know that? I once saw an Indian walk that way."

Hasty generalizing is perhaps the most important of popular vices in thinking. It is interesting to speculate on some of the reasons for this kind of bad thinking. One important factor is prejudice. If we are already prejudiced against unions or businessmen or lawyers or doctors or Jews or Negroes or whites or gentiles, then one or two instances of bad conduct by members of these groups will give us the unshakable conviction that "they're all like that." It is very difficult for a prejudiced person to say, "Some are, and some aren't." A prejudice is a judgment formed *before* examining the evidence.

A psychological reason for asserting "wild" generalizations is exhibitionism: The exhibitionist desires to attract attention to himself. No one pays much attention to such undramatic statements as "Some women are fickle," or "Some politicians are no better than they ought to be." But when one says that "all men are liars," this immediately attracts notice. Goethe once said that it is easy to appear brilliant if one respects nothing, not even the truth.

Let us avoid careless and hasty generalizing. The fault of bad generalizing, however, need not make us take refuge in the opposite error — the refusal to generalize. This error is illustrated in the anecdote concerning the student who wrote an essay on labor relations, in which he argued for equal pay for women. Women, he wrote, word hard; they need the money; they are the foundation of the family; and, above all, they are the mothers of most of the human race! There is an old anecdote about the cautious man whose friend pointed to a flock of sheep with the remark, "Those sheep seem to have been sheared recently." "Yes," said the cautious man, "at least on this side."

Generalizations are dangerous, but we must generalize. To quote Justice Holmes once more: he said that he welcomed "anything that will discourage men from believing general propositions." But, he added, he welcomed that "only less than he welcomed anything that would encourage men to make such propositions"! For generalizations are indispensable guides. One of the values of knowledge lies in its predictive power — its power to predict the future. Such knowledge is stated in generalizations. It is of little help to me to know that water froze at 32° F. yesterday unless this information serves as a warning to put antifreeze in my car radiator before winter comes. History, in the "pure" sense of this term, merely tells us what has happened in the past, but science furnishes us with general laws, and general laws tell us what *always* happens under certain specified conditions.

Science is interested in the general, rather than in the particular or individual. When Newton saw an apple fall from a tree in his orchard — even if this story is a fable, and therefore false in a literal sense, it is true in its insight — he was not interested in the size and shape of the apple. Its fall suggested an abstract law to him, the law of gravity. He framed this law in general terms: Every particle of matter attracts every other particle of matter with a force directly proportional to the product of their masses and inversely proportional to the square of their distances. Chemists seek general laws concerning the behavior of matter. The physician wants to know the general characteristics of the disease called myxedema, so that when he has a case he will recognize it and know exactly how to treat it. The finding of general laws, then, is the aim of all science — including history insofar as it is a science.

The problem of the scientist is one of achieving sound generalizations. The scientist is careful not to make assertions which outrun his evidence, and he refuses to outtalk his information. He generalizes, but recognizes that no generalization can be more than probable, for we can never be certain that *all* the evidence is in, nor can the future be guaranteed absolutely — not even future eclipses of the sun and moon. But the scientist knows that certain laws have a very high degree of probability.

Let us look at the logic involved in forming sound generalizations. The number of cases investigated in the course of formulating a scientific law is a factor in establishing the truth of the law, but it is by no means the most important one. Obviously, if we observed one hundred swans, all of which are white, our generalization that "all swans are white" does not have the same probability it would have if we observed one thousand swans. But no matter how great the number of specimens involved in this type of observation, no more than a moderately high degree of probability is ever established. Countless numbers of white swans were observed throughout the ages (without any exceptions), and then in the nineteenth century black swans were observed in Australia.

The weakness of the method of "induction by simple enumeration of cases" is amusingly illustrated by Bertrand Russell's parable in his *History of Western Philosophy.*:

> There was once upon a time a census officer who had to record the names of all householders in a certain Welsh village. The first that he questioned was called William Williams; so were the second, third, fourth. . . . At last he said to himself: "This is tedious; evidently they are all called William Williams. I shall put them down so and take a holiday." But he was wrong; there was just one whose name was John Jones."

Scientific generalizations based on other types of evidence than simple enumeration often acquire a much higher degree of probability after only a few observations. When a chemist finds that pure sulphur melts at 125° C. in an experiment in which every factor is accurately analyzed and controlled, the law concerning the melting point of sulphur achieves as great a degree of certainty as is humanly attainable. Accurate control of every element of one case, then, is more important in establishing probabilities than is *mere enumeration* of many cases.

A single carefully controlled experiment, such as the sulphur experiment, can give us a much higher degree of probability than the mere observation of thousands of swans. The reason is that we also know that no chemical element thus far observed has a variable melting point under conditions of constant pressure. The chemical law is thus consistent with and is borne out by the rest of chemical knowledge, whereas the "law" holding that all swans are white was based on an "accidental" factor. Or consider the generalization concerning the mortality of mankind. This law is based not merely on the fact that countless numbers of human beings have died in the past, but also on the fact that all living beings must, by reason of physiological limitations, die, and that all matter wears out in time. So the harmony of a particular generalization with the rest of our knowledge is also a factor in giving it a high degree of probability.

So much for the logical analysis of generalizations. Thus far, we have been concerned with "uniform" generalizations, which take the form: "All A's are B's." A generalization, we have seen, is a statement that says something about "all" of a group, the evidence consisting of observations of items in which we always find a single characteristic. The observed cases are taken as a *sample* of the whole group or population with which we are concerned. We observe a number of swans and take these as a sample of all swans, past, present, and future. We find that all are white, and make the inductive leap: Swans are always white, everywhere.

### STUDY QUESTIONS
1. What is the "inductive leap"? What are the possible shortcomings of the "inductive leap"?
2. What is a hasty generalization? What fault is opposite to the hasty generalization?
3. What are the similarities between *generalization* and *hypothesis*? What are the differences? (Review Huxley's discussion of *hypothesis.*)

# Causal Connections

## Irving M. Copi

If we drop a glass on a hard surface it shatters. If we place a plastic utensil too close to a source of intense heat it will melt. In these cases, we know that one event inevitably leads to another. We know that a *cause* (impact on a hard surface) will have an effect (a shattered glass). Our knowledge of causality enables us to avoid an effect by removing its cause, or to bring about an effect by introducing its cause.

From Irving M. Copi, *Introduction to Logic,* Third edition (New York: Macmillan, 1968), pp. 322–25.

Because a knowledge of causality can give us control over events, it is an important tool in problem-solving. Copi's selection introduces some of the specialized terminology necessary for a precise understanding of causal relationships.

## The Meaning of "Cause"

To exercise any measure of control over our environment, we must have some knowledge of causal connections. A physician has more power to cure illnesses if he knows what *causes* them, and he should understand the *effects* of the drugs he administers. Since there are several different meanings of the word "cause," we begin by distinguishing them from one another.

It is a fundamental axiom in the study of nature that events do not just happen, but occur only under certain conditions. It is customary to distinguish between necessary and sufficient conditions for the occurrence of an event. A *necessary* condition for the occurrence of a specified event is a circumstance in whose absence the event cannot occur. For example, the presence of oxygen is a necessary condition for combustion to occur: if combustion occurs, then oxygen must have been present, for in the absence of oxygen there can be no combustion.

Although it is a necessary condition, the presence of oxygen is not a sufficient condition for combustion to occur. A *sufficient* condition for the occurrence of an event is a circumstance in whose presence the event must occur. The presence of oxygen is not a sufficient condition for combustion because oxygen can be present without combustion occurring. On the other hand, for almost any substance there is some range of temperature such that *being in that range of temperature in the presence of oxygen* is a sufficient condition for combustion of that substance. It is obvious that there may be several *necessary* conditions for the occurrence of an event, and that they must all be included in the sufficient condition.

The word "cause" is sometimes used in the sense of necessary condition and sometimes in the sense of sufficient condition. It is most often used in the sense of necessary condition when the problem at hand is the elimination of some undesirable phenomenon. To eliminate it, one need only find some condition which is necessary to its existence, and then eliminate that condition. Thus a physician seeks to discover what kind of germ is the "cause" of a certain illness in order to cure the illness by prescribing a drug which will destroy those germs. The germs are said to be the cause of the disease in the sense of a necessary condition for it, since in their absence the disease cannot occur.

The word "cause" is used in the sense of sufficient condition when we are interested not in the elimination of something undesirable but rather in the production of something desirable. Thus a metallurgist seeks to discover the cause of strength in alloys in order to be able to produce stronger metals. The process of mixing and heating and cooling is said to be the cause of the strengthening in the sense of a sufficient condition, since such processing suffices to produce a stronger alloy.

In certain practical situations, the word "cause" is used in still a different sense. An insurance company might send an investigator to determine the cause of a mysterious fire. If the investigator sent back a report that the fire was caused by the presence of oxygen in the atmosphere, he would not keep his job very long. And yet he would be right — in the sense of necessary condition — for had there been no oxygen present, there would have been no fire. But the insurance company did not have *that* sense in mind when they sent him to investigate. Nor is the company interested in the sufficient condition. If after several weeks the investigator reported that although he had proof that the fire was deliberately ignited by the policyholder, he hadn't as yet been able to learn *all* the necessary conditions, and so hadn't been able to determine the cause (in the sense of sufficient condition), the company would recall the investigator and tell him to stop wasting his time and their money. The insurance company was using the word "cause" in another sense — what they wanted to find out was the incident or action which, in the presence of those conditions that usually prevail, made the difference between the occurrence or nonoccurrence of the event.

We may distinguish between two different subdivisions of this third sense of cause. These are traditionally characterized as the *remote* and the *proximate* causes. Where there is a causal sequence or chain of several events, $A$ causing $B$, $B$ causing $C$, $C$ causing $D$, and $D$ causing $E$, we can regard $E$ as effect of any or all of the preceding events. The nearest of them, $D$, is the proximate cause of $E$, and the others are more and more remote causes, $A$ more remote than $B$, and $B$ more remote than $C$. In this case the proximate cause was the policyholder's lighting the fire. But his action, and thus the fire, may have been caused by his wife's nagging him for more money, and her nagging may have been caused by a neighbor's wife getting a new fur coat, which may have been caused by the neighbor's grain speculations turning out well because of rising food prices which were caused by a crop failure in India. The crop failure was a remote cause of the fire, but the insurance company would not have been interested in hearing that the mysterious fire was caused by an Indian crop failure.

## STUDY QUESTIONS

1. What is a *necessary condition*? Does Copi's example of oxygen help you in understanding the term? Can you think of other examples of necessary conditions?
2. What is a *sufficient condition*?
3. Copi says that the word *cause* "is most often used in the sense of necessary condition when the problem at hand is the elimination of some undesirable phenomenon." Explain what he means. What example does he give? What examples can you give?
4. Copi says that "the word 'cause' is used in the sense of sufficient condition when we are interested not in the elimination of something undesirable but rather in the production of something desirable." Explain what he means. What example does he give? What examples can you give?

5. What third sense of cause does Copi discuss?
6. Why is the insurance company interested in cause in this third sense?
7. Why is the insurance company interested in the *proximate* rather than the *remote* cause?
8. Discuss other examples of remote and proximate causes.

# What the Hedgehog Knows

## Garrett Hardin

Some actions start a chain of causal events that lead to undesirable results. Garrett Hardin's lively and informal discussion of causality illustrates the need to consider the larger system when planning projects.

Among the fragments left us by the Greek poet Archilochus there is a line, dark in meaning, that says:

The fox knows many things; the hedgehog knows one big thing.

Isaiah Berlin, who resurrected this enigmatic utterance, uses it to good effect to divide the great literary figures into two classes. (He is fully aware of the dangers of pressing any dichotomy too far, but he finds this one useful nonetheless.)

There exists [he says] a great chasm between those, on the one side, who relate everything to a single central vision, one system less or more coherent or articulate, in terms of which they understand, think and feel — a single, universal, organizing principle in terms of which alone all that they are and say has significance — and, on the other side, those who pursue many ends, often unrelated . . .

Among the latter — the foxes — he places Aristotle, Montaigne, Molière, Goethe, Balzac, and Joyce. The hedgehogs include Plato, Lucretius, Pascal, Nietzsche, and Proust.

Ecologists in my opinion, are hedgehogs. The one big thing they *say* is this:

*We can never do merely one thing.*

This simple sentence imperfectly mirrors the one big thing ecologists *know* — the idea of a system. So large an idea is best defined ostensively, i.e., by pointing to examples.

From Garrett Hardin, *Exploring New Ethics for Survival* (New York: Viking, 1972), pp. 38–41.

The ostensive work of defining a system can begin with a not entirely sober example, first cited by Charles Darwin:

> The number of humble-bees in any district depends in a great measure upon the number of field-mice which destroy their combs and nests; and Col. Newman who has long attended to the habits of humble-bees, believes that "more than two-thirds of them are thus destroyed all over England." Now the number of mice is largely dependent, as everyone knows, on the number of cats; and Col. Newman says, "Near villages and small towns I have found the nests of humble-bees more numerous than elsewhere, which I attribute to the number of cats that destroy the mice." Hence it is quite credible that the presence of a feline animal in large numbers in a district might determine, through the intervention first of mice and then of bees, the frequency of certain flowers in that district!

Darwin published this story in the *Origin of Species,* in 1859. Others, amused, embroidered on it. It was pointed out, on the one hand, that cats are kept (as is well known) by old maids; and, on the other, that red clover (which requires humble-bees as pollinators) is used to make the hay that nourishes the horses of the cavalry, on which the defense of the British Empire depends. Putting all this into a causal chain, we see:

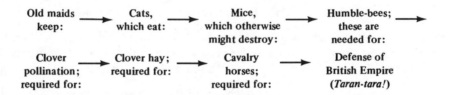

Thus, "it logically follows" that the perpetuation of the British Empire is dependent on a bountiful supply of old maids!

Far-fetched? A bit; but counsel for the defense can offer this fact in evidence: old maids are much less common in England now than they were when Colonel Newman and Darwin started this train of thought. And what has happened to Britain's possessions in India? In East Africa? And what about the Suez Canal? If correlation is causation . . .

Now for a completely serious example. Everyone has heard of the Aswan Dam in Egypt. Actually, a succession of dams has been constructed on the Nile at Aswan during the twentieth century, of which the "High Aswan," built with Russian help, is only the latest. The builders meant to do only one thing — dam the water — and that for two purposes: to generate electricity and to provide a regular flow of water for irrigation of the lower Nile basin. Ecology tells us that we cannot do merely one thing; neither can we do merely two things. What have been the consequences of the Aswan dams?

First, the plain has been deprived of the annual fertilization by flooding that served it so well for five thousand years. Where else in the world can you point to farmland that is as fertile after five millennia of cultivation as it was in the beginning? (Much of the farmland in the southeastern United States was ruined in half a generation.) Now the Egyptians will have to add artificial fertilizer to the former floodplains of the Nile — which will cost money.

Second, controlled irrigation without periodic flushing salinates the soil, bit by bit. There are methods for correcting this, but they too cost money. This problem has not yet been faced by Egypt.

Third, the sardine catch in the eastern Mediterranean has diminished from 18,000 tons a year to 400 tons, a 97 percent loss, because the sea is now deprived of floodborne nutrients. No one has reimbursed the fishermen (who are mostly not Egyptians) for their losses.

Fourth, the rich delta of the Nile is being eroded away by storms on the Mediterranean. In the past, a nearly "steady state" existed between the deposition of silt by the river and the erosion of it by the sea, with a slight positive balance in favor of deposition, which gradually extended the farmlands of Egypt. With deposition brought to a virtual halt, the balance is now negative and Egypt is losing land.

Fifth, schistosomiasis (a fearsomely debilitating disease) has greatly increased in the Nile valley. The disease organism depends on snails, which depend on a steady supply of water, which constant irrigation furnishes but annual flooding does not. Of course, medical control of the disease is possible — but that too costs money.

Is this all? By no means. The first (and perhaps only a temporary) effect of the Aswan Dam has been to bring into being a larger population of Egyptians, of whom a greater proportion than before are chronically ill. What will be the political effects of this demographic fact? This is a most difficult question — but would anyone doubt that there will be many political consequences, for a long time to come, of trying to do "just one thing," such as building dams on the Nile? The effects of any sizable intervention in an "ecosystem" are like ripples spreading out on a pond from a dropped pebble; they go on and on.

Ironically, in the end, the whole wretched game will return to the starting point. All dam-ponds are transient: in the scale of historical time they are soon filled by siltation behind the dam, and then they are useless. Theoretically, a dam-pond could be dredged clean, but engineers, inclined though they are to assume they can do anything, never suggest this. Evidently the cost is, in the strict sense, prohibitive.

So in a short time — perhaps a century, certainly nothing like the fifty centuries during which Nile agriculture prospered before the dams were built at Aswan — in a short time the dams themselves will be uselss and then the silt-laden waters of the Nile will once again enrich the river bottom.

Will history start over again then? Certainly not. Too much will have happened to Egypt in the meantime, most of it bad; and perhaps much to the rest of the world, as the Egyptian people struggle desperately to free themselves from the net in which their well-wishers have unwittingly ensnared them. Things will never again be the same. History does not repeat.

## STUDY QUESTIONS

1. What does Hardin mean by his statement "We can never do merely one thing"?
2. What errors in reasoning can you see in the chain of events that lead from old maids' cats to the defense of the British Empire? Use Copi's terminology to label the errors.
3. Why does Hardin use this "not entirely sober example"?
4. According to Hardin, what does the Aswan Dam illustrate?
5. Hardin uses causality to analyze past events that we cannot now remedy completely. How could the planners of the Aswan Dam have avoided the ill effects that it eventually caused?

# Thinking by Visual Images

## Robert H. McKim

Most of us are probably not aware of just how much of our thinking is visual. McKim alerts us to the role of visual imagery in thinking, and discusses some of the ways that "seeing" helps in problem-solving.

### Visual Thinking is Pervasive

Visual thinking pervades all human activity, from the abstract and theoretical to the down-to-earth and everyday. An astronomer ponders a mysterious cosmic event; a football coach considers a new strategy; a motorist maneuvers his car along an unfamiliar freeway: all are thinking visually. You are in the midst of a dream; you are planning what to wear today; you are making order out of the disarray on your desk: *you* are thinking visually.

Surgeons think visually to perform an operation; chemists to construct molecular models; mathematicians to consider abstract space-time relationships; engineers to design circuits, structures, and mechanisms; businessmen to organize and schedule work; architects to coordinate function with beauty; carpenters and mechanics to translate plans into things.

From Robert H. McKim, *Experiences in Visual Thinking* (Monterey, California: Brooks/ Cole, 1972), pp. 6-11.

Visual thinking, then, is not the exclusive reserve of artists. As Arnheim[1] observes, "Visual thinking is constantly used by everybody. It directs figures on a chessboard and designs global politics on the geographical map. Two dexterous moving-men steering a piano along a winding staircase think visually in an intricate sequence of lifting, shifting, and turning. . . . An inventive housewife transforms an uninviting living room into a room for living by judiciously placing lamps and re-arranging couches and chairs."

## See/Imagine/Draw

Visual thinking is carried on by three kinds of visual imagery:

1. the kind that we *see* "People see images, not things."[2]
2. the kind that we *imagine* in our mind's eye, as when we dream.
3. the kind that we *draw,* doodle, or paint.

Although visual thinking can occur primarily in the context of seeing, or only in imagination, or largely with pencil and paper, the expert visual thinker flexibly utilizes all three kinds of imagery. He finds that seeing, imagining, and drawing are interactive.

## Interactive Imagery

The interactive nature of seeing, imagining, and drawing is shown diagrammatically in Figure 1. The overlapping circles can be taken to represent a wide variety of

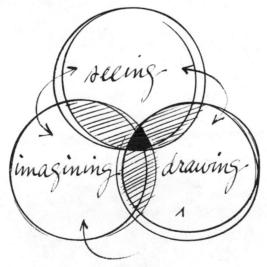

Figure 1.

[1] Arnheim, R. "Visual Thinking," in *Education of Vision* (edited by G. Kepes). Braziller.
[2] Feldman, E. *Art as Image and Idea.* Prentice-Hall.

interactions. Where seeing and drawing overlap, seeing facilitates drawing, while drawing invigorates seeing. Where drawing and imagining overlap, drawing stimulates and expresses imagining, while imagining provides impetus and material for drawing. Where imagining and seeing overlap, imagination directs and filters seeing, while seeing, in turn, provides raw material for imagining. The three overlapping circles symbolize the idea that visual thinking is experienced to the fullest when seeing, imagining, and drawing merge into active interplay.

The visual thinker utilizes seeing, imagining, and drawing in a fluid and dynamic way, moving from one kind of imagery to another. For example, he *sees* a problem from several angles and perhaps even chooses to solve it in the direct context of seeing. Now prepared with a visual understanding of the problem, he *imagines* alternative solutions. Rather than trust to memory, he *draws* a few quick sketches, which he can later evaluate and compare. Cycling between perceptual, inner, and graphic images, he continues until the problem is solved.

Experience this interplay between perceptual, inner, and graphic images for yourself, as you solve this challenging and somewhat difficult classic puzzle:

### 1-1/Pierced Block

Figure 2 shows a solid block that has been pierced with circular, triangular, and square holes. The circle's diameter, the triangle's altitude and base, and the square's sides all have the same dimension. The walls of the three holes are perpendicular to the flat front face of the block. Visualize a single, solid object that will pass *all the way through* each hole and, en route, entirely block the passage of light.

Use seeing, imagining, and drawing to solve this problem, as follows:

1. Simulate the pierced block with a cardboard cut-out. With scissors and cardboard, seek to see a solution by actual "cut-and-try" methods.
2. Close your eyes and seek a solution in your imagination.
3. Make sketches; seek a graphic solution.
4. Consciously alternate between steps 1, 2, and 3.

(An answer to this puzzle is illustrated at the end of this chapter.)

Visual thinking is obviously central to the practice of architecture, design,

Figure 2.

and the visual arts. Less obvious is the importance of visual thinking to other disciplines, such as science and technology. In the next few pages I will present a few brief accounts of seeing, imagining, and drawing in the thinking of scientists and technologists. Interspersed with these, I have placed related problems that will help you to relate these accounts of others to your own experience.

## Seeing and Thinking

Discoveries in the direct context of seeing are common in the history of science. For example, Sir Alexander Fleming "was working with some plate cultures of staphylococci which he had occasion to open several times and, as often happens in such circumstances, they became contaminated. He noticed that the colonies of staphylococci around one particular colony had died. Many bacteriologists would not have thought this particularly remarkable for it has long been known that some bacteria interfere with growth of others. Fleming, however, saw the possible significance of the observation and followed it up to discover penicillin."[3]

Why did Fleming discover penicillin when another scientist saw it and considered it only a nuisance? Because habits of seeing and thinking are intimately related. Fleming, like most creative observers, possessed a habit of mind that permitted him to see things afresh, from new angles. Also he was not burdened by that "inveterate tradition according to which thinking takes place remote from perceptual experience."[1] He didn't look and *then* sit down to think, he used his active eyes and mind *together*.

### 1-2/Cards and Discards[4]

Experience using your eyes and mind together in the following puzzle: "In [the] row of five cards shown below, there is only one card correctly printed, there being some mistake in each of the other four. How quickly can you find the mistakes?"

Figure 3.

[3] Beveridge, W. *The Art of Scientific Investigation.* Random House (Vintage Books).
[4] Kaufman, G. *The Book of Modern Puzzles.* Dover.

Another form of thinking in the context of seeing is described by Nobel Laureate James D. Watson, in *The Double Helix*,[5] a fascinating account of the discovery of the structure of the DNA molecule. Watson and his colleagues visualized this complex structure by interacting directly with a large three-dimensional model. He writes: "Only a little encouragement was needed to get the final soldering accomplished in the next couple of hours. The brightly shining metal plates were then immediately used to make a model in which for the first time all the DNA components were present. In about an hour I had arranged the atoms in positions which satisfied both the x-ray data and the laws of stereochemistry. The resulting helix was right-handed with the two chains running in opposite directions.

> . . . Another fifteen minutes' fiddling by Francis [Crick] failed to find anything wrong, though for brief intervals my stomach felt uneasy when I saw him frowning."

Although a complex structure such as the DNA molecule is difficult to visualize in imagination or on paper, one of Watson's colleagues scorned the model shown in Figure 4. However, Watson observed that his Nobel prize-winning success by this

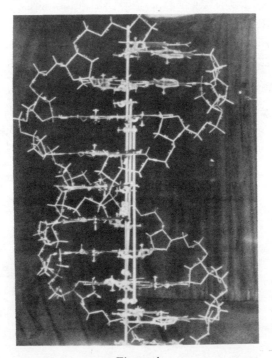

Figure 4.

[5]Watson, J. *The Double Helix*. Atheneum.

method of visual thinking convinced the doubter "that our past hooting about model-building represented a serious approach to science, not the easy resort of slackers who wanted to avoid the hard work necessitated by an honest scientific career."

Watson's account in *The Double Helix* also gives the reader excellent insight into the competitive excitement of science. Ideally, the next problem (an experience in thinking in the direct context of seeing) is given in the spirit of a competition.

### 1-3/Spaghetti Cantilever

With 18 sticks of spaghetti and 24 inches of Scotch tape, construct the longest cantilever structure that you can. Here are three additional constraints:

1. Tape-fasten the base of the structure within a 6-inch-square horizontal area.
2. Don't make drawings. Think directly with the materials.
3. Design and build the structure in 30 minutes.

(Measure length of cantilever from the point on the base nearest to the overhanging end of the cantilever.)

### Imagining and Thinking

Inner imagery of the mind's eye has played a central role in the thought processes of many creative individuals. In rare thinkers, this inner imagery is extremely clear. For example, Nikola Tesla, the technological genius whose list of inventions includes the fluorescent light and the A-C generator, "could project before his eyes a picture, complete in every detail, of every part of the machine. These pictures were more vivid than any blueprint."[6] Tesla's inner imagery was so like perceptual imagery that he was able to build his complex inventions without drawings. Further, he claimed to be able to test his devices in his mind's eye "by having them run for weeks — after which time he would examine them thoroughly for signs of wear."

Although labels lead us to think of the various sensory modes of imagination as though they occur separately, in actuality imagination is polysensory. Albert Einstein,[7] in a famous letter to Jacques Hadamard, described the important role of polysensory (visual and kinesthetic) imagination in his own extremely abstract thinking: "The words or the language, as they are written and spoken, do not seem to play any role in my mechanism of thought. The psychical entities which seem to serve as elements in thought are certain signs and more or less clear images which can be voluntarily reproduced and combined. . . . The above mentioned elements are, in my case, of visual and some of muscular type. Conventional words or other

[6] O'Neil, J. *Prodigal Genius: The Life of Nikola Tesla.* McKay (Tartan Books).
[7] Einstein, A. Quoted by Hadamard, J., in *The Psychology of Invention in the Mathematical Field.* Princeton University Press.

signs have to be sought for laboriously in a secondary stage, when the above mentioned associative play is sufficiently established and can be reproduced at will."

Although Einstein observed that his polysensory imagination could be directed "at will," many important thinkers have obtained imaginative insights more or less spontaneously. For example, the chemist Kekulé[8] came upon one of the most important discoveries of organic chemistry, the structure of the benzene ring, in a dream. Having pondered the problem for some time, he turned his chair to the fire and fell asleep: "Again the atoms were gamboling before my eyes. . . . My mental eye . . . could now distinguish larger structures . . . all twining and twisting in snake-like motion. But look! What was that? One of the snakes had seized hold of its own tail, and the form whirled mockingly before my eyes. As if by a flash of lightning I awoke." The spontaneous inner image of the snake biting its own tail suggested to Kekulé that organic compounds, such as benzene, are not open structures but closed rings.

Those of you who identify high intellectual endeavor exclusively with verbal and mathematical symbols should consider the introspections of Tesla, Einstein, and Kekulé with special care. Has something been overlooked in your education? The following problem, for example, is best solved by inner imagery. Has your education prepared you for this kind of problem-solving?

### 1-4/Painted Cube

Shut your eyes. Think of a wooden cube such as a child's block. It is painted. Now imagine that you take two parallel and vertical cuts through the cube, dividing it into equal thirds. Now take two additional vertical cuts, at 90° to the first ones, dividing the cube into equal ninths. Finally, take two parallel and horizontal cuts through the cube, dividing it into 27 cubes. Now, how many of these small cubes are painted on three sides? On two sides? On one side? How many cubes are unpainted?

Don't be disappointed if you did poorly on this problem. Mental manipulation of mind's-eye imagery improves with practice.

### Drawing and Thinking

Very few people possess the acuity of mind's eye that enabled Tesla to design and build complex machinery without drawings. Most visual thinkers clarify and develop their thinking with sketches. Watson, recollecting the thinking that preceded his discovery of the DNA structure, writes that one important idea "came while I was drawing the fused rings of adenine on paper."[5] An example of a chemical diagram drawn by Watson is shown in Figure 5. As in Watson's experience, drawing and thinking are frequently so simultaneous that the graphic image appears almost

---

[8] Kekulé, F. von. Quoted by Koestler, A., in *The Act of Creation*. Macmillan.

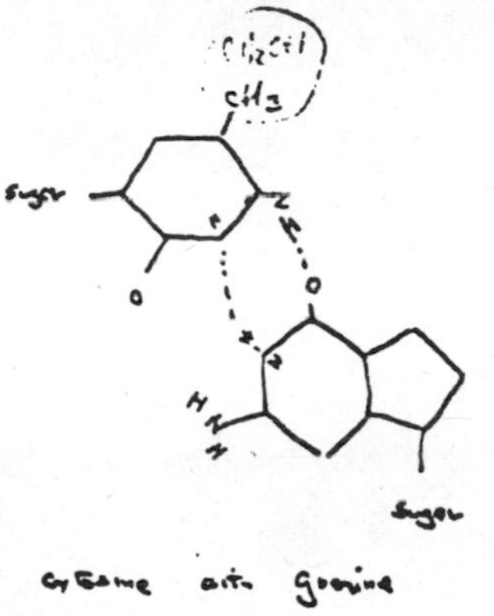

Figure 5.

an organic extension of mental processes. Thus Edward Hill, in *Language of Drawing*,[9] likens drawing to a mirror: "A drawing acts as the reflection of the visual mind. On its surface we can probe, test, and develop the workings of our peculiar vision."

Drawing not only helps to bring vague inner images into focus, it also provides a record of the advancing thought stream. Further, drawing provides a function that memory cannot: the most brilliant imager cannot compare a number of images, side by side in memory, as one can compare a wall of tacked-up idea-sketches.

Two idea-sketches from the notebook of John Houbolt, the engineer who conceived the Lunar Landing Module, are reproduced in Figure 6. Houbolt's drawings show two important attributes of graphic ideation. First, the sketches are relatively "rough." They are not intended to impress or even to communicate; instead, they are a kind of graphic "talking to oneself." Second, one sketch is an abstract schematic of the voyage from earth to moon and back; the other is a relatively more concrete sideview of the landing module. Idea-sketching, like thinking itself, moves fluidly from the abstract to the concrete.

Drawing to extend one's thinking is frequently confused with drawing to communicate a well-formed idea. *Graphic ideation precedes graphic communication;* graphic ideation helps to develop visual ideas *worth communicating.* Because thinking flows quickly, graphic ideation is usually freehand, impressionistic, and rapid. Because communication to others demands clarity, graphic communication is necessarily more formal, explicit, and time-consuming. Education that stresses graphic communication and fails to consider graphic ideation can unwittingly hamper visual thinking.

[9]Hill, E. *The Language of Drawing.* Prentice Hall (Spectrum Books).

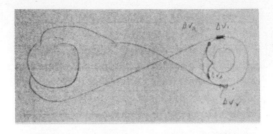

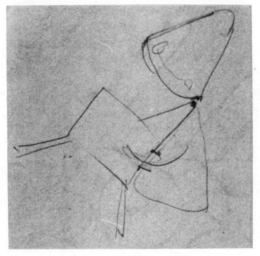

Figure 6.

Some problems are most easily solved by graphic means — for example, this one:

## 1-5/With One Line

With one continuous line that does not retrace itself, draw the pattern shown in Figure 7.

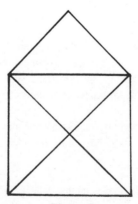

Figure 7.

## From Puzzles to Psychological Tests

The puzzles used to illustrate various modes of visual thinking in this chapter bear a striking resemblance to visual problems posed in psychological tests of intelligence.

### Answers

*Problem 1-1:* Here's one of the infinite number of answers to this problem. Can you generate some of the others? For example, what is the "minimum-volume" solution?

*Problem 1–2:* The fourth card is the correct one.

**Additional Bibliography**

Rudolf Arnheim's *Visual Thinking* (University of California Press) is a readable and erudite treatment of visual thinking, surely destined to be a classic. If you want to delve deeper into the theory of visual thinking, or if you want to be convinced that visual thinking is *the* most important kind of thinking, I highly recommend this book. It is now available in paperback.

**STUDY QUESTIONS**
1. Is visual thinking possible only in "artistic" professions? Explain.
2. When is drawing, sketching, or doodling most constructive?
3. What are the specific ways in which drawing helps us think?
4. What important discoveries have resulted from visual thinking?
5. What are some of the problems in your career field that might be approached with visual thinking?
6. With what kinds of problems could visual thinking not help you?

# Imagining a Lighthouse

## John Smeaton

Until the eighteenth century the lighthouses that English engineers built on the Eddystone Rocks were soon washed away by fierce Atlantic storms. When John Smeaton was commissioned to construct yet another lighthouse on Eddystone, he abandoned the unsuccessful designs and building methods of the past. Instead, he sought to find new methods and designs through analogy. That is, he considered how similar problems had been solved elsewhere, then applied those solutions to the specific circumstances of his problem. Note that he finds one useful analogy in nature and another in carpentry, a craft whose methods are not usually applied to stone construction.

The success of Smeaton's analogical approach is evident in his results. His lighthouse withstood ocean storms for 123 years. It was dismantled in 1882 only because the reef on which it was built was crumbling away.

From John Smeaton, *A Narrative of the Building and a Description of the Construction of the Eddystone Lighthouse with Stone,* 1793.

The natural figure of the waist or bole of a large spreading *Oak,* presented it-self to my imagination. Let us for a moment consider this tree: suppose at twelve or fifteen feet above its base, it branches out in every direction, and forms a large bushy top, as we often observe. This top, when full of leaves, is subject to a very great impulse from the agitation of violent winds; yet party by its elasticity, and partly by the natural strength arising from its figure, it resists them all, even for ages, till the gradual decay of the material diminishes the coherence of the parts, and they suffer piecemeal by the violence; but it is very rare that we hear of such a tree being torn up by the roots. Let us now consider its particular figure. Connected with its roots, which lie hid below ground, it rises from the surface thereof with a large swelling base, which at the height of one diameter is generally reduced by an elegant curve, concave to the eye, to a diameter less by at least one-third, and some-times to half of its original base. From thence its taper diminishing more slow, its side by degrees come into a perpendicular, and for some height form a cylinder. After that a preparation of more circumference becomes necessary, for the strong insertion and establishment of the principal boughs, which produces a swelling of its diameter. Now we can hardly doubt but every section of the tree is nearly of an equal strength in proportion to what it has to resist; and were we to lop off its principal boughs, and expose it in that state to a rapid current of water, we should find it as much capable of resisting the action of the heavier fluid, when divested of the greatest part of its cloathing, as it was that of the lighter when all its spreading ornaments were exposed to the fury of the wind: and hence we may derive an idea of what the proper shape of a column of the *greatest Stability* ought to be, to resist the action of external violence, when the *Quantity of Matter is given* whereof it is to be composed. . . .

The next thing was to consider how the blocks of stone could be *bonded* to the rock, and to one another, in so firm a manner, as that, not only the whole together, but every individual piece, when connected with what preceded, should be proof against the greatest violence of the sea: for, I plainly saw, from the rela-tions I had got, that as every part of the work, even in the most favourable seasons, was liable to be attacked by violent storms; if any thing was left to the mercy of the sea and good fortune, the building of the Edystone Lighthouse with stone would be tantamount to the rolling of the stone of Sisyphus.

On this head I considered the nature of *Cramping;* which, as generally per-formed amounts to no more than a *Bond* upon the upper surface of a course of stone, without having any direct power to hold a stone down, in case of its being lifted upward by an action greater than its own weight; as might be expected fre-quently to happen at the Edystone, whenever the mortar of the ground bed it was set upon was washed out of the joint, when attacked by the sea before it had time to harden; and though upright cramps to confine the stones down to the course below, might in some degree answer this end; yet as this must be done to each individual stone, the quantity of iron, and the great trouble and loss of time that would necessarily attend this method, would in reality render it impracticable. It therefore seemed of the utmost consequence to avoid this, even by any quantity of time and moderate expense, that might be necessary for its performance on shore; provided it prevented hindrance of business upon the rock: because of time upon the rock, there was likely to be a scarcity; but on the shore a very sufficient plenty.

This made me turn my thoughts to what could be done in the way of *dovetailing.* In speaking however of this as a term of art, I must observe that it had been principally applied to works of *Carpentry;* its application in the masonry way had been but very slight and sparing; for in regard to the small pieces of stone that had been let in with a double dovetail, across the joint of larger pieces, and generally to save iron, it was a kind of work even more objectionable than cramping; for though it would not require melted lead, yet being only a superficial bond, and consisting of far more brittle material than iron, it was not likely to answer our end at all. Somewhat more to my purpose I had occasionally observed in many places in the *Streets of London,* that in fixing the *Kirbs* of the walking paths, the long pieces or *Stretchers* were retained between two *Headers* or bond pieces; whose heads being cut dovetail-wise, adapted themselves to and confined in the stretchers; which expedient, though chiefly intended to save iron and lead, nevertheless appeared to me capable of more firmness than any superficial fastening could be; as the *tye* was as good at the bottom as at the top, which was the very thing I wanted; and therefore if the tail of the header was made to have an adequate bond with the interior parts, the work would in itself be perfect.

From these beginnings I was readily led to think, that if the blocks themselves were, both inside and out, all formed into large *dovetails,* they might be managed so as mutually to lock one another together; being primarily engrafted into the rock: and in the round or entire courses, above the top of the rock, they might all proceed from and be locked to one large center-stone.

## STUDY QUESTIONS

1. In the first paragraph, Smeaton recounts the method by which he arrived at the shape appropriate for a lighthouse. Draw the figure that he describes. Does it resemble present day lighthouses? What proof does he have that this shape will resist a storm?
2. Why must he resort to the methods of carpentry to set the stones of the lighthouse?
3. Read the discussion of *analogy* on p. 194. Does Smeaton use *figurative analogies* or *literal analogies*? Would either be appropriate for problem-solving? Why or why not?
4. Discuss other problems that might be solved by analogy.

# The Effective Decision

## Peter Drucker

Decision-making is closely related to problem-solving. The essential difference is that the problem-solver looks for a "right" answer, whereas the decision-maker might have to choose among several "right" answers for the *one that the organization will take action on.* The responsibility for decision-making belongs, of course, to management. But the techniques for decision-making are also important for the engineers, technicians, and scientists who must work at problem-solving, but do not have managerial responsibilities. As Peter Drucker observes, a decision by management has a chance of success only if there is cooperation among all who must work to implement it — including managers and non-managers alike. Although a single person must finally make the decision, he or she needs constructive criticism and suggestions from peers and subordinates. The needed cooperation, support, and constructive criticism are more likely to emerge when key personnel understand that decision-making is a group process, not the boss's whim. It is as a group process that decision-making "mobilizes the vision, energies, and resources of the organization for effective action."

In this selection Peter Drucker discusses some specific approaches to decision-making.

### Facts or Opinions?

A decision is a judgment. It is a choice between alternatives. It is rarely a choice between right and wrong. It is at best a choice between "almost right" and "probably wrong" — but much more often a choice between two courses of action neither of which is probably more nearly right than the other.

Most books on decision-making tell the reader: "First find the facts." But managers who make effective decisions know that one does not start with facts. One starts with opinions. These are, of course, nothing but untested hypotheses and, as such, worthless unless tested against reality. To determine what is a fact requires first a decision on the criteria of relevance, especially on the appropriate measurement. This is the hinge of the effective decision, and usually its most controversial aspect.

But also, the effective decision does not, as so many tests on decision-making proclaim, flow from a "consensus on the facts." The understanding that underlies the right decision grows out of the clash and conflict of divergent opinions and out of the serious consideration of competing alternatives.

From Peter Drucker, *Management: Tasks, Responsibilities, Practices* (New York: Harper and Row, 1974), pp. 470–480.

To get the facts first is impossible. There are no facts unless one has a criterion of relevance. Events by themselves are not facts.

Only by starting out with opinions can the decision-maker find out what the decision is all about. People do, of course, differ in the answers they give. But most differences of opinion reflect an underlying — and usually hidden — difference as to what the decision is actually about. They reflect a difference regarding the question that has to be answered. Thus to identify the alternative questions is the first step in making effective decisions.

Conversely, there are few things as futile — and as damaging — as the right answer to the wrong question.

The effective decision-maker also knows that he starts out with opinions anyhow. The only choice he has is between using opinions as a productive factor in the decision-making process and deceiving himself into false objectivity. People do not start out with the search for facts. They start out with an opinion. There is nothing wrong with this. People experienced in an area should be expected to have an opinion. Not to have an opinion after having been exposed to an area for a good long time would argue an unobservant eye and a sluggish mind.

People inevitably start out with an opinion; to ask them to search for the facts first is even undesirable. They will simply do what everyone is far too prone to do anyhow: look for the facts that fit the conclusion they have already reached. And no one has ever failed to find the facts he is looking for. The good statistician knows this and distrusts all figures — he either knows the fellow who found them or he does not know him; in either case he is suspicious.

The only rigorous method, the only one that enables us to test an opinion against reality, is based on the clear recognition that opinions come first — and that is the way it should be. Then no one can fail to see that we start out with untested hypotheses — in decision-making, as in science, the only starting point. We know what to do with hypotheses. One does not argue them; one tests them. One finds out which hypotheses are tenable, and therefore worthy of serious consideration, and which are eliminated by the first test against observable experience.

The effective decision-maker therefore encourages opinions. But he insists that the people who voice them also think through what it is that the "experiment" — that is, the testing of the opinion against reality — would have to show. The effective executive, therefore, asks, "What do we have to know to test the validity of this hypothesis?" "What would the facts have to be to make this opinion tenable?" And he makes it a habit — in himself and in the people with whom he works — to think through and spell out what needs to be looked at, studied, and tested. He insists that people who voice an opinion also take responsibility for defining what factual findings can be expected and should be looked for.

Perhaps the crucial question here is "What is the measurement appropriate to the matter under discussion and to the decision to be reached?" Whenever one analyzes the way a truly effective, a truly right, decision has been reached, one finds that a great deal of work and thought went into finding the appropriate measurement.

## The Need for Dissent and Alternatives

Unless one has considered alternatives, one has a closed mind. This, above all, explains why the Japanese deliberately disregard the second major command of the textbooks on decision-making and create discussion and dissent as a means to consensus.

Decisions of the kind the executive has to make are not made well by acclamation. They are made well only if based on the clash of conflicting views, the dialogue between different points of view, the choice between different judgments. The first rule in decision-making is that one does not make a decision unless there is disagreement.

Alfred P. Sloan, Jr., is reported to have said at a meeting of one of the GM top committees, "Gentlemen, I take it we are all in complete agreement on the decision here." Everyone around the table nodded assent. "Then," continued Mr. Sloan, "I propose we postpone further discussion of this matter until our next meeting to give ourselves time to develop disagreement and perhaps gain some understanding of what the decision is all about."

Sloan was anything but an "intuitive" decision-maker. He always emphasized the need to test opinions against facts and the need to make absolutely sure that one did not start out with the conclusion and then look for the facts that would support it. But he knew that the right decision demands adequate disagreement.

Every one of the effective presidents in American history had his own method of producing the disagreement he needed in order to make an effective decision. Washington, we know, hated conflicts and quarrels and wanted a united Cabinet. Yet he made quite sure of the necessary differences of opinion on important matters by asking both Hamilton and Jefferson for their opinions.

There are three reasons why dissent is needed. It first safeguards the decision-maker against becoming the prisoner of the organization. Everybody always wants something from the decision-maker. Everybody is a special pleader, trying — often in perfectly good faith — to obtain the decision he favors. This is true whether the decision-maker is the president of the United States or the most junior engineer working on a design modification.

The only way to break out of the prison of special pleading and preconceived notions is to make sure of argued, documented, thought-through disagreements.

Second, disagreement alone can provide alternatives to a decision. And a decision without an alternative is a desperate gambler's throw, no matter how carefully thought through it might be. There is always a high possibility that the decision will prove wrong — either because it was wrong to begin with or because a change in circumstances makes it wrong. If one has thought through alternatives during the decision-making process, one has something to fall back on, something that has already been thought through, studied, understood. Without such an alternative, one is likely to flounder dismally when reality proves a decision to be inoperative.

Both the Schlieffen Plan of the German Army in 1914 and President Franklin D. Roosevelt's original economic program in 1933 were disproved by events at the very moment when they should have taken effect.

The German Army never recovered. It never formulated another strategic concept. It went from one ill-conceived improvisation to the next. But this was inevitable. For twenty-five years no alternatives to the Schlieffen Plan had been considered by the General Staff. All its skills had gone into working out the details of this master plan. When the plan fell to pieces, no one had an alternative to fall back on. All the German generals could do, therefore, was gamble — with the odds against them.

By contrast, President Roosevelt, who, in the months before he took office, had based his whole campaign on the slogan of economic orthodoxy, had a team of able people, the later "Brains Trust," working on an alternative — a radical policy based on the proposals of the old-time Progressives, and aimed at economic and social reform on a grand scale. When the collapse of the banking system made it clear that economic orthodoxy had become political suicide, Roosevelt had his alternative ready. He therefore had a policy.

Above all, disagreement is needed to stimulate the imagination. One may not need imagination to find the *one right* solution to a problem. But then this is of value only in mathematics. In all matters of true uncertainty such as the executive deals with — whether his sphere be political, economic, social, or military — one needs creative solutions which create a new situation. And this means that one needs imagination — a new and different way of perceiving and understanding.

Imagination of the first order is, I admit, not in abundant supply. But neither is it as scarce as is commonly believed. Imagination needs to be challenged and stimulated, however, or else it remains latent and unused. Disagreement, especially if forced to be reasoned, thought through, documented, is the most effective stimulus we know.

The effective decision-maker, therefore, organizes dissent. This protects him against being taken in by the plausible but false or incomplete. It gives him the alternatives so that he can choose and make a decision, but also ensures that he is not lost in the fog when his decision proves deficient or wrong in execution. And it forces the imagination — his own and that of his associates. Dissent converts the plausible into the right and the right into the good decision.

## The Trap of "Being Right"

The effective decision-maker does not start out with the assumption that one proposed course of action is right and that all others must be wrong. Nor does he start out with the assumption "I am right and he is wrong." He starts out with the commitment to find out why people disagree.

Effective executives know, of course, that there are fools around and that there are mischief-makers. But they do not assume that the man who disagrees with what they themselves see as clear and obvious is, therefore, either a fool or a knave. They know that unless proven otherwise, the dissenter has to be assumed to be reasonably intelligent and reasonably fair-minded. Therefore, it has to be assumed that he has reached his so obviously wrong conclusion because he sees a different reality and is concerned with a different problem. The effective executive, therefore, always asks, "What does this fellow have to see if his position were, after all, tenable, rational, intelligent?" The effective executive is concerned first with *understanding*. Only then does he even think about who is right and who is wrong.

Needless to say, this is not done by a great many people, whether executives or not. Most people start out with the certainty that how they see is the only way to see at all. As a result, they never understand what the decision — and indeed the whole argument — is really all about.

The American steel executives have never asked the question "Why do these union people get so terribly upset every time we mention the word 'featherbedding'?" The union people in turn have never asked themselves why steel managements make such a fuss over featherbedding when every single instance thereof they have ever produced has proved to be petty, and irrelevant to boot. Instead, both sides have worked mightily to prove each other wrong. If either side had tried to understand what the other one sees and why, both would be a great deal stronger, and labor relations in the steel industry, if not in U.S. industry, might be a good deal healthier.

No matter how high his emotions run, no matter how certain he is that the other side is completely wrong and has no case at all, the executive who wants to make the right decision forces himself to see opposition as *his* means to think through the alternatives. He uses conflict of opinion as his tool to make sure all major aspects of an important matter are looked at carefully.

### Is a Decision Necessary?

There is one question the effective decision-maker asks: "Is a decision really necessary?" *One* alternative is always the alternative of doing nothing.

One has to make a decision when a condition is likely to degenerate if nothing is done. This also applies with respect to opportunity. If the opportunity is important and is likely to vanish unless one acts with dispatch, one acts — and one makes a radical change.

Theodore Vail's contemporaries agreed with him as to the degenerative danger of government ownership; but they wanted to fight it by fighting symptoms — fighting this or that bill in the legislature, opposing this or that candidate and supporting another, and so on. Vail alone understood that this is the ineffectual way to fight a degenerative condition. Even if one wins every battle, one can never win the war. He saw that drastic action was needed to create a new situation. He alone saw that private business had to make public regulation into an effective alternative to nationalization.

At the opposite end there are those conditions with respect to which one can, without being unduly optimistic, expect that they will take care of themselves even if nothing is done. If the answer to the question "What will happen if we do nothing?" is "It will take care of itself," one does not interfere. Nor does one interfere if the condition, while annoying, is of no importance and unlikely to make much difference.

It is a rare executive who understands this. The controller who in a financial crisis preaches cost reduction is seldom capable of leaving alone minor blemishes, elimination of which will achieve nothing. He may know, for instance, that the significant costs are in the sales organization and in physical distribution. And he will work hard and brilliantly at getting them under control. But then he will discredit himself and the whole effort by making a big fuss about the "unnecessary" employment of two or three old men in an otherwise efficient and well-run plant. And he will dismiss as immoral the argument that eliminating these few semipensioners will not make any difference anyhow. "Other people are making sacrifices," he will argue. "Why should the plant people get away with inefficiency?"

When it is all over, the organization will forget that he saved the business. They will remember, though, his vendetta against the two or three poor devils in the plant – and rightly so. *De minimis non curat praetor* (The magistrate does not consider trifles) said the Roman law almost two thousand years ago – but many decision-makers still need to learn it.

The great majority of decisions will lie between these extremes. The problem is not going to take care of itself; but it is unlikely to turn into degenerative malignancy either. The opportunity is only for improvement rather than for real change and innovation; but it is still quite considerable. If we do not act, in other words, we will in all probability survive. But if we do act, we may be better off.

In this situation the effective decision-maker compares effort and risk of action to risk of inaction. There is no formula for the right decision here. But the guidelines are so clear that decision in the concrete case is rarely difficult. They are:

— act if on balance the benefits greatly outweigh cost and risk; and
— act or do not act; but do not "hedge" or compromise.

The surgeon who takes out only half the tonsils or half the appendix risks as much infection and shock as if he did the whole job. And he has not cured the condition, has indeed made it worse. He either operates or he doesn't. Similarly, the effective decision-maker either acts or he doesn't act. He does not take half-action. This is the one thing that is always wrong.

## Who Has to Do the Work?

When they reach this point, most decision-makers in the West think they can make an effective decision. But, as the Japanese example shows, one essential element is still missing. An effective decision is a commitment to action and results. If it has to be "sold" *after* it has been made, there will be no action and no results – and, in effect, no decision. At the least, there may be so much delay as to obsolete the decision before it has become truly effective.

The first rule is to make sure that everyone who will have to do something to make the decision effective — or who could sabotage it — has been forced to participate responsibly in the discussion. This is not "democracy." It is salesmanship.

But it is equally important to build the action commitments into the decision from the start. In fact, no decision has been made unless carrying it out in specific steps has become someone's work assignment and responsibility. Until then, there are only good intentions.

This is the trouble with so many policy statements, especially of business: they contain no action commitment. To carry them out is no one's specific work and responsibility. No wonder that the people in the organization tend to view these statements cynically if not as declarations of what top management is really not going to do.

Converting a decision into action requires answering several distinct questions: "Who has to know of this decision?" "What action has to be taken?" "Who is to take it?" "And what does the action have to be so that the people who have to do it *can* do it?" The first and the last of these are too often overlooked — with dire results.

A story that has become a legend among management scientists illustrates the importance of the question "Who has to know?" A major manufacturer of industrial equipment decided to discontinue one model. For years it had been standard equipment on a line of machine tools, many of which were still in use. It was decided, therefore, to sell the model to present owners of the old equipment for another three years as a replacement, and then to stop making and selling it. Orders for this particular model had been going down for a good many years. But they shot up temporarily as former customers reordered against the day when the model would no longer be available. No one had, however, asked, "Who needs to know of this decision?" Therefore nobody informed the clerk in the purchasing department who was in charge of buying the parts from which the model itself was being assembled. His instructions were to buy parts in a given ratio to current sales — and the instructions remained unchanged. When the time came to discontinue further production of the model, the company had in its warehouse enough parts for another eight to ten years of production, parts that had to be written off at a considerable loss.

Above all, the action must be appropriate to the capacities of the people who have to carry it out.

A chemical company found itself, in the early sixties, with fairly large amounts of blocked currency in two West African countries. To protect this money, it decided to invest in local businesses which would contribute to the local economy, would not require imports from abroad, and would, if successful, be the kind that could be sold to local investors if and when currency remittances became possible again. To establish these businesses, the company developed a simple chemical process to preserve a tropical fruit which is a staple crop in both countries and which, up until then, had suffered serious spoilage in transit to its markets.

The business was a success in both countries. But in one country the local manager set the business up in such a manner that it required highly skilled and, above all, technically trained management of the kind not easily available in West Africa. In the other country the local manager thought through the capacities of the people who would eventually have to run the business and worked hard at making both process and business simple and at staffing from the start with nationals of the country right up to the top.

A few years later it became possible again to transfer currency from these two countries. But though the business flourished, no buyer could be found for it in the first country. No one available locally had the necessary managerial and technical skills. The business had to be liquidated at a loss. In the other country so many local entrepreneurs were eager to buy the business that the company repatriated its original investment with a substantial profit.

The process and the business built on it were essentially the same in both places. But in the first country no one had asked, "What kind of people do we have available to make this decision effective? And what can they do?" As a result, the decision itself became frustrated.

All this becomes doubly important when people have to change behavior, habits, or attitudes if a decision is to become effective action. Here one has to make sure not only that responsibility for the action is clearly assigned and that the people responsible are capable of doing the needful. One has to make sure that their measurements, their standards for accomplishment, and their incentives are changed simultaneously. Otherwise, the people will get caught in a paralyzing internal emotional conflict.

Theodore Vail's decision that the business of the Bell System was service might have remained dead letter but for the yardsticks of service performance which he designed to measure managerial performance. Bell managers were used to being measured by the profitability of their units, or at the least, by cost. The new yardsticks made them accept rapidly the new objectives.

If the greatest rewards are given for behavior contrary to that which the new course of action requires, then everyone will conclude that this contrary behavior is what the people at the top really want and are going to reward.

Not everyone can do what Vail did and build the execution of his decisions into the decision itself. But everyone can think through what action commitments a specific decision requires, what work assignment follows from it, and what people are available to carry it out.

## The Right and the Wrong Compromise

The decision is now ready to be made. The specifications have been thought through, the alternatives explored, the risks and gains weighed. Who will have to do what is understood. At this point it is indeed reasonably clear what course of action should be taken. At this point the decision does indeed almost "make itself."

And it is at this point that most decisions are lost. It becomes suddenly quite obvious that the decision is not going to be pleasant, is not going to be popular, is not going to be easy. It becomes clear that a decision requires courage as much as it requires judgment. There is no inherent reason why medicines should taste horrible — but effective ones usually do. Similarly, there is no inherent reason why decisions should be distasteful — but most effective ones are.

The reason is always the same: there is no "perfect" decision. One always has to pay a price. One has always to subordinate one set of *desiderata*. One always has to balance conflicting objectives, conflicting opinions, and conflicting priorities. The best decision is only an approximation — and a risk. And there is always the pressure to compromise to gain acceptance, to placate strong opponents of the proposed course of action or to hedge risks.

To make effective decisions under such circumstances requires starting out with a firm commitment to what is right rather than with the question "Who is right?" One has to compromise in the end. But unless one starts out with the closest one can come to the decision that will truly satisfy objective requirements, one ends up with the wrong compromise — the compromise that abandons essentials.

For there are two different kinds of compromise. One kind is expressed in the old proverb "Half a loaf is better than no bread." The other kind is expressed in the story of the Judgment of Solomon, which was clearly based on the realization that "half a baby is worse than no baby at all." In the first instance, objective requirements are still being satisfied. The purpose of bread is to provide food, and half a loaf is still food. Half a baby, however, is not half of a living and growing child. It is a corpse in two pieces.

It is, above all, fruitless and a waste of time to worry about what is acceptable and what one had better not say so as not to evoke resistance. The things one worries about never happen. And objections and difficulties no one thought about suddenly turn out to be almost insurmountable obstacles. One gains nothing, in other words, by starting out with the question "What is acceptable?" And in the process of answering it, one loses any chance to come up with an effective, let alone with the right, answer.

## The Feedback

A feedback has to be built into the decision to provide continuous testing, against actual events, of the expectations that underlie the decision. Few decisions work out the way they are intended to. Even the best decision usually runs into snags, unexpected obstacles, and all kinds of surprises. Even the most effective decision eventually becomes obsolete. Unless there is feedback from the results of a decision, it is unlikely to produce the desired results.

This requires first that the expectations be spelled out clearly — and in writing. Second, it requires an organized effort to follow up. And this feedback is part of the decision and has to be worked out in the decision process.

When General Eisenhower was elected president, his predecessor, Harry Truman, said: "Poor Ike; when he was a general, he gave an order and it was carried out. Now he is going to sit in that big office and he'll give an order and not a damn thing is going to happen."

The reason why "not a damn thing is going to happen" is, however, not that generals have more authority than presidents. It is that military organizations learned long ago that futility is the lot of most orders and organized the feedback to check on the execution of the order. They learned long ago that to go oneself and look is the only reliable feedback.[1] Reports — all an American president is normally able to mobilize — are not much help. All military services have long ago learned that the officer who has given an order goes out and sees for himself whether it has been carried out. At the least he sends one of his own aides — he never relies on what he is told by the subordinate to whom the order was given. Not that he distrusts the subordinate; he has learned from experience to distrust communications.

One needs organized information for the feedback. One needs reports and figures. But unless one builds one's feedback around direct exposure to reality — unless one disciplines oneself to go out and look — one condemns oneself to sterile dogmatism and with it to ineffectiveness.

In sum: decision-making is not a mechanical job. It is risk-taking and a challenge to judgment. The "right answer" (which usually cannot be found anyway) is not central. Central is understanding of the problem. Decision-making, further, is not an intellectual exercise. It mobilizes the vision, energies, and resources of the organization for effective action.

## STUDY QUESTIONS
1. According to Drucker, why should the decision-maker start with opinions rather than facts?
2. How are unsound opinions eliminated?
3. Why are dissent and alternatives important in decision-making?
4. Under what circumstances might we decide to make no decision?
5. Why must a "feedback" be built into a decision? How do military commanders insure that decisions are carried out?

---

[1] This was certainly established military practice in very ancient times — Thucydides and Xenophon both take it for granted, as do the earliest Chinese texts on war we have — and so did Caesar.

# Part Two

---

# Developing the Topic

# Developing the Topic

When we speak of *rhetoric,* we are speaking of strategy. Successful writing, like successful engineering, successful business, successful tennis or football, requires strategy — thought, planning, and a method of attacking the problem at hand. The writing in science, business, and industry — the writing that does the world's work — probably requires more strategical thinking than other kinds of writing, because technical and business writing *must* communicate clearly if information is to be transmitted effectively.

Studying strategy, then, involves studying rhetoric, and rhetoric is often studied by examining examples of *methods of development* or *rhetorical techniques.* In the following pages you will find examples of the methods of development most commonly used in business and technical writing: *description, narration, process, example/illustration, comparison/contrast, definition, classification/division, cause-effect,* and *analogy.* You should be aware that these are not necessarily all the writing strategies that might be employed. Some rhetoricians might add several more strategies to this list of important rhetorical techniques; some rhetoricians, on the other hand, might delete some of the strategies from this list, arguing that there were really only three or four truly major rhetorical techniques. This disagreement exists not because rhetoricians have failed to study this problem carefully, but because of the very nature of writing itself. To understand this fully let us compare for a moment two very different things — a river and a brick wall.

A well-written essay develops smoothly as each sentence flows into the next, each paragraph blends with the whole. In a well-written article words merge and blend, flowing like a river. When a certain writing pattern is repeated frequently it may be identified and labelled. Imagine for a moment that you are the pilot of a riverboat. Think of the many wave patterns, current patterns, and ripple patterns that you might distinguish as you travelled down a river. Some ripples might be seen as patterns in themselves. Other ripples might instead be seen as parts of larger waves or current patterns. Methods of development are like these patterns in the water. Identifying them and labelling them is not a precise activity.

Students, however, sometimes form the opinion that identifying and discussing methods of development is a precise activity, because instructors and textbooks (such as this one) usually present these techniques as though they are precise and distinct entities — like the bricks that make up a brick wall. Rhetorical techniques, however, seldom exist in what we might call a pure state. They are not precise building blocks — distinct, separate, easily handled and managed — they are more like ripples and wave patterns — blended, fluid, sometimes difficult to distinguish and identify. If you as a student of technical and business writing can come to understand methods of development in this way, you should not become perplexed when these strategies fail to emerge as neatly and precisely in your own prose as they do in textbook examples.

On the following pages you will find samples of business and technical writing that have been chosen as representative illustrations of rhetorical techniques. You should notice that in almost all cases these samples are excerpts from longer reports and articles. They are excerpts because we do not write complete reports using only one strategy — *narration,* or *comparison/contrast,* or *analogy* — just as we do not play football relying on only one strategy — the on-side kick, or the halfback pass. We must combine several strategical elements in order to perform successfully in football, tennis, or writing reports, and although the following examples may present samples of pure *definition* or *classification/division,* the complete works they have been collected from combine many methods of development effectively, just as your own articles and reports will.

As one of your prewriting activities you should stop to examine the information you are attempting to transmit and ask yourself what the most effective way of presenting this material may be. You might then consciously decide to rely upon *definition, example/illustration,* or *analogy* (as this introduction does) to make your point clear. Do not necessarily be disturbed, however, if you intend to begin a report using narration and suddenly find yourself creating an extended analogy. You have arrived at this situation naturally and spontaneously. The method of development you have selected, whether you have selected it consciously or not, is quite probably exactly the right strategy for the occasion.

# Description

*Description* gives a verbal account of a physical object, a process, or a condition. Even visual illustration cannot completely replace description, because in a verbal account the writer can readily focus on the important details. Used for manuals, specifications, technical proposals, catalogues, and sales literature, description is one of the report writer's most commonly employed tools. We emphasize physical description in this discussion, although the principles involved will apply to describing processes and conditions.

Description may be either *technical* or *artistic.* In technical description, the writer gives objective, quantifiable information about objects — length, width, height, weight, number, and so on. In artistic description, the writer allows his subjective interpretation to color his account.

Although the report writer will most often use technical description, there are times when the artistic approach is appropriate. The purpose of the report should determine the form of description used. If you are informing the reader on how to fix a carburetor, only objective information about the physical aspects of the mechanism will be useful to him. The person who must fix the carburetor is not likely to be receptive to your subjective commentary on design features. On the other hand, if you are preparing sales literature for a new model of carburetor, you might use artistic description to enhance its features. In this case, purely technical description would not be appropriate.

Description must be systematic, otherwise the reader will not be able to visualize the object. Establish an orderly pattern, then stick with it. In technical description, use a pattern that helps you in accounting for the individual parts of a mechanism, and the relation of each part to the whole. Left to right, front to back, and top to bottom are only the obvious patterns: they do not begin to exhaust the possibilities. Always allow the mechanism itself to dictate its own pattern, and keep your mind open to creative strategies. Artists do not have sole possession of creativity.

The two examples that follow describe new products. Dave Bode describes the features of a drillship, a product with a highly specialized industrial use. The *Motor Trend* article presents a consumer product, the Oldsmobile Toronado. Note that in each case the approach to the description is appropriate for the intended audience.

# New Class of Global Marine Drillship

## Dave Bode

The first of a new class of dynamically-positioned vessels began operation last year when *Glomar Pacific*, a drillship owned and operated by Global Marine Development, Inc., went into action under long-term contract to Exxon Corp. Currently the *Glomar Pacific* is working 95 miles (153 km) east of Atlantic City, New Jersey,

Dave Bode, "New Class of Global Marine Drillship," *Diesel and Gas Turbine Progress,* May 1978, pp. 41–42.

Drillship Glomar Pacific is now working east of Atlantic City, New Jersey, in the Baltimore Canyon trough area.

in the Baltimore Canyon trough area, where it will drill through the ocean floor to a depth of 14,000′ (4300 m). A sister ship, the *Glomar Atlantic,* is now under construction with delivery expected about the middle of this year. Global Marine has 13 drillships of various classes in its fleet operating worldwide.

The *Glomar Pacific* has an overall length of 452′ (138 m), 72′ (22 m) beam, 35′ 10″ (11 m) depth and a loaded draft of 23′6″ (7 m). Displacement is 14, 751 long tons (15 million kg) loaded and 8189 long tons (8.3 million kg) lightship. Center-well is 26′ (8 m) square.

The vessel's four main powerplants are GE model 7HPD16CC2 16-cylinder Diesel Electric Marine Modules, each rated at 2750 kW, 600 volt, 3-phase, 60 Hz, that provide power for propulsion, ship's service, positioning and drilling. Speed of the turbocharged engines, equipped with Woodward governors, is 1200 rpm. The GE Power Modules combine on a single skid GE's custom 8000 self-ventilated marine synchronous a.c. generator having brushless exciter and static voltage regulator, with the company's marine-type diesel engine, plus auxiliary equipment. A seawater heat exchanger meets engine cooling requirements, and instrument panels feature all necessary alarm and shutdown functions for control and safety.

The GE Power Modules feed a common a.c. bus that routes power to nine GE SCR units, each rated 750 volts, d.c., 1800 amps, for d.c. power for vessel propulsion, dynamic positioning and on-site drilling, as well as other d.c. equipment. The bus also routes 600 volts a.c. through two 1000 kva transformers where voltage is stepped down to 480 volts a.c. for ship's service.

Six GE 1600 hp (1200kW) d.c. propulsion motors drive the twin 11' (3.4 m) diameter main screws; three motors power each screw. These screws give the *Glomar Pacific* a cruisng speed of 13 knots and maximum speed of 14 knots. Five Schottel 102" (2.6 m) fixed pitch thrusters, three forward and two astern, powered by GE 1675 hp (1250 kW) d.c. motors, provide dynamic positioning when used with the main propulsion system.

Emergency power is provided by one Detroit Diesel 12V-149 generator set rated at 600 kW, 480 volts, 3-phase, 60 Hz, which powers all safety equipment such as fire main pumps, bilge pump and ballast pump, as well as the dual Delco computer system that controls the dynamic positioning thrusters.

Two Triton water distillation units, having a total capacity of 14,400 gal. (54,500 l) per day operate using waste heat from the engines' heat exchangers, various air compressors and any other onboard equipment that generates heat. The vessel's two Mariner No. 300 and one Bucyrus-Erie MK-35 cranes are driven by Detroit Diesel model 3-71 engines and Detroit Diesel model 8V71N engine, respectively.

Crew capacity of the *Glomar Pacific* is 101; accommodations include a sick bay for six. She stores 6562 long tons (6.7 million kg) net of cargo, 15, 751 barrels (2.5 million l) of fuel oil, 407 barrels (64,700 l) of lube oil, 19,163 barrels (3 million l) of drill water, 1140 barrels (181,000 l) of potable water, 2748 barrels (437,000 l) wash water, 594 barrels (94,000 l) active mud, 3044 barrels (484,000 l) reserve mud, 16,890 cu. ft. (478 m$^3$) bulk mud and 7415 cu. ft (209 m$^3$) bulk cement. Her automatic drill pipe racker carries over 23,000' (7000 m) of drill pipe.

The vessel's derrick is a 142' X 61' X 38' (43 m X 19 m X 12 m) GMI design having one million pounds (454,000 kg) hook load capacity. Three GE 752 1000 hp (750 kW) d.c. drawworks drilling motors drive the National 1625 DE drawworks, with Elmagco eddy current brake, 6500' (2000 m) of 1-1/2" (4 cm) drill line and 20,500' (6200 m) of 9/16" (14 mm) sand line capacity. The vessel's National C-495 rotary table is driven by one GE 752 drilling motor.

Other equipment includes: two National 12P-160 triplex mud pumps, each driven by two GE 752 drilling motors; two 6" X 5" (15 cm X 13 cm), 75 hp (56 kW) mud mixing pumps rated 800 gpm (3600 1/m) at 105' (32 m) TDH each; Halliburton twin HT-400 cementing unit with two GE 752 drilling motors; and Halliburton wire line measuring equipment driven by a Lister two-cylinder diesel engine rated at 15.2 hp (11.3 kW).

A GE electrical control system is used for maneuvering controls for the main propulsion system as well as the dynamic positioning thrusters. The control stations are located in both the fore and aft pilot houses.

## STUDY QUESTIONS

1. Has the author used any artistic description in his account of the drillship? Discuss.
2. The article is taken from *Diesel and Gas Turbine Progress: The Engine Drive Systems Magazine,* a publication for specialists on drive systems. Is the form of description used appropriate for the audience of the magazine? What is the purpose of the description?
3. Discuss the descriptive pattern that the author employs.
4. *Writing Assignment:* Select a portion of the article and rewrite it using an artistic approach.

# ' 79 Oldsmobile Toronado

## *Motor Trend*

### GM's Original Front-drive Sedan Continues in Its Best Form to Date

Toronado was the lone front-wheel-drive entry from General Motors back in 1966, when it first appeared on the scene. The front-drive configuration was recognized for its superior traction in snow and rain and the fact that it removed the huge hump from the floor of the passenger compartment. Other than that it was looked upon as little more than a novelty by a public that was swimming in cheap gasoline.

Thirteen years later we find the new-generation Toronado emerging into a world where front-drive cars are rapidly becoming the norm. A much more sophisticated car-buying public sees front-wheel drive as not only a way to plod through inclement weather with less difficulty but also as a means of reducing vehicle weight, which translates into better fuel mileage.

The stylist's pen and engineer's ingenuity combined to trim more than 900 pounds off the Toronado for '79. The wheelbase was reduced from 122 inches to 114 inches, and overall length dropped from a mammoth 227.5 inches to a more efficient 205.6 inches. Width was reduced by 8.7 inches.

Hip room suffered somewhat but both head and leg room were improved over last year's model. The car does have a smaller appearance in all respects, but inside there is more than ample room for four adult passengers.

With its long nose, high rear deck and formal roofline the Toronado looks a bit awkward from certain angles. The choice of color is critical to the overall

"'79 Oldsmobile Toronado," *Motor Trend,* October 1978, p. 88.

appearance of the car. It looks best in solid colors and worst when ordered with a vinyl roof that contrasts sharply with the body color. Such a contrast tends to off-set the roof too greatly, giving the car a very choppy look. Vinyl toppings of the same color as the body look fine.

The new Toronado is much more manageable in all respects. The lighter weight and reduced size allow it to slip through city traffic with ease and tackle twisting country roads with new-found agility. The turning radius has been reduced by 3 feet over last year's model, making it much easier to squeeze into a tight spot.

Overall riding comfort hasn't suffered a bit. In fact, the new automatic load-leveling system and independent rear suspension greatly improve ride under all conditions. The most noticeable benefit of the independent rear suspension is superior handling on rough road surfaces. The tendency for the rear to hop in tight bumpy turns also has been eliminated.

A 5.7-liter (350cid) V-8 gasoline engine is the base for the Toronado. Its 170 horsepower pulls the car along quite nicely and has plenty of reserve for passing at high speeds. Preliminary figures for the engine rate fuel mileage at 16 mpg, which is a 3-mpg improvement over the city mileage figure for the '78 Toronado.

The only optional engine offered is the 5.7-liter (350cid) diesel V-8 which features Oldsmobile's new fast-start system. At 0° Fahrenheit the glow plug requires a warmup period of only 6 seconds. Most diesels would require about a full minute at the same temperature.

The engine has the usual diesel clatter, but heavy layers of insulation up front have effectively isolated the passenger compartment from any noticeable level of noise.

Oldsmobile has found an 8-cylinder solution to the age-old diesel problem of sluggish performance. This smooth diesel will leave all but the Mercedes turbo diesel clattering at the starting line. Acceleration is brisk and constant with the automatic transmission. Fuel mileage is estimated to be 21 mpg in the city, and a special 27.8-gallon fuel tank allows for many miles in between fill ups.

The word *plush* pretty much sums up the Toronado interior; creature comforts abound. Standard features include air conditioning, split front seat with 6-way power adjustment on the driver's side, power windows, digital clock, *AM/FM* stereo radio, power steering, side window defogger and power brakes.

The wide vertical dashboard is cleanly styled, with all vital instrumentation and controls placed within sight and reach of the driver. A fiber optic system lights the center-mounted Toronado nameplate, light switch, ashtray and lighter, cruise-control switch and rear window defogger control. European influence is evident in the steering-column-mounted headlight dimmer switch which is incorporated into the turn-signal stalk.

Notable options for the Toronado include leather upholstery, reclining front seats, heavy-duty suspension, and door lock and interior illumination system that lights when the door handle is lifted.

Drivers who like to buy their automobiles by the pound may lament the passing of the old, heavy Toronado design, but those who can appreciate an automobile for the way it rides and handles will love this new front-drive sedan. It offers personalized, luxurious transportation and outstanding fuel mileage when equipped with the optional diesel V-8.

The Toronado's character is of a different flavor than that of the Riviera. It is a softer car which is most at home cruising the open highway. It offers a level of driving pleasure that is akin to that of the Eldorado and proves to be a match for it in every respect except name.

**STUDY QUESTIONS**
1. What is the purpose of the article? (Consider its source: *Motor Trend.*)
2. Which form of description is dominant, technical or artistic? Is this appropriate for the purpose of the article?
3. Could the article have the same purpose if it were purely technical?
4. Does the writer use a rigid pattern in his description? Discuss.
5. *Writing Assignment:* Rewrite the article using only technical description.

# Narration

*Narration* is concerned with telling a story, with relating a *narrative,* a sequence of events. The easiest way to tell a story is to begin at the beginning and relate events in the sequence in which they happened, a chronological sequence. This is the standard narrative method, and it is the narrative method most commonly employed in business and technical writing. We often find such writing at the beginning of reports or studies, when the history of a process or action is recounted, or in the body of a report, when the actions presently being proposed are examined.

In any kind of narration the writer must decide upon his narrative stance, his position relative to the material being presented. He might, for example, write a first-person narrative and put himself into the action, either as an observer or a central participant. Or he might choose to write a third-person narrative, a more detached and objective point of view that remains totally outside the action being related.

In the following examples we can see that Edward C. Lambert's history of hexachlorophene is a rather objective third-person account. Lowell Foster's article on metric studies employs a third-person point of view, even though the author is writing a history of his own work. The excerpt from James Watson's classic work on DNA, *The Double Helix,* uses a far more personal tone in its first-person account. Watson is of course attempting to create a book that will be interesting to the general reading public. He succeeded by writing a book on genetics that was also a best seller.

# Hexachlorophene

## Edward C. Lambert

The most recent of the laboratory mistakes and accidents is the tragedy occurring after the use of baby talcum powder containing hexachlorophene. This effective, non-irritating skin antiseptic has been used in most countries since its introduction thirty-three years ago. For several years it has been an ingredient of many creams, ointments, powders, cosmetics, antiperspirants, mouth washes, feminine deodorant sprays, toothpastes, and soaps. Its important use as Phisohex has been as a pre-operative skin preparation in surgery and in newborn nurseries, where it is effective in preventing epidemics of staphylococcal infections.

Ever since it was synthesized, it has been known to be toxic and potentially fatal if taken by mouth. In 1968 toxicity and deaths were reported from the use of hexachlorophene preparations on a few patients who had lost large areas of skin because of burns. Subsequently, researchers found that prolonged application to the skin on rats and monkeys could lead to central nervous system damage. Because of this finding, in December 1971 the FDA and the Committee on Fetus and Newborn of the American Academy of Pediatrics warned against its routine use in nurseries.

In the United States, many hospitals stopped total body bathing of newborns with hexachlorophene as a result of the warning by the FDA and the Academy of Pediatrics. Within three months there had been twenty-four outbreaks of staphylococcal infection related to discontinuance of the preparation. (There was no increase of such infections in fifty-eight other hospitals where such bathing had been continued.)

During the spring of 1972, 120 babies became seriously ill in various provinces of France, and 40 of them died. All had had a rash, which was followed by loss of appetite, irritability, drowsiness, twitching, and often convulsions. When death occurred, it arrived a few days after the onset of the illness. The only common factor which could be implicated was the use of "Talc Mohange," a baby talcum powder. In August 1972 the French Minister of Health announced that by accident a concentration of over 6 per cent hexachlorophene had been included in certain batches of the powder (3 percent is the usual amount).

Lessons of caution have been learned. An essential drug known to be toxic in higher than usual doses should be handled with caution and respect, but should not be discontinued unless it can be suitably replaced — a delicate balance between risk and benefit.

At present, the usual practice is to use it on only full-term infants, and to rinse off the residue thoroughly with water. Premature babies are washed with it only if threatened by an epidemic of staphylococcal infection.

Edward C. Lambert, "Hexachlorophene," *Modern Medical Mistakes* (Bloomington: Indiana Univ. Press, 1978), pp. 64–65.

**STUDY QUESTIONS**
1.  Lambert presents a fairly straightforward narrative. What is his relationship to his subject matter? Would you say that he is a detached observer or a sub-jective participant?
2.  *Writing Assignment:*  Pretend that you are the author of this selection. Rewrite it using a first-person point of view.

# The Double Helix

## James D. Watson

When I got to our still empty office the following morning, I quickly cleared away the papers from my desk top so that I would have a large, flat surface on which to form pairs of bases held together by hydrogen bonds. Though I initially went back to my like-with-like prejudices, I saw all too well that they led nowhere. When Jerry came in I looked up, saw that it was not Francis, and began shifting the bases in and out of various other pairing possibilities. Suddenly I became aware that an adenine-thymine pair held together by two hydrogen bonds was identical in shape to a guanine-cytosine pair held together by at least two hydrogen bonds. All the hydrogen bonds seemed to form naturally; no fudging was required to make the two types of base pairs identical in shape. Quickly I called Jerry over to ask him whether this time he had any objection to my new base pairs.

When he said no, my morale skyrocketed, for I suspected that we now had the answer to the riddle of why the number of purine residues exactly equaled the number of pyrimidine residues. Two irregular sequences of bases could be regularly packed in the center of a helix if a purine always hydrogen-bonded to a pyrimidine. Furthermore, the hydrogen-bonding requirement meant that adenine would always pair with thymine, while guanine could pair only with cytosine. Chargaff's rules then suddenly stood out as a consequence of a double-helical structure for DNA. Even more exciting, this type of double helix suggested a replication scheme much more satisfactory than my briefly considered like-with-like pairing. Always pairing adenine with thymine and guanine with cytosine meant that the base sequences of the two intertwined chains were complementary to each other. Given the base sequence of one chain, that of its partner was automatically determined. Con-ceptually, it was thus very easy to visualize how a single chain could be the tem-plate for the synthesis of a chain with the complementary sequence.

Upon his arrival Francis did not get more than halfway through the door before I let loose that the answer to everything was in our hands. Though as a

From James D. Watson, *The Double Helix* (New York: Atheneum Publishers, 1968), pp. 194–197.

matter of principle he maintained skepticism for a few moments, the similarly shaped A-T and G-C pairs had their expected impact. His quickly pushing the bases together in a number of different ways did not reveal any other way to satisfy Chargaff's rules. A few minutes later he spotted the fact that the two glycosidic bonds (joining base and sugar) of each base pair were systematically related by a diad axis perpendicular to the helical axis. Thus, both pairs could be flipflopped over and still have their glycosidic bonds facing in the same direction. This had the important consequence that a given chain could contain both purines and pyrimidines. At the same time, it strongly suggested that the backbones of the two chains must run in opposite directions.

The question then became whether the A-T and G-C base pairs would easily fit the backbone configuration devised during the previous two weeks. At first glance this looked like a good bet, since I had left free in the center a large vacant area for the bases. However, we both knew that we would not be home until a complete model was built in which all the stereochemical contacts were satisfactory. There was also the obvious fact that the implications of its existence were far too important to risk crying wolf. Thus I felt slightly queasy when at lunch Francis winged into the Eagle to tell everyone within hearing distance that we had found the secret of life.

### Study Questions

1. What personal elements does Watson incorporate into his narrative? Do these elements make the selection more, or less, readable?
2. Watson's narrative builds to a climax in this selection. How does he maintain an air of excitement throughout this piece?
3. *Writing Assignment:* Write a narrative account of an important discovery you have made.

# Honeywell Metric Studies: 1962 - 1971

## Lowell Foster

Also of particular value and interest are some company metric studies undertaken in the 1960s. Each of these events can be counted as significant to Honeywell's positive posture and ready acceptance of metrication.

In 1962, the first action concerned with the possible formal metric conver-

Lowell Foster, "Honeywell Metric Studies: 1962–1971," *Managing Metrication in Business and Industry* (New York: Marcel Dekker, 1976), pp. 12–16.

sion within the company (and the U.S.) was undertaken. At the request of executive management, a metrication committee of six individuals was charged with studying the possible implications of metric conversion; that is, if it were to occur in the "sometime" future. This committee was composed of corporate staff personnel and residential division (automatic controls division). After three months of spasmodic efforts, a report of a very general nature was submitted to upper management. The report removed some of the fears about conversion, clarified some of the issues, set somewhat of an optimistic tone, but left the impression that more in-depth study was required.

In early 1963, the author was assigned the task of conducting a more in-depth metric study, but within a limited time and resource constraint. It was to be done "solo" and within approximately 40 working hours. The job was ultimately accomplished within the foregoing dimensions, but with the voluntary involvement of numerous resource persons at various positions within the company. A brief report of impressions, methods, objectives, and results of the 1963 study is of interest.

After initial consideration, it was decided that the report would:

1. Be cost-and result-oriented
2. Be based on a 10-year conversion period
3. Assess the major functions impacted, especially the affects on design and production
4. Establish a company position on the subject
5. Be based on available or speculated "facts" and written in terms of rations or percentage figures so that a sliding scale criteria could serve in future assessment or up-dating

The first step was to gather data, to study the impact and possibilities of metric manufacture (could it be done with existing equipment?), look at the raw material and supply problems (e.g., dual inventories and dual documentation), talk to knowledgeable and concerned people to ascertain the psychological effects, etc. Needless to say, this was a pioneering effort undertaken with fear and trepidation, a challenge to be sure. However, the results of this study were both surprising and encouraging.

Surprising, because of some 50 people contacted and burdened somewhat by the uniqueness and weight of the questions asked, *no one* seemed opposed to the idea of metric conversion. Remembering that this was 1963, when metrication was far from the buzz word it is today, is even more surprising today than it was then. Granted, some had an "it can't happen here" attitude. Nevertheless, only interest, optimism, and curiosity seemed evident in their reactions. However, there was a great concern on the part of some as to when this great event was to be expected. Surely, they said, you do not encourage the change now. The assurance was given that it was only a study, but the participants were told that speculation as to the future was serious.

The cooperation received from everyone involved in the study as resource persons was excellent. Facts and figures were available on such things as metric conversion of machines and equipment. Other areas, such as the costs of dual documentation in the transition period and the training of people, required an "educated guess."

In the categories of design and documentation, modification of equipment, training, etc. (shown in Figure 1), individual studies of cause and effect, time, and costs were determined on the basis of a 10-year conversion plan. It was said then, and it is repeated loudly now, that the cost results were (and are) somewhat crude. Any price tag on metrication made before the fact must be crude. However, a later (1971) study, which is discussed below, corroborated the original estimates. The early study did isolate the areas of concern, set some priorities, and gave a "ball park" evaluation.

Figure 1 reveals the cost impact results of the 1963 study. Each of the areas used as key concerns is listed with a percentage and dollar value assigned. Totaling 3.63% or $3.63 million, cost over the 10-year period represents a "then assumed" optimistic figure. Only one or two other companies to our knowledge had made any similar studies at that time (1963), and their figures were in the 15% to 20% range, with a very rubbery base (obviously).

Note that the tabulated results of the 1963 study shown in Figure 1 are based upon a hypothetical figure of $100 million gross sales. This permitted the sponsoring division to plug in its true sales figures at any time in the future (within approximately a 5-year period) and derive a metric cost figure for the speculated 10-year conversion period. The 10-year period cycle was envisioned as shown in Figure 2.

**Company "X"[a]**

|  |  | % of<br>Gross Sales |
| --- | --- | --- |
| Modification or replacement of<br>  machines and equipment | ($1 380 000) | 1.38 |
| Maintaining dual supply tools<br>  and facilities | (750 000) | .75 |
| Modification or replacement of<br>  gages and testing equipment | ($580 000) | .58 |
| Redocumentation | ($500 000) | .5 |
| Education and Training | ($20 000) | .42 |
| Totals | $3 630 000 | 3.63% |

[a]This study is based on $100 million gross annual sales year. Total cost for 10-year transition to metric use on basis of any one point in time, e.g. start 10-year transition program in 1972 and end 1982 (as based on 1972 $ value), est cost = $3,630,000.

Figure 1. Results of 1963 Honeywell metrication cost study.

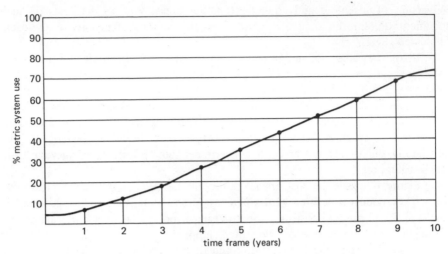

Figure 2. Honeywell's 10 year conversion cycle.

The report satisfied management needs for setting some direction, evaluating the impact, and determining company posture on metric conversion.

The years from 1963 to 1968 passed with monitoring of metric progress, both within the company (then drawing in more U.S. divisions in its concerns), in the U.S., and worldwide. Participation by the author and others in early national metric activities, such as the ANSI Metric Advisory Committee, supplemented Honeywell's activities.

Progress and interest from 1968 through 1971 were accelerated by participation, as a company, in the NBS Metric Study and increased involvement with multinational goals and worldwide markets. For Honeywell, metrication was no longer a question of "if"; it was a matter of "how" and "when."

In 1971, a detailed status report and policy and plan guidelines were developed for submittal to executive management. Also, as corroborating support, a restudy was made of the cost effect of metrication. The 1971 results, using the same base ingredients as the 1963 report, are shown in Figure 3. Note that with the advantage of some experience, "hindsight," and company divisional experiences in U.K., the margins were narrowed to a cost effect of 1.12% of gross sales for one year, stretched over a speculated 10-year transition.

These latter figures (Figure 3) are admittedly crude; there are many implications to conversion and their cost is not predictable to any degree of accuracy. However, the figures served a purpose in helping to identify concerns and set priorities.

The popular thought today is to avoid attempting to isolate costs; it is thought to be an exercise in futility. We agree with this thought in principle. However, as good managers, some cost data seem essential to create the assurance that we are in control. In actuality, the more germane question, not appropriately approached as yet, is: what are the cost advantages of metrication? We are convinced that they are impressive to the point of overshadowing costs.

Cost Estimate Table Division "X"[a]

|  | Costs 10 Year Total | % of Annual Gross Sales |
|---|---|---|
| Design and Documentation | ($250 000) | .25 |
| Modification or replacement of machines and equipment | ($150 000) | .15 |
| Maintaining dual supply of tools, parts, and materials | ($250 000) | .25 |
| Modification or replacment of testing equipment and gages | ($210 000) | .21 |
| Education and training | ($160 000) | .16 |
| Miscellaneous (marketing, sales, legal, tax) | ($100 000) | .10 |
| Total | $1 120 000 | 1.12% |

[a]Based on $100 million gross annual sales over 10 year period.

Figure 3. Results of 1971 Honeywell metrication cost study.

Honeywell's optimism toward metrication is a product of a general feeling that conversion can be accomplished within tolerable cost limits. The company believes metrication offers significant advantages in product expansion, new market-place horizons, and better management of a worldwide corporation as a result of a worldwide language of measurement.

### STUDY QUESTIONS
1. Compare this selection with the selection from James D. Watson's *The Double Helix*. Would you say that Foster is more, or less, objective than Watson?
2. Foster, like Watson, is writing about a project in which he was personally involved. How do you account for the different tone of the article?
3. Foster frequently juxtaposes previous ideas and approaches with present concepts and methods, reminding his readers of the changes in attitudes that have taken place. What do these references add to the narrative? Would it have been better for him to move directly from 1962 to 1971 without these comparisons?

# Process

Our world is committed to knowing and communicating how things work. The technologist and the engineer must know how mechanisms work, the scientist must know how natural laws work, and the manager must know how business systems work. Each requires a knowledge of *process.*

Simply defined, a process is an orderly series of actions by which something is accomplished. The operation of the combine, a farm machine, is a process. The formation of stars out of interstellar gas is a process. A medical procedure is a process. As a method of development, process involves the principles of *narration* and *description* (themselves methods of development) and *logical analysis.* Process is narrative in that it establishes the time relationships of a sequence of actions. The sequence, however, must be orderly and predictable. Unlike fictive, biographical, or historical narrative, process doesn't account for random or inexplicable events. Process uses description to indicate place relationships. For example, in relating the process of an internal combustion engine, the writer must account for the parts that move as well as their movement. Description allows the writer to translate the complex dynamics of a process into a series of frozen time frames that the reader can easily understand.

Further, the writer must use logical analysis in marking off the steps or phases of a process. By their nature, processes tend to be continuous and unbroken. It is up to the writer to determine the number of steps in the sequence, and where a step begins and ends. Analysis of cause-effect relations, a logical operation discussed in Copi's "Causal Connections" (pp. 98–100) and Hardin's "What the Hedgehog Knows" (pp. 101–104), many times provides the key to identifying the steps.

Process writing is a mainstay of instruction manuals, textbooks, and reference works. It also has an important role in many proposals, evaluations, and feasibility studies.

Although process is often used to introduce a set of instructions, it should not be confused with the instructions themselves. A detailed account of a manufacturing process may give the reader enough information to perform the process itself. But process focuses on the operation, not the role of the operator. Instructions (discussed in Part III, section 3), must be written solely from the operator's point of view.

The following selections illustrate process writing in several fields. Roy C. Selby, Jr., narrates a medical procedure. "Stars: Birth, Life, and Death" relates the life-cycle of stars. "Combine Harvester and Thresher" outlines a mechanical process. Jim Mann marks off the basic steps in assessing the effectiveness of product design.

# A Delicate Operation

## Roy C. Selby, Jr.

In the autumn of 1973 a woman in her early fifties noticed, upon closing one eye while reading, that she was unable to see clearly. Her eyesight grew slowly worse. Changing her eyeglasses did not help. She saw an ophthalmologist, who found that her vision was seriously impaired in both eyes. She then saw a neurologist, who confirmed the finding and obtained X-rays of the skull and an EMI scan — a photograph of the patient's head. The latter revealed a tumor growing between the optic nerves at the base of the brain. The woman was admitted to the hospital by a neurosurgeon.

Further diagnosis, based on angiography, a detailed X-ray study of the circulatory system, showed the tumor to be about two inches in diameter and supplied by many small blood vessels. It rested beneath the brain, just above the pituitary gland, stretching the optic nerves to either side and intimately close to the major blood vessels supplying the brain. Removing it would pose many technical problems. Probably benign and slow-growing, it may have been present for several years. If left alone it would continue to grow and produce blindness and might become impossible to remove completely. Removing it, however, might not improve the patient's vision and could make it worse. A major blood vessel could be damaged, causing a stroke. Damage to the undersurface of the brain could cause impairment of memory and changes in mood and personality. The hypothalamus, a most important structure of the brain, could be injured, causing coma, high fever, bleeding from the stomach, and death.

The neurosurgeon met with the patient and her husband and discussed the various possibilities. The common decision was to operate.

The patient's hair was shampooed for two nights before surgery. She was given a cortisonelike drug to reduce the risk of damage to the brain during surgery. Five units of blood were cross-matched, as a contingency against hemorrhage. At 1:00 P.M. the operation began. After the patient was anesthetized her hair was completely clipped and shaved from the scalp. Her head was prepped with an organic iodine solution for ten minutes. Drapes were placed over her, leaving exposed only the forehead and crown of the skull. All the routine instruments were brought up — the electrocautery used to coagulate areas of bleeding, bipolar coagulation forceps to arrest bleeding from individual blood vessels without damaging adjacent tissues, and small suction tubes to remove blood and cerebrospinal fluid from the head, thus giving the surgeon a better view of the tumor and surrounding areas.

A curved incision was made behind the hairline so it would be concealed when the hair grew back. It extended almost from ear to ear. Plastic clips were

Roy C. Selby, Jr., "A Delicate Operation," *Harper's Magazine,* December 1975, pp. 118–119.

applied to the cut edges of the scalp to arrest bleeding. The scalp was folded back to the level of the eyebrows. Incisions were made in the muscle of the right temple, and three sets of holes were drilled near the temple and the top of the head because the tumor had to be approached from directly in front. The drill, powered by nitrogen, was replaced with a fluted steel blade, and the holes were connected. The incised piece of skull was pried loose and held out of the way by a large sponge.

Beneath the bone is a yellowish leatherlike membrane, the dura, that surrounds the brain. Down the middle of the head the dura carries a large vein, but in the area near the nose the vein is small. At that point the vein and dura were cut, and clips made of tantalum, a hard metal, were applied to arrest and prevent bleeding. Sutures were put into the dura and tied to the scalp to keep the dura open and retracted. A malleable silver retractor, resembling the blade of a butter knife, was inserted between the brain and skull. The anesthesiologist began to administer a drug to relax the brain by removing some of its water, making it easier for the surgeon to manipulate the retractor, hold the brain back, and see the tumor. The nerve tracts for smell were cut on both sides to provide additional room. The tumor was seen approximately two-and-one-half inches behind the base of the nose. It was pink in color. On touching it, it proved to be very fibrous and tough. A special retractor was attached to the skull, enabling the other retractor blades to be held automatically and freeing the surgeon's hands. With further displacement of the frontal lobes of the brain, the tumor could be seen better, but no normal structures — the carotid arteries, their branches, and the optic nerves — were visible. The tumor obscured them.

A surgical microscope was placed above the wound. The surgeon had selected the lenses and focal length prior to the operation. Looking through the microscope, he could see some of the small vessels supplying the tumor and he coagulated them. He incised the tumor to attempt to remove its core and thus collapse it, but the substance of the tumor was too firm to be removed in this fashion. He then began to slowly dissect the tumor from the adjacent brain tissue and from where he believed the normal structures to be.

Using small squares of cotton, he began to separate the tumor from very loose fibrous bands connecting it to the brain and to the right side of the part of the skull where the pituitary gland lies. The right optic nerve and carotid artery came into view, both displaced considerably to the right. The optic nerve had a normal appearance. He protected these structures with cotton compresses placed between them and the tumor. He began to raise the tumor from the skull and slowly to reach the point of its origin and attachment — just in front of the pituitary gland and medial to the left optic nerve, which still could not be seen. The small blood vessels entering the tumor were cauterized. The upper portion of the tumor was gradually separated from the brain, and the branches of the carotid arteries and the branches to the tumor were coagulated. The tumor was slowly and gently lifted from its bed, and for the first time the left carotid artery and optic nerve could be seen. Part of the tumor adhered to this nerve. The bulk of the tumor was amputated, leaving a small bit attached to the nerve. Very slowly and carefully the tumor fragment was resected.

The tumor now removed, a most impressive sight came into view — the pituitary gland and its stalk of attachment to the hypothalamus, the hypothalamus itself, and the brainstem, which conveys nerve impulses between the body and the brain. As far as could be determined, no damage had been done to these structures or other vital centers, but the left optic nerve, from chronic pressure of the tumor, appeared gray and thin. Probably it would not completely recover its function.

After making certain there was no bleeding, the surgeon closed the wounds and placed wire mesh over the holes in the skull to prevent dimpling of the scalp over the points that had been drilled. A gauze dressing was applied to the patient's head. She was awakened and sent to the recovery room.

Even with the microscope, damage might still have occurred to the cerebral cortex and hypothalamus. It would require at least a day to be reasonably certain there was none, and about seventy-two hours to monitor for the major postoperative dangers — swelling of the brain and blood clots forming over the surface of the brain. The surgeon explained this to the patient's husband, and both of them waited anxiously. The operation had required seven hours. A glass of orange juice had given the surgeon some additional energy during the closure of the wound. Though exhausted, he could not fall asleep until after two in the morning, momentarily expecting a call from the nurse in the intensive care unit announcing deterioration of the patient's condition.

At 8:00 A.M. the surgeon saw the patient in the intensive care unit. She was alert, oriented, and showed no sign of additional damage to the optic nerves or the brain. She appeared to be in better shape than the surgeon or her husband.

## STUDY QUESTIONS

1. What process is described? What are the steps of the process? What principle orders the steps?
2. Are all the narrative elements essential to understanding the process? What information could be eliminated without interfering with our understanding of the process?
3. Is it the author's purpose to relate only a process? Is the process secondary in importance to the overall narrative? Why or why not?
4. Are we told how to perform brain surgery, or how brain surgery is performed? Discuss.
5. *Writing Assignment:* Rewrite a portion of this selection, using only the information that relates to the process.

# Stars: Birth, Life, and Death

## Editors of *Time*

In our galaxy and in galaxies yet to be discovered, stars are going through a continuous cycle of birth, life and death. Indeed, there are places where the observer who knows what to look for can practically see stars forming before his eyes. These star wombs are great clouds of gas and dust floating in interstellar space. Like the clouds that formed in the expanding primordial fireball shortly after the big bang, they consist mostly of nature's simplest molecule, hydrogen. A star is born when some force, perhaps a shock wave, drives enough of the hydrogen molecules in a cloud sufficiently close to one another that they are held together by their mutual gravity. As a result, a huge pocket of condensed gas, trillions of miles across, is formed at the edge of the larger cloud. In a model proposed by Astronomers Bruce Elmegreen and Charles Lada of the Harvard-Smithsonian Center for Astrophysics, shock waves from the ignition of earlier massive stars help create the conditions for the birth of other stars from the same cloud.

Under the force of their own gravity, the great clouds of gas slowly begin to contract, raising the pressures and temperatures at their centers. They have become embryonic stars. The process continues for some 10 million years, during which the clouds shrink to globes more than a million miles in diameter. At this point, temperatures near the centers of the great gas balls have reached the critical level of 20 million degrees F., hot enough to cause fusion — the awesome process that occurs in a detonating hydrogen bomb.

Long since stripped of their electrons by the high temperatures, the nuclei of the hydrogen atoms slam together at tremendous speeds, fusing to form helium and releasing huge amounts of energy. Though the nuclear fires have been lit, the actual ignition is hidden deep within the interstellar clouds. "Nature very discreetly pulls the curtain over the act of birth," says Thaddeus. But the infant star soon makes its presence known, shining through and illuminating the obscuring cloud. This process is occurring in the Orion Nebula, the illuminated portion of a gigantic cloud of gas and dust that is giving birth to new stars. Some of the stars spawned by the nebula have been formed as recently as the time when the human species first stood upright; the newest offspring are only about 100,000 years old — mere infants by stellar standards.

The fusion of hydrogen to form helium marks the beginning of a long and stable period in the evolution of the star — a combination of adolescence and middle age that constitutes 99% of the lifespan of a sun-size star. During this period, the tremendous energy radiating from the star's center neutralizes its gravitational force, and the great glowing orb shrinks no further. But as it must to all stars, death

From "Stars: Where Life Begins," *Time*, December 27, 1976, pp. 31–32. Title selected by editors.

eventually comes. How long a star lives depends on its mass. Generally, the more massive a star is, the shorter its life is. Stars with a mass significantly greater than that of the sun burn their fuel in a profligate manner and die young; a star ten times as massive as the sun, for example, burns 1,000 times faster and survives only 100 million years. The sun, which is some 5 billion years old, is only at the mid-point in life. Smaller stars, on the other hand, are the Methuselahs of the celestial community. A star with one-tenth the mass of the sun can burn for a trillion years.

> Overhead, without any fuss, the stars were going out.
> — Arthur Clarke, *The Nine Billion Names of God*

In Clarke's haunting story, the stars switch off — like lights in an office building at closing time — when mankind has fulfilled its purpose and determined all the names of God. In fact, stars do go out, but for reasons that are much more complex, and in a variety of ways: some end with a whimper, others with a bang.

The beginning of the end comes when the star has exhausted much of the hydrogen near its core and starts to burn the hydrogen in its outer layers. This process causes the star gradually to turn red and swell to 100 times its previous size, pouring out prodigious amounts of energy. Betelgeuse, in the constellation Orion, is such a "red giant," visible to the naked eye. When the sun undergoes a similar metamorphosis, it will envelop Mercury and Venus and vaporize the earth. By that time, 5 billion years from now, man's descendants may have found a new home in an outer planet or beyond.

A star's red-giant phase lasts until the hydrogen in the layer around the core is exhausted, perhaps as long as a billion years. The stage that follows is short-lived. Its fires banked, the star is deprived of the outward radiation pressure. It contracts violently, driving the core temperature up again, until it reaches 200 million degrees. That is hot enough to ignite the helium, which fuses into a still heavier element: carbon. Its radiation energy restored, the star zooms back toward red-giant status 100 times faster than it took to get there the first time.

What happens after the helium is consumed depends on the size of the star. If a star's mass is no more than about four times that of the sun, its second red-giant stage may be its death rattle. As the star contracts again, its gravitational energy cannot produce enough heat to fuse carbon into heavier elements. But as its internal temperature rises, the outer envelope expands and cools. Held loosely by gravity, the outer layers then slough off into space in a billow of gas. All that is left behind is the core, which continues contracting into a ball a few thousand miles in diameter with a density of tons per cubic inch. The result is a "white dwarf," hotter than the surface of the sun but only about the size of the earth and ready to enter a long period of stellar senility. As the millenniums pass, the white dwarf gradually loses its heat, turning first yellow, then red; eventually, its fires burn out entirely, leaving behind a "black dwarf," a cold cinder in the graveyard of space.

Many large stars manage to lose much of their mass as they evolve, shedding their matter as gas and dust. If they manage to shed sufficient mass, in fact, they can die quietly as white dwarfs. But for stars with a mass greater than four times that of the sun, the end may be far more dramatic. In these giant stars, fusion does not end when all the helium has been converted into carbon. In some of the massive stars, in a catastrophic event known as a supernova, the carbon core explodes, dispersing most of the elements it has produced into space. Stars of more than eight solar masses may go through several more contracting and expanding cycles, forming elements such as magnesium, silicon, sulfur, cobalt, nickel and ultimately iron. When the star has formed an iron core, its fate is sealed. It begins to contract again, but does not have enough gravity to cause fusion of the densely packed nuclei of iron. Instead of being suspended again by the energy of a rekindled nuclear fire, the great mass of the star continues to fall toward the core, unable to resist the pull of its own gravity.

This event is also catastrophic. In a matter of seconds, a star that has lived several million years caves in with a devastating crash, most of its material crushing into an incredibly dense and small sphere at the center. Then, like a giant spring, the star rebounds from this collapse in a massive explosion. The result is another kind of supernova, a fantastic explosion that blows the star to smithereens, dispersing into space most of the remaining elements that it has manufactured during its lifetime. So brilliant is the light from the exploding star that it briefly outshines all of the galaxy's other billions of stars combined. The last supernova observed in the Milky Way Galaxy was seen by Johannes Kepler in 1604.

What remains after this explosion again depends on the size of the star. Its death throes may leave behind a rapidly spinning, incredibly dense sphere (about ten miles in diameter), consisting only of tightly packed neutrons. Such an object, called a neutron star, or pulsar, has been located in the center of the Crab Nebula, a glowing cloud that is still expanding from a supernova reported by the Chinese in A.D. 1054.

A very massive star may have an even stranger fate. Driven by its own immense gravitation, it collapses through its neutron star stage, crushing its matter into a volume so small that it virtually ceases to exist. The gravity of its tiny remnant is so great that nothing, not even light, can escape from it. All external evidence of its presence disappears, and the star, like the Cheshire cat, vanishes, leaving behind only the grin of its disembodied gravity. Anything that fell into such a "black hole" would quite literally be crushed out of existence.

## STUDY QUESTIONS

1. What ordering principle underlies the process described?
2. Although stars are not living organisms, the article uses terms such as "womb," "embryonic," and "spawned" to describe their development. Is the author justified in using biological terminology to account for an astrophysical process? Discuss.
3. Would the process be easier to follow if the author had numbered the steps of the process? Why or why not?
4. For what audience is the piece intended? Support your choice with evidence from the article.

# Combine Harvester and Thresher

## Editors of *The Way Things Work*

The combine harvester is a combination of a grain harvesting machine and a threshing machine. The grain is cut, threshed and cleaned in one operation. The machine may be self-propelled or be towed by a tractor.

The wheatstalks are cut by an oscillating knife while the revolving reel pushes them back towards the knife and auger (feed screw). Grain flattened by wind or rain is raised by the spring prongs on the reel. The cut wheat is conveyed into the machine by the auger and reaches the threshing cylinder which rubs the grain out of the heads against a "concave". The grains of wheat, together with the chaff and short fragments of straw fall through the interstices between the bars of the concave and into the cleaning "shoe". Some of the grain is carried along with the straw, is stopped by check flaps, and is shaken out of it on shaking screens on the straw rack of the machine. The straw drops out of the back of the machine and is left in a windrow for later baling, or is baled directly by a baling attachment or press, or is scattered over the ground by a fan-like straw spreader. The grain shaken out of the straw is also delivered to the cleaning shoe. In the shoe the grain is separated from the chaff and cleaned by sieves and a blast of air. The chaff and fragments of straw are thrown out from the back of the machine. The grains of wheat fall through the sieves and into the cleangrain auger (screw conveyor) which conveys them to an elevator and on into the storage tank or into bags. Any heads of wheat which fail to go through the sieves and are thrown backwards by the air blast fall short in comparison with the lighter chaff and drop into a return auger which, via an elevator, returns them to the threshing cylinder. Correct adjustment of the air blast — by

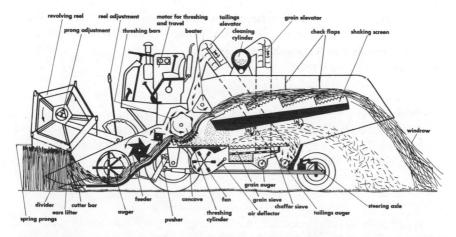

"Combine Harvester and Thresher," *The Way Things Work,* Vol. I (New York: Simon and Schuster, 1967), pp. 432–433.

throttling the intake of the fan and by altering the setting of baffles — is important in determining the degree of cleaning of the grain and the magnitude of the grain losses that occur.

## STUDY QUESTIONS

1. Would numbered steps make the process easier to follow? Why or why not?
2. Is the visual illustration *necessary* for understanding the process? Discuss.
3. *Writing Assignment:* Rewrite this selection using graphics and layout to make the process easier to follow. You might use the selection that follows — Jim Mann's "Assessing Design Effectiveness" — as a guide.

# Assessing Design Effectiveness

## Jim Mann

The biggest mistake a marketing executive can make is to withhold judgment regarding a new design on the ground that he isn't a designer or has no formal design training. You don't have to be a designer to spot a design that won't work any more than you have to be a chicken to smell a rotten egg.

The quickest systematic way to evaluate a design's marketing effectiveness — whether the design of the product or its package — is to consider how it will work in each step of the purchasing decision.

Behavioral scientists see the average purchasing decision as a six-step process:

1. **Awareness.** The customer becomes aware that the product exists. This is the first step in any buying process. No matter how great the prospect's need, unless he or she knows a product is available there can be no response. A large part of today's advertising dollars are spent solely to create awareness.
2. **Interest.** The prospect decides that the product may solve one or more of his or her problems. Awareness is the passive state of knowing; interest is an active step beyond. It relates the product to personal desire or need, and includes the urge to investigate further to make a decision.
3. **Evaluation.** The interested prospect proceeds to actively gathering information about the product (and, usually, competitive or alternate products). In this step the product is tested in the mind.

From "Assessing Effectiveness Before A Design is Produced," *Industrial Design,* November/December 1978, p. 46. Title selected by editors.

4. **Trial.** Evaluation goes beyond mental test to real test. Sampling is the most common marketing technique directed to facilitating this step in the buying process.
5. **Decision.** The step every marketer has been aiming at. The prospect uses his or her evaluation or trial to decide whether to buy. From the salesman's view, this is "the close."
6. **Confirmation.** The final step — the most neglected in market planning — comes after the customer has decided in your favor and bought your product. Whether he intends to buy again or not, he always reviews his decision in some way. This review and its outcome are important to the marketer, not only in creating and holding repeat customers but in developing new ones. Nothing can launch a product faster than word-of-mouth advertising, and nothing can scuttle it more quickly than widespread badmouthing.

It is highly unlikely that a marketing executive will make a serious design mistake if he is satisfied with the way the new design will function at each of these six steps in the buying process.

## STUDY QUESTIONS
1. What ordering principle underlies the steps of this process?
2. Are the steps in this process as rigid as the steps in the formation of stars or the operation of a thresher? Discuss.

# Illustration / Example

When somebody makes a generalization that we don't understand, we commonly ask, "Can you give me an example?" or "Can you illustrate that?" And in writing we almost instinctively use *examples* and *illustrations* to support or clarify points. They are perhaps the most used of the methods of development.

As terms, *example* and *illustration* are used loosely and often interchangeably. There are, however, important distinctions between them. Strictly speaking, illustration does just that — it illustrates, and, as such, might clarify a point or aid in making a distinction. But an illustration does not stand as proof. The "Too Many People" selection below is a case in point. Ehrlich observes that if the population continues to increase at the current rate, the world population in 900 years will be sixty million billion people. His purpose, however, is not to prove that such an incredible increase will actually come about. By illustrating that drastic increases in population will adversely affect living conditions, he warns us that we must think seriously about population control now.

An example, on the other hand, might serve to illustrate, but it essentially stands as a specific instance in support of a generalization. A sufficient number of examples might even prove the generalization. In "Collision Damage," Feld contends that a collision between a "moving barge or ship and a structure usually ends as damage to the structure." He then follows with numerous examples as supporting evidence for this generalization.

The report writer should have an awareness of when it is sufficient to only illustrate a point and when it is necessary to include supporting examples. In addition to "Too Many People" and "Collision Damage," we include three other short selections for introducing illustration and example. Be prepared to discuss whether they clarify or support.

# Too Many People

## Paul Ehrlich

In a book about population there is a temptation to stun the reader with an avalanche of statistics. I'll spare you most, but not all, of that. After all, no matter how you slice it, population is a numbers game. Perhaps the best way to impress you with numbers is to tell you about the "doubling time" — the time necessary for the population to double in size.

It has been estimated that the human population of 8000 B.C. was about five million people, taking perhaps one million years to get there from two and a half million. The population did not reach 500 million until almost 10,000 years later — about 1650 A.D. This means it doubled roughly once every thousand years or so. It reached a billion people around 1850, doubling in some 200 years. It took only 80 years or so for the next doubling, as the population reached two billion around 1930. We have not completed the next doubling to four billion yet, but we now have well over three and a half billion people. The doubling time at present seems to be about 35 years. Quite a reduction in doubling times: 1,000,000 years, 1,000 years, 200 years, 80 years, 35 years. Perhaps the meaning of a doubling time of around 35 years is best brought home by a theoretical exercise. Let's examine what might happen on the absurd assumption that the population continued to double every 35 years into the indefinite future.

If growth continued at that rate for about 900 years, there would be some 60,000,000,000,000,000 people on the face of the earth. Sixty million billion people. This is about 100 persons for each square yard of the Earth's surface, land and sea. A British physicist, J. H. Fremlin, guessed that such a multitude might be housed in a continuous 2,000-story building covering our entire planet. The upper 1,000 stories would contain only the apparatus for running this gigantic warren. Ducts, pipes, wires, elevator shafts, etc., would occupy about half of the space in

From Paul Ehrlich, *The Population Bomb* (Sierra Club Edition, 1969), pp. 3–5.

the bottom 1,000 stories. This would leave three or four yards of floor space for each person. I will leave to your imagination the physical details of existence in this ant heap, except to point out that all would not be black. Probably each person would be limited in his travel. Perhaps he could take elevators through all 1,000 residential stories but could travel only within a circle of a few hundred yards' radius on any floor. This would permit, however, each person to choose his friends from among some ten million people! And, as Fremlin points out, entertainment on the worldwide TV should be excellent, for at any time "one could expect some ten million Shakespeares and rather more Beatles to be alive."

## STUDY QUESTIONS

1. Discuss the effectiveness of the illustration. Can you think of a better way to illustrate the dangers of over-population?
2. Discuss in general the value of hypothetical illustrations. For what kinds of reports is the hypothetical illustration appropriate?
3. *Writing Assignment:* Develop a hypothetical illustration of another current trend (the number of automobiles manufactured each year, the inflation rate, the fuel shortage, or the weapons race), and where it may lead in the future.

# The Gifted Child

## Dorothy A. Sisk

Barbara, aged 12, gobbles up learning. She shifts a stack of papers from the floor beside her cluttered desk, riffles through them for her outline, and then dashes to the school library for yet one more source. She keeps a "to do" list of things she wants to accomplish, books to be read, and what she calls "poem starters," words that spur the imagination. Barbara is also a compulsive reader. "I read at least a book every two or three days. It's like a big thirst. There's so much to read that I just can't waste time."

Barbara is a sixth grader now, but she was adding simple sums at age three and reading before she was four. Her parents viewed these achievements as unusual, but they did not see any reason to refer Barbara to a psychologist or testing expert. Barbara's first-grade teacher was introduced to the youngster's special abilities on the very first day of school when Barbara plowed through the preprimer, primer and first-grade reading book. Fortunately, the teacher didn't mind that she would

From Dorothy A. Sisk, "What If Your Child Is Gifted?" *American Education,* October 1977. Title selected by editors.

have one less beginning reading student; instead, she arranged for Barbara to read independently during the regular reading period.

Over the next two years Barbara encountered some resistance from teachers who found — within the crowded classrooms — that Barbara's advanced reading skills and endless questions were too taxing on their time and attention. When a program for gifted children was started during Barbara's fourth year, her teachers nominated her for the program, and she took a series of intelligence tests under the observant eye of the school psychologist. Barbara scored in the top two percent of her age group, registering an IQ score of 140 on the Wechsler Intelligence Test for Children. This meant that, although Barbara was then ten years old, her mental abilities were comparable to those of the average 15-year-old child. After Barbara was admitted to the gifted program, she attended special enrichment classes for one period each day.

Barbara's special intellectual abilities put her among what is often called the "gifted." Although a high IQ score or high grades are frequently used to indicate "giftedness," the concept of gifted children is now expanding to include other areas of talent such as leadership, performing and visual arts, creativity, and psychomotor.

### Understanding Giftedness

Almost every child has something that he or she can do better than most children of the same age. . . .

### STUDY QUESTIONS
1. Is this selection illustration or example? Explain.
2. This excerpt introduces the article on gifted children from which it was taken. Does it make an effective opening?
3. Toward what audience level does this selection seem to be directed?

# Pin - Making

## Adam Smith

To take an example, therefore, from a very trifling manufacture; but one in which the division of labour has been very often taken notice of, the trade of the pin-maker; a workman not educated to this business (which the division of labour has rendered a distinct trade), nor acquainted with the use of the machinery employed

From Adam Smith, *The Wealth of Nations,* 1776. Title selected by editors.

in it (to the invention of which the same division of labour has probably given occasion), could scarce, perhaps, with his utmost industry, make one pin in a day, and certainly could not make twenty. But in the way in which this business is now carried on, not only the whole work is a peculiar trade, but it is divided into a number of branches, of which the greater part are likewise peculiar trades. One man draws out the wire, another straightens it, a third cuts it, a fourth points it, a fifth grinds it at the top for receiving the head: to make the head requires two or three distinct operations; to put it on is a peculiar business; to whiten the pins is another; it is even a trade by itself to put them into the paper; and the important business of making a pin is, in this manner, divided into about eighteen distinct operations, which, in some manufactories, are all performed by distinct hands, though in others the same man will sometimes perform two or three of them. I have seen a small manufactory of this kind where ten men only were employed, and where some of them consequently performed two or three distinct operations. But though they were very poor, and therefore but indifferently accommodated with the necessary machinery, they could, when they exerted themselves, make among them about twelve pounds of pins in a day. There are in a pound upwards of four thousand pins of a middling size. Those ten persons, therefore, could make among them upwards of forty-eight thousand pins in a day. Each person, therefore, making a tenth part of forty-eight thousand pins, might be considered as making four thousand eight hundred pins in a day. But if they had all wrought separately and independently, and without any of them having been educated to this peculiar business, they certainly could not each of them have made twenty, perhaps not one pin in a day; that is, certainly, not the two hundred and fortieth, perhaps not the four thousand eight hundredth part of what they are at present capable of performing, in consequence of a proper division and combination of their different operations.

In every other art and manufacture, the effects of the division of labour are similar to what they are in this very trifling one . . . .

## STUDY QUESTIONS

1. Could Smith's selection be an example as well as an illustration? Discuss.
2. Does the selection give a detailed account of all the steps of the pin-making process? What element does he emphasize?

# Collision Damage.

## Jacob Feld

The contact between a moving barge or ship and a structure usually ends as damage to the structure. The Pontchartrain Causeway in Louisiana had been struck by barges, seven times between 1956 and 1966. Replacement items of the precast elements are kept available to permit rapid replacement. In 1958, a 150-ft concrete span at Topeka, Kan., was put out of commission by a barge. The next year the Hood Canal Bridge in Washington was damaged and some of its precast boxes sunk by a similar collision, with consequent serious delay in completion of the project. The fender around the pier of the Mackinac Bridge was dented by a ship's prow and as a result it was discovered that the mortar intrusion of preplaced concrete aggregate had not provided a solid concrete. In 1964 the Maracaibo Bridge in Venezuela was damaged by a seagoing vessel that missed the main channel and hit a pier.

In 1966 the Raritan River railroad bridge was hit by a barge with serious effect on New Jersey commuter traffic, and the Poplar Street Bridge over the Mississippi River, under construction, had some of its falsework carried away by a string of barges. Engineering construction talent is used to maximum value in the repair and reconstruction of such bridge accidents. A recent example well illustrates this know-how. The Murphy Pacific Corp. built a 2500-ft multispan bridge in 1958 crossing the Sacramento River at Rio Vista, Calif. In 1967, a 10,000-ton freighter knocked down the 108-ft suspended span and an adjacent 36-ft cantilever section, cracked the concrete supporting piers, and damaged an adjoining truss. The bridge was back in temporary service almost immediately with wooden trestles and steel decking. New trusses were fabricated, assembled in full spans with reinforced-concrete deck, loaded on a 560-ton derrick barge, and moved 45 miles upstream from the assembly yard to the bridge site. Meanwhile the damaged piers and truss were repaired. The temporary trestles were removed and the new 36-ft span lifted into position and bolted securely. The sunken lift span was removed from the river bed, loaded on the barge and the replacement span placed into position. Fill-in strips of concrete pavement were placed while the steel was permanently connected. The total job required only 84 hours of continuous work by Murphy Pacific after that company was called in by the California Division of Highways, and for most of that time traffic was using the temporary trestle.

Equipment sometimes gets involved in these physical contacts with damage to both structure and moving item. In 1959 a 525-ft grain ship bumped into the 200-ft lift bridge over the Buffalo River, wrecking the span. Falling water after the crash lowered the stern of the boat, which became enmeshed with the bridge steel. Fuel and cargo were removed to lighten the boat so that it could be pulled free.

Trucks crossing the Muscantine Bridge over the Mississippi River connecting

From Jacob Feld, *Construction Failure* (New York: Wiley, 1968), pp. 145–147.

to U.S. Highway 92 on the Illinois side in 1956 smashed into the steel trusses and one span fell into the river, with one truck hanging by its rear wheels over the edge of the adjacent upright span. A tractor being hauled on a flatcar in a 44-car Illinois Central freight train came loose during the crossing of the single track bridge at Fort Dodge River, Ia. The 196-ft span bridge fell into the river 25 ft below, carrying six railroad cars with it. In the same year, 1964, a 23-yd scraper working on an embankment in Seattle went out of control and ran backwards down three blocks, demolishing two automobiles and ending in the corner of a hotel. Four stories of masonry collapsed into the street exposing the full height of the hotel rooms to street view. The wood framed floors deflected somewhat but stayed in place, even with the loss of part of the return wall where seven beams had been supported.

## STUDY QUESTIONS

1. The book from which this selection is taken is often used as a text in construction engineering courses. What textbook features can you detect in this short excerpt?

# Supership Breakdown

## Noël Mostert

Ships have always been liable to breakdown, either from the shattering force of the weather upon their hulls or means of propulsion, or because of mechanical failure, but probably no class of ship since the age of steam has been more systematically prone to breakdown than supertankers. Nor has any class of vessel, whether in the age of sail or steam, been less able in the face of disaster to make do, mend itself a little, and perhaps get to where it is going through patchwork or improvisation.

In the days of sail the weather could be the only real cause of breakdown of a ship's physical equipment, unless it was all so aged and rotted that it collapsed of its own accord; even then, sailor's artifice could always make something new of what was available. The sailmaker and the carpenter were the most indispensable men aboard after the master and the mate. The sailmaker had to restitch and remake the sails blown and shredded by the winds. On a bad voyage a ship might have had to stitch an entirely new suit of sails for itself, and just one of these sails of brutally heavy canvas could hold as many as two million stitches.

If a sailing vessel was dismasted and otherwise damaged, a carpenter could

From Noël Mostert, *Supership* (New York: Warner Edition, 1976), pp. 185-187. Title selected by editors.

hammer a manageable rig or steering gear from splinters, so to speak; improvisation was navigation's complementary art. It had to be, for such a vessel was entirely cut off, alone in its own unbeseechable eternity.

The age of steam and its mechanical breakdowns often required even greater resourcefulness from sailors. Lacking radio, early steamships often drifted for weeks, even months, while repairs were made to engines or steering gear. When, for example, the 5,000-ton British tramp steamer *Titania* lost her propeller during heavy gales in the South Atlantic in 1900, the ship's head was held to the wind and sea by a sea-anchor, the forward holds were flooded to raise the stern and, lurching and pounding in this dangerous position, the spare propeller was swung outboard by cargo winches and then guided onto the wildly gyrating tail shaft, threaded onto it, and then sealed into place by a locknut. The whole business took several days and was supervised by the ship's master, who was suspended outboard in a bo'sun's chair which, with the motion of the ship, alternately swung him across the skies, then plunged him into the crests. Lloyd's awarded him its medal for Meritorious Services.

The sagas of make-do which have in this manner brought steamships and their crews to port against apparently hopeless circumstances are beyond counting. It still often happens during the North Atlantic winter: some flag of convenience World War Two-built freighter, perhaps a Liberty, whose cargo has shifted, whose power is faltering or gone, and which is lying almost on its beam ends, nonetheless manages to hold its head to the sea and, eventually, to make port. The two world wars of course provided scores of instances of skillful handling of disabled ships. When a torpedo in 1942 blew off the troopship *Llangibby Castle's* stern and took with it the ship's rudder, she sailed 3,400 miles steering with her twin screws. Many of the greatest survival sagas of the last war involved tankers. A notable episode was that of the 8,000-ton British tanker *San Demetrio,* which was shelled and set on fire by the German pocket battleship *Admiral Scheer* when it raided an Atlantic convoy. The *San Demetrio* was so badly damaged, her fires so uncontrollably raging, that she was abandoned. Twenty hours later, however, one of the lifeboats sighted the ship, still afire, and, since no other succor was in sight, its occupants maneuvered alongside. They boarded the tanker with great difficulty, fought for two days to extinguish the fires, got the engines going and, although the bridge was one of the areas demolished by the flames, they conned the ship from aft. The *San Demetrio* finally arrived with most of its precious gasoline cargo still intact.

One was always inclined to believe that nothing was impossible at sea, but it would be difficult to imagine too many such feats, or even opportunities for resourcefulness, in the case of supertankers. That of course doesn't preclude feats of seamanship or courage with them. It was seamanship of the highest order, notably calm maneuvering of the ship, that kept flames from engulfing the 206,000-ton *Mactra* after her tank explosion. Her master, Captain J. E. Palmer, turned the ship to keep the flames and thick smoke away from the accommodation area which, if it caught fire, would have meant abandoning the vessel. Her chief engineer, E. H. Edmondson, directed fire-fighting on deck for nine hours. But although much of her deck was blasted open, the working spaces of the ship were unharmed; no windows even were broken because the force of the explosion went straight up. Had her accommodation and bridge been burned out, however, and the ship been abandoned, it is hard to conceive that a *San Demetrio* type of salvage operation could have been carried out by her crew. Reboarding her sheer towering sides alone would have been a ferocious problem.

The mere size of the ship always emerges as one of the main restrictions upon the ingenuity of any supertanker sailor, however good he may be in an old-fashioned sense. The distance can be too great or access too awkward for practical solutions. If a supertanker loses a propeller it is virtually inconceivable that the spare that is carried right forward on the main deck could be maneuvered aft as it was on the *Titania* and set in place. It weighs between thirty and sixty tons. There is no practical means of handling such a weight for a distance of close to a quarter of a mile outboard along the ship's side, or even over the decks. It is there simply to avoid having to wait in a dockyard should a replacement be necessary. Anyway, on a loaded VLCC the propeller shaft is about forty feet under the water. Diving gear not a bo'sun's chair would be in order for any master wishing to supervise the task.

**STUDY QUESTIONS**
1. What is the central purpose of this selection?
2. Does the author illustrate or support?

# Comparison/Contrast

*Comparison/contrast* is a natural method of development, for we often find ourselves placing two things — friends, automobiles, movies, opinions — side by side and noting their similarities and differences. Such a thought pattern allows us to understand or evaluate in a manner that is both clear and effective, so comparison/contrast is a technique that frequently appears in technical communication.

It is possible to use comparison/contrast in two different manners. Let us assume that mechanisms A and B are to be examined together. One could first write a long analysis of A, pointing out its features and purpose, then write an analysis of B, doing essentially the same thing. If you are presenting a relatively short or simple analysis, this would really not be a bad way to develop a comparison. It does have its weaknesses, however, particularly when you are examining rather complex mechanisms at length. The chief weakness of this method is that after the reader concludes his examination of your analysis of component A, and then proceeds to examine your analysis of component B, certain facets of A may be forgotten. Also, some details may not seem as important as they should, simply because they have appeared earlier in the report and are no longer uppermost in the reader's mind. The effectiveness of this comparison can be limited in extended analyses, because in effect the writer does not really create a comparison. He simply presents the two mechanisms, A and B, and leaves it up to the reader to establish parallels between the two.

For this reason a point by point comparison, in which features of the mechanisms are compared one at a time, is generally preferable. It may at times be necessary, however, for a writer first to lay out his comparison in the form described above — all of component A, then all of component B. Many of us just cannot juggle all the points of an elaborate comparison without first delineating clearly to ourselves the things to be compared. Then when we do have our items clearly in focus, we can begin the dove-tailing, the paralleling, the comparing, the contrasting. Consequently, our finished report will take on the preferable form, contrasting components one feature at a time, even though it may have begun with the rather simple format, all of mechanism A juxtaposed with all of mechanism B.

Lastly, you should remember that there is no point in comparing things only to be comparing them. Your comparison must have a point and it must be meaningful, for any two things in the world can be set side by side and their differences and similarities noted. We can compare a ballpoint pen with a Mercedes-Benz. We can compare the same pen with the planet Venus, a stuffed coyote, a treatise on nuclear physics, and a dandelion growing in our yard. Any two things can be compared, and probably some similarities (and certainly some differences) can be noted when we place any two objects or ideas together. The important question to keep in mind is "What is the point"? Do we have anything significant to say about the similarities or differences we find in our comparison? Does this comparison inform and enlighten our reader? Or are we just examining objects and commenting upon what we see?

In the following selections, Joseph Weizenbaum, a noted computer expert, compares two kinds of computer programmers, and Robert Gorman, writing for *Popular Science,* compares two kinds of smoke detectors. Clinton J. McGirr's "Guidelines For Abstracting" presents an example of comparison/contrast while it explains the differences between abstracts, summaries, and introductions — formats that all writers in business, science, and technology will be exposed to sometime in their careers.

# Computer Programmers

## Joseph Weizenbaum

Wherever computer centers have become established, that is to say, in countless places in the United States, as well as in virtually all other industrial regions of the world, bright young men of disheveled appearance, often with sunken glowing eyes, can be seen sitting at computer consoles, their arms tensed and waiting to fire their fingers, already poised to strike, at the buttons and keys on which their attention seems to be as riveted as a gambler's on the rolling dice. When not so transfixed, they often sit at tables strewn with computer printouts over which they pore like possessed students of a cabalistic text. They work until they nearly drop, twenty, thirty hours at a time. Their food, if they arrange it, is brought to them: coffee, Cokes, sandwiches. If possible, they sleep on cots near the computer. But only for a few hours — then back to the console or the printouts. Their rumpled clothes, their unwashed and unshaven faces, and their uncombed hair all testify that they are oblivious to their bodies and to the world in which they move. They exist, at least when so engaged, only through and for the computers. These are computer bums, compulsive programmers. They are an international phenomenon.

How may the compulsive programmer be distinguished from a merely dedicated, hard-working professional programmer? First, by the fact that the ordinary professional programmer addresses himself to the problem to be solved, whereas the compulsive programmer sees the problem mainly as an opportunity to interact with the computer. The ordinary computer programmer will usually discuss both his substantive and his technical programming problem with others. He will generally do lengthy preparatory work, such as writing and flow diagramming, before beginning work with the computer itself. His sessions with the computer may be comparatively short. He may even let others do the actual console work. He develops his program slowly and systematically. When something doesn't work, he may spend considerable time away from the computer, framing careful hypotheses to account for the malfunction and designing crucial experiments to test them. Again, he may leave the actual running of the computer to others. He is able, while waiting for results from the computer, to attend to other aspects of his work, such as documenting what he has already done. When he has finally composed the program he set out to produce, he is able to complete a sensible description of it and to turn his attention to other things. The professional regards programming as a means toward an end, not as an end in itself. His satisfaction comes from having solved a substantive problem, not from having bent a computer to his will.

The compulsive programmer is usually a superb technician, moreover, one who knows every detail of the computer he works on, its peripheral equipment, the

From Joseph Weizenbaum, "Science and the Compulsive Programmer," *Computer Power and Human Reason* (San Francisco: W. H. Freeman, 1976), pp. 116–118. Title selected by editors.

computer's operating system, etc. He is often tolerated around computer centers because of his knowledge of the system and because he can write small subsystem programs quickly, that is, in one or two sessions of, say, twenty hours each. After a time, the center may in fact be using a number of his programs. But because the compulsive programmer can hardly be motivated to do anything but program, he will almost never document his programs once he stops working on them. A center may therefore come to depend on him to teach the use of, and to maintain, the programs that he wrote and whose structure only he, if anyone, understands. His position is rather like that of a bank employee who doesn't do much for the bank, but who is kept on because only he knows the combination to the safe. His main interest is, in any case, not in small programs, but in very large, very ambitious systems of programs. Usually the systems he undertakes to build, and on which he works feverishly for perhaps a month or two or three, have very grandiose but extremely imprecisely stated goals. Some examples of these ambitions are: new computer languages to facilitate man-machine communication; a general system that can be taught to play any board game; a system to make it easier for computer experts to write super-systems (this last is a favorite). It is characteristic of many such projects that the programmer can long continue in the conviction that they demand knowledge about nothing but computers, programming, etc. And that knowledge he, of course, commands in abundance. Indeed, the point at which such work is often abandoned is precisely when it ceases to be purely incestuous, i.e., when programming would have to be interrupted in order that knowledge from outside the computer world may be acquired.

Unlike the professional, the compulsive programmer cannot attend to other tasks, not even to tasks closely related to his program, during periods when he is not actually operating the computer. He can barely tolerate being away from the machine. But when he is nevertheless forced by circumstances to be separated from it, at least he has his computer printouts with him. He studies them, he talks about them to anyone who will listen — though, of course, no one else can understand them. Indeed, while in the grip of his compulsion, he can talk of nothing but his program. But the only time he is, so to say, happy is when he is at the computer console. Then he will not converse with anyone but the computer. We will soon see what they converse about.

The compulsive programmer spends all the time he can working on one of his big projects. "Working" is not the word he uses; he calls what he does "hacking." To hack is, according to the dictionary, "to cut irregularly, without skill or definite purpose; to mangle by or as if by repeated strokes of a cutting instrument." I have already said that the compulsive programmer, or hacker as he calls himself, is usually a superb technician. It seems therefore that he is not "without skill" as the definition would have it. But the definition fits in the deeper sense that the hacker is "without definite purpose": he cannot set before himself a clearly defined long-term goal and a plan for achieving it, for he has only technique, not knowledge. He has nothing he can analyze or synthesize; in short, he has nothing to form theories about. His skill is therefore aimless, even disembodied. It is simply not connected with anything other than the instrument on which it may be exercised. His skill is like that of a monastic copyist who, though illiterate, is a first-rate calligrapher. His grandiose projects must therefore necessarily have the quality of illusions, indeed, of illusions of grandeur. He will construct the one grand system in which all other experts will soon write their systems.

## STUDY QUESTIONS

1. Weizenbaum is of course comparing and contrasting computer programmers here, but can we make similar distinctions in other careers? Are there, for instance, "professional workers" and "compulsive workers" in all fields? How might one recognize these types?
2. It is sometimes said that flaws are easier to point out than virtues. Does that generalization seem to apply to Weizenbaum's comparison?
3. Weizenbaum does not actually create a point by point comparison. How would you characterize the format he does employ? Is it more effective than a point by point comparison? Less effective?
4. *Writing Assignment:* Write a comparison of professional and compulsive workers in your own field.

# Guidelines for Abstracting

## Clinton J. McGirr

### Abstract Versus Summary

Are an abstract and summary identical? The answer to that question often depends upon the semantics of the publisher. Some publishers interchange the two words in their instructions to authors, thus implying that abstracts and summaries are the same. However, though the words *abstract* and *summary* may be sometimes considered to be synonymous, they are not so viewed by many authors and publishers. To those who differentiate, a summary is generally longer than an abstract. It is included as part of a report and is not published separately in a secondary publication.

The American National Standards Institute, for example, defines a summary as a restatement within a document (usually at the end) of its salient findings and conclusions. The summary is intended to complete the orientation of a reader who has studied the preceding text.

From "Guidelines for Abstracting," *Technical Communication,* Second Quarter, 1978, pp. 3–4.

Here at GEND, on the other hand, the summary of each internal laboratory report is the first page of that report. The laboratory report summary is made up of these topics: subject, objective, abstract of results, conclusions, and recommendations. In these reports, the summary provides busy managers with the salient facts about a study without the necessity of their reading the entire report, which may be long and complex.

Westinghouse Research Laboratories also requires that a Technical Document Summary accompany all technical documents emanating from the labs. The Westinghouse report summary includes these elements: purpose, scope, approach, results, conclusions, and significance.

As Souther and White point out, for many readers the summary *is* the report — the part of the report that they read, anyway. For that reason, summaries are usually longer and more detailed than abstracts. As can be seen by the GEND and Westinghouse requirements, the summary usually includes a description of the problem, the approach taken to solve it, results, final conclusions, and recommendations. A summary is always informative.

While an abstract rarely, if ever, is illustrated, a summary may include an illustration if the author thinks it is desirable as an aid in covering the subject. A summary may also include tables.

## Abstract Versus Introduction

Some writers, when faced with preparing an abstract, take an easy path. They duplicate the introduction of the report, labeling it "abstract." However, most publishers' instructions to authors delineate a functional difference between the abstract and the introduction. Whereas the abstract provides a brief synopsis of the entire report, the introduction introduces the subject so that the reader will more easily understand the report. The American National Standard Institute recommends that the introduction make clear the aim of the paper, the author's purpose in undertaking the work described, and the relationship of the work to a larger field of inquiry. Often an author may assume that the purpose of the work is obvious to experts in the field, but he should, nevertheless, make the purpose explicit (or at least evident) and intelligible to those who are not specialists. Readers should learn from the introduction exactly what question(s) the author is to address. The introduction to a long paper may include an outline of the rest of the paper or an indication of its content.

Some of the information in the introduction is obviously going to be duplicated in the abstract. However, the abstract should not be devoted largely to introductory material rather than to the substance of the report.

## STUDY QUESTIONS

1. McGirr actually creates two comparisons in this selection, first comparing the abstract with the summary, then comparing the abstract with the introduction.

Using the guidelines he has provided, can you then go on to compare the summary with the introduction? Is such a comparison meaningful?

# Smoke Detectors

## Robert Gorman

Last Christmas, a Fairfield, Conn., family received a prophetically timed present — a battery-powered smoke detector. Unwrapped but left under the tree, it sounded an alarm in the small hours of the following morning after a fire started in the basement. Five people escaped the rapidly spreading blaze that a fire chief later said could otherwise have been fatal to them all.

It's a true story and, as the experience indicates, smoke detectors can be life-savers. But there are several different kinds on the market with lots of conflicting claims. Which is best for you? Where should you put it in your home for greatest effectiveness? How can you be sure it is working properly a year or two after you've installed it? And finally, how about all that flap over the type that contains a radioactive substance? Is it really dangerous? To get the answers, I went to the experts. But first, I acquainted myself with the inner workings of the two types of detectors now used for the home: photoelectric and ionization.

### Inside the Detector

Photoelectrics have been used professionally for years as part of an overall system. Recently they have become a significant factor as a single-station unit. The reason is the development of battery-operated models with improved case geometry and sophisticated electronics. They use low-drain LED light sources instead of incandescent bulbs, which need more power and have a much more limited life.

The operating principle is simple. The LED light source and a photosensitive cell are placed inside a dark, shielded chamber. A barrier is between them so the photocell cannot see the light. When visible smoke particles (larger than about 1.0 micron) enter the chamber, they scatter and deflect the light, allowing some of it to get around the barrier and be detected by the photocell. The detector's circuitry responds to the resulting current change and sounds the alarm.

An ionization detector contains one or two tiny, foil-wrapped grains of radioactive materal. This material, now almost exclusively an oxide of americium 241, ionizes the air in a sensing chamber, making it conductive. This permits a small current to flow between two electrodes. On some units two chambers are used — the

From "Smoke Detectors," *Popular Science*, October 1978, pp. 50–52.

second being a reference of clean air. Others — single-chamber units — use an electronic circuit to maintain a level of equalization. When ions and other small particles produced by a fire enter the detector, they reduce the conductivity of the air. The unit senses the decreased current flow and sets off the alarm.

Both types of detectors have strengths and weaknesses. Ionization units are inherently most sensitive to subvisible products of combustion 0.01 to 1.0 micron in size. These particles are most numerous close to flaming fires; they are also released in quantity by such fires and — being very light — are accelerated upward by thermal lift toward ceiling or high-wall positions where detectors are normally located.

At the same time, the small-particle sensitivity of ion detectors makes them react easily to a variety of non-hostile products of combustion (from cooking, for example) that are common in homes. They can also be affected by air currents, steam, humidity, and the like.

In a photoelectric type, particles too small to reflect light cannot set off the system, so it responds slowly to close-in flaming fires as well as to other particles and currents that could cause nuisance or false alarms. These units are most sensitive to "gray" smoke. It is a fairly consistent signature of smoldering fires and very likely to be produced within a short time by flaming fires because smaller particles tend to cluster (or agglomerate) as they travel away from their source. (You can see this effect by looking at a flame. You may notice a smoke-free gap just above the flame. A foot or so higher, visible smoke appears.)

## Which Type is Better?

Probably the most comprehensive data so far available on detector performance in simulated residential fires have come out of a study made for the National Bureau of Standards by Underwriters' Laboratories and IIT Research Institute. Most previous studies had been done in fire-test rooms, built especially for that purpose. But researchers for NBS found two brick houses slated for demolition at the Indiana Dunes National Lakeshore. They set some 40 fires in the houses, exposing 20 different detectors mounted in many different locations to a wide variety of conditions.

The result: The photoelectric type responded faster to smoldering fires, the ion type to flaming fires, just as previous research indicated they should. But more important, the researchers concluded that *both* types generally responded well to all fires. Either type, said the authors, could "provide more than adequate lifesaving potential under most real residential-fire conditions when properly installed."

**STUDY QUESTIONS**

**1.** What form does Gorman's comparison of smoke detectors employ? Is it a juxtaposition of complete components, a point by point comparison, or a combination of the two?

2. It is easier to create a comparison when the items being compared share some major common elements. What important characteristics do the two types of smoke detectors have in common?

# Definition

To define a term is to give its meaning. In technical, business, and scientific fields definition is an important activity because of the many new terms that are generated.

You owe it to your reader to define your use of a term under one of two conditions: 1) if you are using a term that your reader is unlikely to be familiar with, and 2) if you are using a term that has many possible meanings.

Of the many methods of definition we will discuss the ones that are commonly used:

1. *Formal definition.* This is the dictionary approach, and works quite well when a terse, one sentence definition is necessary to orient the reader. The formal definition has three elements: a) the *term* to be defined, b) the larger *class* (or *genus*) to which the term belongs, and c) the specific features (sometimes called the *differentia*) that distinguish your term from all others Included in the same class. For example:

   term                class              distinguishing feature
   An ammeter is an instrument for measuring electrical current.

   Note the effectiveness of this scheme for informing your reader. By placing the term in a familiar class, you bring the reader into the realm of the known. Although he perhaps doesn't know what an ammeter is, he likely knows what instrument means. The *differentia* then indicates what the ammeter specifically does as an instrument.
2. *Definition by synonym.* A synonym is a word having much the same meaning as the term you are defining. Although no two words ever have exactly the same meaning, the synonym might serve as a useful approximation. For example, a *gazebo* is a *summerhouse.* There is more to be explained about a gazebo, but a synonym is a place to start. The synonym is especially useful in translating specialized terminology: an *autoclave* is a *sterilizer* (an equivalent term that the layman is more likely to understand.)
3. *Definition by word origin.* The origin, or etymology, of a word can provide a valuable clue to its current usage. For example, *taxonomy* is formed from the Greek words *taxis* (arrangement) and *nomos* (law). "Law of arrangement" is a good beginning for explaining the meaning of taxonomy.
4. *Negative definition.* Telling the reader what the term is *not* might be useful for cases in which there has been past confusion. For example, a *parameter* is not a *perimeter.*

5. *Definition by example.* A suspension bridge might be defined by citing some of the well known examples of suspension bridges, such as the Brooklyn Bridge and the Golden Gate Bridge.
6. *Stipulative definition.* The writer states, briefly, the precise way in which he is going to use a term in the text. He might stipulate that he is using the term science to mean a *systematized body of factual knowledge* — which is but one of the many possible meanings.
7. *Operational definition.* The writer emphasizes what the term does. If you state that a generator *converts mechanical power into electrical energy,* you have given an operational definition. You might even include information on the process involved. In fact, operational definition is often similar to process description, although it is not as complete.

These methods can be used singly, or several might be combined for an extended definition.

You might analyze each of the following extended definitions to determine the methods used. *Technology,* an encyclopedia entry, illustrates the systematic and straightforward approach to definition that is common in reference works. *Magnetism* employs several techniques to aid the beginning technician in understanding a scientific principle. In *But What Do the Numbers Mean,* a slightly irreverent view of the business world, "Adam Smith" presents some slightly unorthodox definitions of business terms.

# Technology

## Robert S. Sherwood and Harold B. Maynard

Systematic knowledge and action, usually of industrial processes but applicable to any recurrent activity. Technology is closely related to science and to engineering. Science deals with man's understanding of the real world about him — the inherent properties of space, matter, energy, and their interactions. Engineering is the application of objective knowledge to the creation of plans, designs, and means for achieving desired objectives. Technology deals with the tools and techniques for carrying out the plans.

For example, certain manufactured parts may need to be thoroughly clean. The technological approach is to use more detergent and softener in the wash water, to use more wash cycles, to rinse and rerinse, and to blow the parts dry with

From *McGraw-Hill Encyclopedia of Science and Technology,* Vol. 13 (New York: McGraw-Hill, 1972), pp. 428–429.

a stronger, warmer air blast. Often such refinements provide an adequate action. However, if they do not suffice, the basic technique may need to be changed. Thus, in this example, science might contribute the knowledge that ultrasonically produced cavitation counteracts surface tension between immiscible liquids and adhesion between clinging dirt and the surface to be cleaned, and thereby produces emulsions. Engineering could then plan an ultrasonic generator and a conveyor to carry the parts through a bath tank in which the ultrasonic energy could clean them. The scientist may use ultrasonic techniques to determine properties of materials. The engineer may design other types of devices that employ ultrasonics to perform other functions. These specialists enlarge their knowledge of ultrasonics and their skill in using this technique not for its own sake but rather for its value in their work. The technologist is the specialist who carries out the technique for the purpose of accomplishing a specified function. He extends his knowledge and skill of ultrasonic cleaning by refinement and perfection of the technique for use on various materials soiled in different ways. Technological advances improve and extend the application to cleaning other parts under other conditions. *See* AERO-NAUTICAL ENGINEERING; CHEMICAL ENGINEERING; CIVIL ENGINEERING; ELECTRICAL ENGINEERING; ENGINEERING; ENGINEERING, SOCIAL IMPLICATIONS OF; ENGINEERING DESIGN; FOOD ENGINEERING; HUMAN-FACTORS ENGINEERING; INDUSTRIAL ENGINEERING; MARINE ENGINEERING; MECHANICAL ENGINEERING; NUCLEAR ENGINEERING; PRODUCTION ENGINEERING; SCIENCE; SPACE TECHNOLOGY.

**STUDY QUESTIONS**
1. Do the authors use formal definition? Where?
2. Why do the authors define engineering and science?
3. Is the example used for definition well chosen? Explain.
4. Why do the authors list related terms at the end of the article?

# Magnetism

## Editors of *Automotive Alternators*

Just what is magnetism? Strangely enough, scientists cannot give us a precise answer. We really don't know what creates this strange force or even why it exists at all. The dictionary defines it as ". . . that properly possessed by various bodies,

From *Automotive Alternators* (Indianapolis: Howard W. Sams and Co., 1977), pp. 11–12. Title selected by editors.

such as iron or steel, of attracting or repelling similar substances according to certain physical laws." Although we cannot say exactly what magnetism is, we do know how it behaves, how to produce it, and how to harness its strange powers for useful purposes. Interestingly enough, most of what we know about magnetism has been discovered only in the past 150 years.

However, magnetism, or rather magnetic effects, were known in ancient times. In fact, it is believed that the word "magnetism" is derived from a particular form of iron ore, known as *lodestone,* that originally was found near the ancient city of Magnesia. A lodestone is a natural magnet that, as early experimenters discovered, had the power to attract pieces of iron (Figure 1). The lodestone remained merely a scientific curiosity until about the 11th or 12th century when it was discovered that an iron needle, made magnetic by rubbing against a lodestone, pointed toward the earth's north pole. Hence, the magnetic compass was invented.

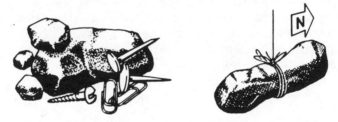

Figure 1. Lodestone, or natural magnet, attracts pieces of iron.

From this it was learned that a magnet, either natural or artificial, had two *poles* — a north pole and a south pole. The end of a magnetized needle that pointed to the earth's north pole was, logically enough, called the *north pole* of the needle. Conversely, the other end was named the *south pole.* This phenomenon also led to the conclusion that the earth itself was a giant magnet, because it was known that unlike poles attracted each other whereas like poles repelled (Figure 2). This behavior of a magnet illustrates an unusual and little-known fact: The earth's geographical north pole must be, in reality, a *south* magnetic pole. If it were not, the north pole of the compass needle would not point toward the earth's geographical north pole.

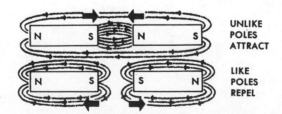

Figure 2. Like magnets repel and unlike magnets attract.

This was about all that was known about magnetism until the year 1819 when it was discovered that there is a relationship between electricity and magnetism. This was demonstrated by placing a compass near a wire, as shown in Figure 3, and then passing a current through that wire. The compass needle deflected, proving that a magnetic field existed around a wire carrying an electrical current. When the current ceased to flow, the compass needle return to its normal position. After this discovery, major strides were made in magnetic theory and its practical application, principally by Joseph Henry, a great American scientist.

Figure 3. Demonstrating the magnetic field surrounding a current-carrying conductor.

One of Henry's most notable discoveries was that not only could magnetism be produced electrically, but that electricity could be produced magnetically. It is this unique characteristic that has made possible our present-day utilization of electricity; without it, virtually all the electrical products we know and use today would be nonexistent. And the few electrical devices that could exist without employing magnetism in one way or another — a battery-operated light, for example — would simply be expensive novelties or merely laboratory curiosities.

Now, let's take a closer look at the more important principles of magnetism and electromagnetism to help us better understand the way an alternator works. First, what is the difference between a magnetized and an unmagnetized piece of iron or other magnetic material? Again, we really don't know that much about magnetism, but the generally accepted theory states that a magnetized piece of material consists of a great number of small, magnetized particles all aligned in the same direction. This is shown in Figure 4. All the north poles point in one direction and the south poles point in the other direction. When the material is unmagnetized,

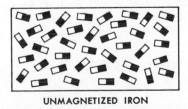

UNMAGNETIZED IRON

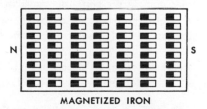

MAGNETIZED IRON

Figure 4. Alignment of magnetic particles in unmagnetized and magnetized iron.

the individual magnetic particles form a random pattern with no predominate magnetic polarity.

Whether or not this is the way magnetism really works is not really important for our purposes. The important point is that we can magnetize and demagnetize iron at will. It is due to this principle that we are able to generate and *regulate* the current in an alternator. If we were unable to control the magnetism or magnetic field in an alternator, the current we generate would be of little use to us. We will cover the regulation phase of alternators a little later. For now, let's look at the way in which magnetism is produced electrically.

## STUDY QUESTIONS

1. Why does the author quote a dictionary definition of magnetism? Should he have stated the definition in his own words? Why or why not?
2. Is the quoted definition an example of formal definition? If so, identify the class and *differentia.*
3. Why must the author define magnetism by its behavior?
4. Does the source of the word "magnetism" add anything to your knowledge of magnetic principles? Why does the author include this information?
5. Does the discussion of discoveries in magnetism add to your knowledge of magnetic principles?
6. What features indicate that the definition is written for the layman?
7. *Writing Assignment:* Using as your audience the layman reader, write an extended definition of a scientific or technical term.

# But What do the Numbers Mean ?

## "Adam Smith"

You can see that there are a lot of numbers floating around Wall Street, that the Game is played with numbers, and that with computers more people can play with more numbers in more combinations than anyone would have dreamed possible in the old, archaic pre-computer days BC in 1960. But what are the base numbers? They are the figures reported by the subject companies as sales and earnings, and earnings, in anybody's systems, are one of the most important factors.

But what are earnings?

It really ought to be easy. You pick up the paper, and Zilch Consolidated says its net profit for the year just ended was $1 million or $1 a share. When Zilch Con-

From *The Money Game* (New York: Random House, 1968), pp. 181–184.

solidated puts out its annual report, the report will say the company earned $1 million or $1 a share. The report will be signed by an accounting firm, which says that it has examined the records of Zilch and "in our opinion, the accompanying balance sheet and statement of income and retained earnings present fairly the financial position of Zilch. Our examination of these statements was made in accordance with generally accepted accounting principles."

The last four words are the key. The translation of "generally accepted accounting principles" is "Zilch could have earned anywhere from fifty cents a share to $1.25 a share. If you will look at our notes 1 through 16 in the back, you will see that Zilch's earnings can be played like a guitar, depending on what we count or don't count. We picked $1. That is consistent with what other accountants are doing this year. We'll let next year take care of itself."

Numbers imply precision, so it is a bit hard to get used to the idea that a company's net profit could vary by 100 percent depending on which bunch of accountants you call in, especially when the market is going to take that earnings number and create trends, growth rates, and little flashing lights in computers from it. And all this without any kind of skulduggery you could get sent to jail for.

How can this be?

Let's say you are an airline, and you buy a brand-new, freshly painted Boeing 727. Let's say the airplane costs you $5 million. At some point in the future the airplane is going to be worth 0, because its useful life will be over. So you must charge your income each year with a fraction of the cost of your airplane. What is the life of your airplane? You say the useful life of the airplane is ten years, so on a straight-line basis you will charge your income $500,000, or 10 percent of the cost, this year. If your net income from ferrying passengers and cargo is $1 million, it will drop by half when you apply this depreciation charge. Obviously the year you buy the airplane your earnings are going to look worse than they are next year, when you have the full use of the airplane and it is shuttling back and forth all the time. Your profits will certainly look better if you are still running that airplane in eleven years, because that year there will be no charge at all for depreciation; it will have been written off.

But that is only the beginning of the complications. Right next door at the airport is another airline. It has also bought a brand-new, freshly painted Boeing 727. So you and your competitor can be compared side by side when you both report your earnings on the same day. Right?

Hardly. The airline next door says it can run an airplane twelve years. So it is depreciating its airplane over twelve years, and its depreciation charge this year is 1/12, not 1/10, so it has only penalized its earnings $416,666 instead of $500,000, and for this year on that basis it has made more money than you have.

Don't the accountants make everybody charge the same thing for the same airplane? No, they don't. It just makes another little bit of work for the security analysts, who have to adjust the varying depreciation rates to constants. Accountants are not some kind of super-authority, they are professionals employed by clients. If you say the life of your airplane is twelve years, you must know your business; the life is twelve years. Delta Airlines depreciates a 727 in ten years; United in sixteen.

The airplane example is, of course, a very simple one. But what about two second-generation computers, say a Honeywell H200 and something in the IBM 1400 series? Do they have the same life? They may, as far as usage is concerned, but if you are going to sell or trade up it may be easier on the IBM. Then there is an investment credit available on new equipment, a tax assist passed to encourage capital expenditures. Is the investment credit "flowed through," as the jargon says, right to the earnings the first year? Or is the investment credit spread through the whole life of the equipment?

If everybody used the same depreciation method but with different periods of use, life would be tough enough. But equipment is not always depreciated straight-line, an equal percentage for each year. Some companies use heavy charges at the beginning, say 150 percent declining. Some use a method with the charming appellation "sum-of-the years-digits." If you really want to go into details, call up your accountant and ask *him* for definitions.

## STUDY QUESTIONS

1. What conventional financial terms does "Smith" introduce? Does he increase your understanding of these terms?
2. Are the terms explicitly defined, or must we figure out the meanings from the context?
3. Why does "Smith" use a hypothetical company — Zilch Consolidated — to introduce the term "generally accepted accounting principles"?
4. What method does he use to define "depreciation"?
5. Why does "Smith" suggest you ask an accountant if you want a definition for "sum-of-the-years-digits"?
6. What is "Smith's" attitude toward his subject? Is this appropriate for the conventional business report? Why or why not?
7. For what audience would his attitude be appropriate?
8. *Writing Assignment:* Rewrite the definition of "depreciation" (or any other term in the selection) in the form that would be appropriate for a reference work. You might use the "Technology" selection as a guide.

# Classification/Division

*Classification* involves grouping small things together into larger categories; *division* involves breaking large things down into smaller, individual parts. Classification, then, is essentially the opposite of division. When a small boy disassembles an alarm clock into its individual pieces he is engaged in division. When he groups together those pieces that seem similar — gears, springs, levers, wheels — he is engaged in classification. Or, to cite another example, we may divide a simple ballpoint pen by separating and labeling its various individual parts; we classify the same pen by identifying it as part of the larger category *writing instruments,* or the still larger category *manufactured items.*

Classification/division is frequently used in technical communication, for one of the central tasks of business and technical writing is organizing information in an orderly fashion. You may find yourself being asked to describe a certain mechanism, and you will find it necessary to do this by dividing the mechanism into its component parts, classifying them according to their appearance or function. Or, because classification/division is not only used when writing about objects, you may employ it when you are explaining the various steps in a process, or when you are sorting out the most important factors in an experiment from those that are less important, and those that are least important. Similarly, you will use classification/division when you are asked to bring meaning out of pages of raw data, when, for example, you are deciding which parts of your research belong to the planning and design stage, which to the construction stage, and which to the testing stage. When you separate the theoretical advantages from the practical advantages, separate the procedure from the data and the data from the analysis, when you create categories by placing together items that share some common similarity, or separate larger entities into smaller bits and pieces, you are using classification/division.

It is important to remember that when we classify or divide we must maintain a consistent principle of classification or division for the material we are organizing, and it is a good idea to jot this principle down in our notes before we begin writing. If we were classifying alarm clocks, for example, and we chose the categories *mechanical, electronic,* and *digital,* we would be in for some difficulties. Some digital clocks are electronic clocks that use lighted diodes and contain no moving parts; some digital clocks are basically mechanical clocks using revolving numbers on moving cylinders or panels. Then there are some digital clocks that employ both revolving mechanisms and lighted diodes; these clocks are both mechanical and electronic and belong to both categories in some respects. If we attempt to work with this faulty set of categories our classification will fail; we will not be certain which category some clocks belong in, because we have created overlapping categories. The overlapping categories exist because we did not maintain a consistent principle of classification in creating our categories.

Two of our categories, *mechanical* and *electronic,* describe modes of operation, the ways in which these clocks function. The third category, *digital,* describes the appearance of the clock as much as anything else, for it refers to the manner in which the time is displayed — directly as numbers — and has nothing to do with the clock's mechanism or the way it functions. By momentarily overlooking the basis or principle of classification we have created confusion rather than clarity; although it is likely we will discover such a problem while we are still writing our report, we may waste valuable time trying to resolve difficulties that could have been prevented if we had clearly stated and identified our basic principle of classification in the beginning.

In the following selections, Robert Pirsig's excerpt from his best seller *Zen and the Art of Motorcycle Maintenance* demonstrates rather thoroughly the principle of division by showing how a motorcycle might be divided into its many components. A different kind of division is presented in the piece entitled "Tavist," an advertisement taken from a medical journal. In this selection we see something that would seem to be rather simple — a pill — analyzed in a number of different ways. "Engineering Mistakes" and "Types and Limitations of Respiratory Protective Devices" demonstrate the principle of classification, illustrating both fairly simple and very complex classifications.

# The Motorcycle

## Robert Pirsig

A motorcycle may be divided for purposes of classical rational analysis by means of its component assemblies and by means of its functions.

If divided by means of its component assemblies, its most basic division is into a power assembly and a running assembly.

The power assembly may be divided into the engine and the power-delivery system. The engine will be taken up first.

The engine consists of a housing containing a power train, a fuel-air system, an ignition system, a feedback system and a lubrication system.

The power train consists of cylinders, pistons, connecting rods, a crankshaft and a flywheel.

The fuel-air system components, which are part of the engine, consist of a gas tank and filter, an air cleaner, a carburetor, valves and exhaust pipes.

The ignition system consists of an alternator, a rectifier, a battery, a high-voltage coil and spark plugs.

The feedback system consists of a cam chain, a camshaft, tappets and a distributor.

The lubrication system consists of an oil pump and channels throughout the housing for distribution of the oil.

The power-delivery system accompanying the engine consists of a clutch, a transmission and a chain.

The supporting assembly accompanying the power assembly consists of a frame, including foot pegs, seat and fenders; a steering assembly; front and rear shock absorbers; wheels; control levers and cables; lights and horn; and speed and mileage indicators.

From Robert Pirsig, *Zen and the Art of Motorcycle Maintenance* (New York: Morrow, 1974), pp. 76–78. Title selected by editors.

That's a motorcycle divided according to its components. To know what the components are for, a division according to functions is necessary:

A motorcycle may be divided into normal running functions and special, operator-controlled functions.

Normal running functions may be divided into functions during the intake cycle, functions during the compression cycle, functions during the power cycle and functions during the exhaust cycle.

And so on. I could go on about which functions occur in their proper sequence during each of the four cycles, then go on to the operator-controlled functions and that would be a very summary description of the underlying form of a motorcycle. It would be extremely short and rudimentary, as descriptions of this sort go. Almost any one of the components mentioned can be expanded on indefinitely. I've read an entire engineering volume on contact points alone, which are just a small but vital part of the distributor. There are other types of engines than the single-cylinder Otto engine described here; two-cycle engines, multiple-cylinder engines, diesel engines, Wankel engines — but this example is enough.

This description would cover the "what" of the motorcycle in terms of components, and the "how" of the engine in terms of functions. It would badly need a "where" analysis in the form of an illustration, and also a "why" analysis in the form of engineering principles that led to this particular conformation of parts. But the purpose here isn't exhaustively to analyze the motorcycle. It's to provide a starting point, an example of a mode of understanding of things which will itself become an object of analysis.

## STUDY QUESTIONS

1. Pirsig first divides a motorcycle according to its component assemblies and then divides it according to its functions. *Writing Assignment:* Write a paper that performs a similar analysis by dividing a simpler object — a bicycle — according to the same criteria. Do you find any overlapping within these systems?
2. What other ways of analyzing a bicycle or a motorcycle can you think of? Can you, for example, provide the "why" analysis Pirsig describes?

# Types and Limitations of Respiratory Protective Devices

## Walter E. Ruch and Bruce J. Held

### Introduction

Respiratory protective devices can be divided into two broad categories — air-purifying and atmosphere-supplying. The air-purifying devices filter out harmful particles in the air or remove toxic gases or vapors from the air. The atmosphere-supplying devices provide clean air, containing sufficient oxygen to sustain the wearer. The clean air comes from a source separated from the contaminated or oxygen-deficient air where the wearer is working. A combination air-purifying and atmosphere-supplying respirator is also available for use in special circumstances. Further information on classification of devices can be found in ANSI Standard Z88.2 and the Respirator Manual published by the American Industrial Hygiene Association.

### Face and Head Coverings

Facepieces fall into three types: the full-face mask which covers the entire face including the eyes, the half-mask which covers only the nose and mouth and fits under the chin, and the quarter-mask which covers only the nose and mouth and fits on the chin.

All facepieces have a body and harness or set of straps to hold the body on the face. All respirators have an exhaust valve and exhaust valve cover except the single-use, throw-away type in which the body is also the filter. The exhaust valve opens on exhalation to allow the air from the lungs to escape the face-piece. It must instantly close when inhaling to prevent contaminated air from seeping into the mask and into the lungs. The exhaust valve cover plays an important role in the operation of the mask. When exhaling, a small amount of uncontaminated air from the lungs is trapped between the exhaust valve and the exhaust valve cover. On inhalation, this uncontaminated air is drawn into the mask while the exhaust valve closes and properly seats.

The facepiece body has one or two air inlets. These air inlets also have valves to prevent moist, exhaled air from blowing back into the air inlet source. The inlet air may come from one or two cartridges or canisters containing particulate filters, chemical sorbents, or a combination of both, or from an air-line from a clean air source. In addition to facepieces, the entire head might be covered with a hood or helmet. Generally the hood or helmet are the supplied-air type rather than the air-purifying type. Full head coverings are generally used when the skin must also be

From Walter E. Ruch and Bruce J. Held, *Respiratory Protection* (Ann Arbor, Michigan: Ann Arbor Science Publishers, Inc., 1975), pp. 45–53.

protected or in situations where the supplied air is also used for cooling or heating the head area.

## Air-Purifying Devices

Air-purifying devices use a mechanical filter to remove particulate matter such as dust, fumes, or mists and a chemical such as a type of charcoal to remove gas or vapor from the air, or a combination of the two. The filter or chemical is contained in a cartridge or canister. The cartridge generally refers to a small cylindrically-shaped holder used on half- and quarter-facepieces, whereas a canister is the term usually used for the large holder used on full-facepiece respirators and gas masks.

### Mechanical Filters

These devices are usually made of a fibrous material and physically screen out particulates such as silica dust, lead fumes, or chromic acid mists. A dust filter does not necessarily protect against fumes or mists. Furthermore, dust filters vary according to the size particle they can capture and the efficiency of collection. A high efficiency type filter removes 99.97% of particles 0.3 microns or larger in size (a human hair is about 50 microns in diameter). This type filter must be used against highly hazardous particles such as beryllium or radionuclides.

### Chemical Cartridges

Unlike gas masks, these devices may not be used in atmospheres containing gas or vapor concentrations immediately dangerous to life or health. Whereas mechanical filters become more efficient as the dust particles are collected in the filter's fibers, chemical cartridges become less efficient with sorbent saturation. The contaminating gas or vapor enters the facepiece and then the lungs unless the cartridge is changed before breakthrough occurs. Chemical cartridges should be used only once, whereas filter cartridges are reusable.

Chemical cartridges offer protection against organic vapors and certain acid gases. NIOSH will also approve chemical cartridges for a specific gas or vapor if a manufacturer submits the device and it passes certain tests. Organic vapor cartridges are tested by NIOSH only against carbon tetrachloride. However, the absorption efficiency of the charcoal used in these cartridges varies between different organic solvents. Thus, some knowledge of the solvent to be protected against and the sorption characteristics of the cartridge sorbent is essential. Every chemical cartridge approved by NIOSH states on the label what it protects against and the maximum concentration in which it may be used. The user of the respirator must be very sure that the cartridge is the proper one and that the concentration in which it is to be used is within the limits of its capabilities. Generally, chemical cartridge respirators are quite unsatisfactory, and their use should be limited to people very knowledgeable in the field.

### Combination Air-Purifying Respirators

Respirators are available which combine a filter and chemical cartridge for protection in atmospheres where both particulates and gases or vapors are present. Selection of these devices must be made only when the contaminants present and their concentrations have been identified, and the respirator cartridge capabilities are known.

### Gas Masks

Canister-type gas masks may be subdivided into three categories:

1. Front- or back-mounted masks in which the canister is hung in a harness on the wearer's chest or back and connected to the full-facepiece by a hose.
2. Chin-style masks in which the canister is mounted directly onto the full-facepiece.
3. Escape masks, designed for emergency escape and using a half-mask facepiece or mouthpiece with an attached canister.

Front- or back-mounted gas masks are approved for concentrations of 2% by volume or less of certain acid gases, carbon monoxide, and certain organic vapors. They are also available for protection against up to 3% ammonia concentrations. The chin-style gas masks may be used for 0.5% by volume of certain acid gases, certain organic vapors, or ammonia.

The escape masks are approved for all of the above-mentioned chemicals in concentrations of 0.1% for certain acid gases, 0.5% for ammonia and certain acid gases, and 1% for carbon monoxide. Escape masks are not to be used for working in a re-entry to areas containing unsafe concentrations of these gases or vapors. These masks use only a half-mask facepiece or a mouthpiece and are for escape purposes only.

One other type of front- or back-mounted canister gas mask should be mentioned. It is commonly known as a Type N or universal gas mask as it contains sorbents to protect against certain acid gases, ammonia, carbon monoxide, and certain organic vapors. A high-efficiency filter for particulate protection is also enclosed in the canister. These devices were commonly used by fire-fighters until recent years. Since they did not offer protection in oxygen-deficient atmospheres, several injuries and fatalities resulted by misuse and led to decreasing use. Manufacturers of Type N masks have decreased production and some have ceased production.

### Powered Air-Purifying Devices

A new device, the powered air-purifying device, was so named because contaminated air is pulled through a filter or chemical cartridge by a small battery-operated blower, and delivered to a facepiece or head covering. This type of device has a distinct advantage because it puts the facepiece or head covering under positive pressure with respect to the outside contaminated atmosphere. Other air-purifying devices are under negative pressure when the wearer is inhaling. Thus, outside contaminated air can enter the facepiece through any leaks in the mask itself and, more significantly, through leaks in the seal between the facepiece and the face. The powered device maintains the facepiece or head covering under positive pressure at all times, therefore insuring an outward flow of air through any leaks that exist, even during inhalation.

The device consists of a small pack which can be hung on the belt or carried and contains cartridges, blower and battery. The contaminated air is pulled through the cartridges and forced up a hose to the facepiece or head covering. Excess air is released through the exhalation valve or from under the hood or helmet.

## STUDY QUESTIONS
1. Ruch and Held create a very complex (and orderly) system as they analyze air-purifying devices. Diagram or outline their series of classifications. Would you have organized these items any differently? Why or why not?
2. Only half of this chapter — the category of air-purifying devices — is reprinted here. The discussion of face and head coverings applies to both types of devices. What are the advantages or disadvantages of placing this discussion where it is?

## Tavist

## Dorsey Laboratories

# TAVIST® (clemastine fumarate) Tablets

## Description:

Tavist (clemastine fumarate) belongs to the benzhydryl ether group of antihistaminic compounds. The chemical name is (+)-2-[-2-[(p-chloro-α-methyl-α-phenyl benzyl)oxy]ethyl]-1-methylpyrrolidine* hydrogen fumarate and the structural formula is:

From Dorsey Laboratories, Division of Sandoz, Inc., Lincoln, Nebraska.
*U.S. Patent No. 3,097,212

## Actions:

*Tavist* is an antihistamine with anticholinergic (drying) and sedative side effects. Antihistamines appear to compete with histamine for cell receptor sites on effector cells. The inherently long duration of antihistaminic effects of Tavist (clemastine fumarate) has been demonstrated in wheal and flare studies. In normal human subjects who received histamine injections over a 24-hour period, the antihistaminic activity of *Tavist* reached a peak at 5-7 hours, persisted for 10-12 hours and, in some cases, for as long as 24 hours. Pharmacokinetic studies in man utilizing $^3$H and $^{14}$C labeled compound demonstrate that: *Tavist* is rapidly and nearly completely absorbed from the gastrointestinal tract, peak plasma concentrations are attained in 2–4 hours, and urinary excretion is the major mode of elimination.

## Indications:

Tavist (clemastine fumarate) is indicated for the relief of symptoms associated with seasonal allergic rhinitis such as sneezing, rhinorrhea, pruritus, and lacrimation.

*Tavist* is indicated for the relief of mild, uncomplicated allergic skin manifestations of urticaria and angioedema.

## Contraindications:

**Use in Nursing Mothers:** Because of the higher risk of antihistamines for infants generally and for newborns and prematures in particular, antihistamine therapy is contraindicated in nursing mothers.

**Use in Lower Respiratory Disease:** Antihistamines *should not* be used to treat lower respiratory tract symptoms including asthma.

Antihistamines are also contraindicated in the following conditions:

Hypersensitivity to Tavist (clemastine fumarate) or other antihistamines of similar chemical structure.

Monamine oxidase inhibitor therapy (see Drug Interaction Section).

## Warnings:

Antihistamines should be used with considerable caution in patients with:

Narrow angle glaucoma
Stenosing peptic ulcer

Pyloroduodenal obstruction
Symptomatic prostatic hypertrophy
Bladder neck obstruction

**Use in Children**: Safety and efficacy of *Tavist* have not been established in children under the age of 12.

**Use in Pregnancy**: Experience with this drug in pregnant women is inadequate to determine whether there exists a potential for harm to the developing fetus.

**Use with CNS Depressants**: *Tavist* has additive effects with alcohol and other CNS depressants (hypnotics, sedatives, tranquilizers, etc.).

**Use in Activities Requiring Mental Alertness**: Patients should be warned about engaging in activities requiring mental alertness such as driving a car or operating appliances, machinery, etc.

**Use in the Elderly (approximately 60 years or older)**: Antihistamines are more likely to cause dizziness, sedation, and hypotension in elderly patients.

## Precautions:

Tavist (clemastine fumarate) should be used with caution in patients with:

History of bronchial asthma
Increased intraocular pressure
Hyperthyroidism
Cardiovascular disease
Hypertension

**Drug Interactions**: MAO inhibitors prolong and intensify the anticholinergic (drying) effects of antihistamines.

## Adverse Reactions:

Transient drowsiness, the most common adverse reaction associated with Tavist (clemastine fumarate), occurs relatively frequently and may require discontinuation of therapy in some instances.

**Antihistaminic Compounds**: It should be noted that the following reactions have occurred with one or more antihistamines and, therefore, should be kept in mind when prescribing drugs belonging to this class, including *Tavist*. The most frequent adverse reactions are underlined.

1. **General**: Urticaria, drug rash, anaphylactic shock, photosensitivity, excessive perspiration, chills, dryness of mouth, nose, and throat.
2. **Cardiovascular System**: Hypotension, headache, palpitations, tachycardia, extrasystoles.
3. **Hematologic System**: Hemolytic anemia, thrombocytopenia, agranulocytosis.

4. **Nervous System**: *Sedation, sleepiness, dizziness, disturbed coordination,* fatigue, confusion, restlessness, excitation, nervousness, tremor, irritability, insomnia, euphoria, paresthesias, blurred vision, diplopia, vertigo, tinnitus, acute labyrinthitis, hysteria, neuritis, convulsions.
5. **GI System**: *Epigastric distress,* anorexia, nausea, vomiting, diarrhea, constipation.
6. **GU System**: Urinary frequency, difficult urination, urinary retention, early menses.
7. **Respiratory System**: Thickening of bronchial secretions, tightness of chest and wheezing, nasal stuffiness.

## Overdosage:

Antihistamine overdosage reactions may vary from central nervous system depression to stimulation. Stimulation is particularly likely in children. Atropine-like signs and symptoms: dry mouth; fixed, dilated pupils; flushing; and gastrointestinal symptoms may also occur.

**If vomiting has not occurred spontaneously** the conscious patient should be induced to vomit. This is best done by having him drink a glass of water or milk after which he should be made to gag. Precautions against aspiration must be taken, especially in infants and children.

**If vomiting is unsuccessful** gastric lavage is indicated within 3 hours after ingestion and even later if large amounts of milk or cream were given beforehand. Isotonic and ½ isotonic saline is the lavage solution of choice.

**Saline cathartics,** such as milk of magnesia, by osmosis draw water into the bowel and therefore, are valuable for their action in rapid dilution of bowel content.

**Stimulants** should *not* be used.

Vasopressors may be used to treat hypotension.

## Dosage and Administration:

DOSAGE SHOULD BE INDIVIDUALIZED ACCORDING TO THE NEEDS AND RESPONSE OF THE PATIENT.

The maximum recommended dosage of Tavist (clemastine fumarate) is one tablet three times daily. Many patients respond favorably to a single tablet dose, which may be repeated as required but not to exceed three tablets daily.

## How Supplied:

Tablets: 2.68 mg clemastine fumarate. White, round, compressed tablet, embossed "43" over "70" and scored on one side, "Tavist" on the other. Packages of 100.

**STUDY QUESTIONS**

1. Dorsey Laboratories, the company that supplies Tavist, has provided physicians with a great deal of information, dividing the drug into many components through the use of several analytical matrices or systems. What aspects of this drug would you, as a patient, be most interested in? Is there any information or system of division that is omitted here?
2. *Writing Assignment:* Using the Tavist advertisement as a model, construct a similar analysis for another simple object, such as a piece of chalk, employing multiple analytical systems. What categories might your chalk share with Tavist?

# Engineering Mistakes

## Samuel C. Florman

As I have already suggested, the current crisis in engineering should be the occasion for calm reflection. Instead of indulging in an orgy of penitence, we should be thinking soberly about our failures and about the lessons to be learned from them.

The first thought that comes to mind is that if we have failed in our endeavors, perhaps we should retire from the field. Have we made so many mistakes, and damaged the environment so badly that technological enterprise should cease? Can technology simply be abandoned, the way fascism is abandoned when the people overthrow a totalitarian regime? Hardly anybody thinks so. The proposition is not even worth debating. It would be fun, perhaps, to recount in sordid detail the ways in which society would begin to disintegrate within hours after the technologists walked away from their jobs. But that would be proving a point that needs no proving. Distasteful as it may be to some people, it is clear that our survival, and the salvaging of our environment, are dependent upon more technology, not less. In *The Limits to Growth,* the Club of Rome's report which so distressed the public when it was presented at international meetings in the summer of 1971, it was noted that technology was a necessary factor in solving the critical problems inherent in exponential population growth. "Technological advance," states the report, "would be both necessary and welcome in the equilibrium state." Indeed, conservation itself is a "technique," as are forestry, horticulture, sanitary engineering, and other endeavors dear to the heart of environmentalists.

Since technology cannot be abandoned, the next logical step is to see what can be done to avoid repetition of technological mistakes that have been made in

From Samuel C. Florman, *The Existential Pleasures of Engineering* (New York: St. Martin's Press, 1976), pp. 31–33. Title selected by editors.

the past. Toward this end, let us consider the types of mistakes that engineers have made, and the reasons for them.

First, there are the mistakes that have been made by carelessness or error in calculation. Occasionally a decimal point is misplaced, the mistake is not picked up on review, and a structure collapses or a machine explodes. Human error. It happens rarely, and much less frequently now than in times past.

More often, failure results from lack of imagination. The Quebec Bridge collapsed while under construction in 1907 because large steel members under compression behaved differently than the smaller members that had been tested time and again. The Tacoma Narrows Bridge failed in 1940 because the dynamic effect of wind load was not taken into account. Although designed to withstand a static wind load of fifty pounds per square foot, the bridge was destroyed by harmonic oscillations resulting from a wind pressure of a mere five pounds per square foot. We do not have to be too concerned about bridge failures anymore. (In 1869 American bridges were failing at the rate of 25 or more annually!) But the problem of reasoning from small to large, and from static to dynamic, is symbolic of the difficulties we face in designing anything in a complex, interdependent, technological society. The Aswan Dam is an example. As a structure it is a success. But in its effect on the ecology of the Nile Basin — most of which could have been predicted — it is a failure.

Finally, there are the mistakes that result from pure and absolute ignorance. We use asbestos to fireproof steel, with no way of knowing that it is any more dangerous to health than cement or gypsum or a dozen other common materials. Years later, we find that the workers who handled it are developing cancer.

Human error, lack of imagination, and blind ignorance. The practice of engineering is in large measure a continuing struggle to avoid making mistakes for these reasons. No engineer would quarrel with the objective of being more cautious and farseeing than we have been before, of establishing ever stricter controls, testing as extensively as possible, and striving for perfection. It is beside the point to say that everyone makes mistakes, that there is no such thing as perfection, nothing ventured, nothing gained, and so forth. In such matters the engineer is willing to be held to the strictest standards of accountability. Of course, the elimination of error, depending as it does upon additional man-hours of study, has a direct relationship to cost. The public must be willing to pay such cost. Sometimes the expense is not warranted. We know that every once in a while an old water main will explode in one of our cities, but we do not want to spend the money to replace all old mains. Yet, on balance, it would seem that, knowing what we now know, both the engineer and the public would be willing to invest substantially in added protection, particularly where the environment is concerned.

## STUDY QUESTIONS

1. Do Florman's categories cover all the possibilities? Can you think of a type of mistake that does not properly belong to one of these categories?

2. It is possible that Florman does not have enough categories, but it is also possible that he has too many. Can any of these engineering mistakes be eliminated or combined with another one?

## Cause-Effect

The report writer uses cause-effect development to answer questions like "Why is vehicle A getting poor mileage?" or "What will be the effect of the new credit policy?" The first question assumes that something causes the poor mileage. The second assumes that the credit policy, a cause, will result in an effect. If we know the causes for the first, or wish to speculate on the possible effects of the second, then our reporting task is simply to write a clear and orderly record of the information. In organizing cause-effect development, we might first state the cause, then go to the effects that lead from that cause. Or we might start with the effect, then back up to explain its cause or causes. But in either case, we must make it clear to the reader which is cause and which is effect.

Cause-effect is also a method of logical analysis. We have dealt with cause-effect as a logical method in Part I, Problem Solving (specifically Copi's "Causal Connections," pp. 98–100, and Hardin's "What the Hedgehog Knows," pp. 101–104).

The three short selections that follow illustrate patterns of cause-effect development. Sir James Jeans moves from cause to effect in explaining why the sky is the color it is. Richard Nelson Bolles uses cause-effect development to encourage job-hunters to take a more positive approach in searching for employment. Aaron Sussman uses cause-effect in a highly functional document — a trouble-shooting guide for photographers.

# The Color of the Sky

## Sir James Jeans

Imagine that we stand on any ordinary seaside pier, and watch the waves rolling in and striking against the iron columns of the pier. Large waves pay very little attention to the columns — they divide right and left and re-unite after passing each column, much as a regiment of soldiers would if a tree stood in their road; it is almost as though the columns had not been there. But the short waves and ripples find the columns of the pier a much more formidable obstacle. When the short

From Sir James Jeans, *The Stars in Their Courses* (New York: Cambridge Univ. Press, 1954), pp. 25–27. Title selected by editors.

waves impinge on the columns, they are reflected back and spread as new ripples in all directions. To use the technical term, they are "scattered." The obstacle provided by the iron columns hardly affects the long waves at all, but scatters the short ripples.

We have been watching a sort of working model of the way in which sunlight struggles through the earth's atmosphere. Between us on earth and outer space the atmosphere interposes innumerable obstacles in the form of molecules of air, tiny droplets of water, and small particles of dust. These are represented by the columns of the pier.

The waves of the sea represent the sunlight. We know that sunlight is a blend of lights of many colours — as we can prove for ourselves by passing it through a prism, or even through a jug of water, or as Nature demonstrates to us when she passes it through the raindrops of a summer shower and produces a rainbow. We also know that light consists of waves, and that the different colours of light are produced by waves of different lengths, red light by long waves and blue light by short waves. The mixture of waves which constitutes sunlight has to struggle through the obstacles it meets in the atmosphere, just as the mixture of waves at the seaside has to struggle past the columns of the pier. And these obstacles treat the light-waves much as the columns of the pier treat the sea-waves. The long waves which constitute red light are hardly affected, but the short waves which constitute blue light are scattered in all directions.

Thus, the different constituents of sunlight are treated in different ways as they struggle through the earth's atmosphere. A wave of blue light may be scattered by a dust particle, and turned out of its course. After a time a second dust particle again turns it out of its course, and so on, until finally it enters our eyes by a path as zigzag as that of a flash of lightning. Consequently the blue waves of the sunlight enter our eyes from all directions. And that is why the sky looks blue. But the red waves come straight at us, undeterred by atmospheric obstacles, and enter our eyes directly. When we look towards the sun, we see it mainly by these red rays. They are not the whole light of the sun; they are what remains after a good deal of blue has already been filtered out by atmospheric obstacles. This filtering of course makes the sunlight redder than it was before it entered our atmosphere. The more obstacles the sunlight meets, the more the blue is extracted from it, and so the redder the sun looks. This explains why the sun looks unusually red when we see it through a fog or a cloud of steam. It also explains why the sun looks specially red at sunrise or sunset — the sun's light, coming to us in a very slantwise direction, has to thread its way past a great number of obstacles to reach us. It also explains the magnificent sunsets which are often seen through the smoky and dusty air of a city — or even better after a volcanic eruption, when the whole atmosphere of the world may be full of minute particles of volcanic dust.

## STUDY QUESTIONS

1. Jeans uses elaborate comparisons to clarify the cause-effect relation. Would our understanding of the topic be as full if Jeans didn't use comparisons? Why or why not?

2. Jeans presents the cause before stating the effect. This order is logically appropriate, but is it rhetorically effective? Could the order be reversed?
3. Does his explanation have any purpose other than informing the reader? Discuss.


# Assumptions in Job Hunting

## Richard Nelson Bolles

What, then, is it that makes the present job-hunting system in this country so disastrous? That was the question which the creative minority, wherever they were, first asked themselves. What are the fatal *assumptions* that are so casually made, taught, propagated, and reproduced, by some of our best business schools and job counselors, without ever being critically questioned? To the creative minority, the fatal assumptions seemed to be these:

*Fatal Assumption No.1: The job-hunter should remain somewhat loose (i.e., vague) about what he wants to do, so that he is free to take advantage of whatever vacancies may be available.* Good grief, said the creative minority, this is why we have so great a percentage (80 or whatever) of under-employment in this country. If you don't state just exactly what you want to do, first of all to yourself, and then to others, you are (in effect) handing over that decision to others. And others, vested with such awesome responsibility, are either going to dodge the decision or else make a very safe one, which is to define you as capable of doing only such and such a level of work (safe, no risk diagnosis).

*Fatal Assumption No.2: The job-hunter should spend a good deal of time identifying the organizations that might be interested in him (no matter in what part of the country they may be), since employers have the initiative and upper hand in this whole process.* Nonsense, said the creative minority. This isn't a high school prom, where the job-hunters are sitting around the edge of the dance-floor, like some shy wall-flower, while the employers are whirling around out in the center of the floor, and enjoying all the initiative. In many cases, those employers are stuck with partners (if we may pursue the metaphor) who are stepping on their toes, constantly. As a result, although the employer in theory has all the initiative as to whom he chooses to dance with, in actuality he is often praying *someone will pay no attention to this silly rule, and come to his rescue by cutting in.* And indeed, when someone takes the initiative with the employer, rather than just sitting on the sidelines with *I'll-be-very-lucky-if-you-choose-me* written all over their demeanor, the employer cannot help thinking *I-am-very-lucky-that-this-one-has-chosen-me.* People who cut in are usually pretty good dancers.

From Richard Nelson Bolles, *What Color Is Your Parachute?*, Revised Edition (Berkeley: Ten Speed Press, 1975 and 1979), pp. 38–39. Title selected by editors.

*Fatal Assumption No.3: Employers see only people who can write well.* Pretty ridiculous, when it's put that way, But, say the creative minority, isn't that just exactly what our present job-hunting system is based on? To get hired, you must get an interview. To get an interview, you must let the personnel department see your resume first. Your resume will be screened out (and the interview never granted), if it doesn't make you sound good. But the resume is only as good as your writing ability (or someone else's) makes it. If you write poorly, your resume is (in effect) a Fun House mirror, which distorts you out of all proportion, so that it is impossible to tell what you really look like. *But no allowance is made for this possibility, by personnel departments, except maybe one out of a thousand.* Your resume is assumed to be an accurate mirror of you. You could be Einstein, but if you don't write well (i.e., if you write a terrible resume) you will not get an interview. Employers only see people who can write well. Ridiculous? You bet it is. And, say the creative minority, this is an assumption which is long overdue for a rest. It just doesn't have to be this way.

### STUDY QUESTIONS

1. What is the effect that the author accounts for?
2. What are its causes?
3. Why has the author developed the cause and effect relation in this order?
4. Is the author's purpose only to inform you about the job market? What other purpose might stand behind his cause-effect analysis?

# Negative Faults

## Aaron Sussman

Cause-effect relationships are often explained in trouble-shooting guides. Many people, it seems, are interested in knowing more than just how to correct an error; they want to know what caused the error in the first place. The following guide, taken from a handbook of photography, informs the reader on the causes of unsatisfactory negative development.

Here are the most common negative defects and a few ideas on what you can do to avoid or correct them.

From Aaron Sussman, *The Amateur Photographer's Handbook,* 8th edition (New York: Crowell, 1973), pp. 365–366.

## STREAKS OR OTHER IRREGULARITIES

1. *Cause:* Insufficient or improper agitation. *Cure:* Constant agitation during development, changing motion at frequent and regular intervals.

## AIR BELLS

2. *Cause:* Air bubbles clinging to emulsion surface and preventing action of developer. *Cure:* Place film *into* solution instead of pouring solution into tank. Agitate. Knock tank against table to dislodge bubbles.

## BLACK SPOTS

3. *Cause:* Undissolved developing agent particles settling on emulsion surface. *Cure:* Better preparation of solution, and filtering.

## UNEVEN DEVELOPMENT

4. *Cause:* Developer poured into tank too slowly. *Cure:* Pour solution into tank quickly and continuously.

## PINHOLES

5. *Cause:* Gas bubbles formed when film is placed in acid rinse or acid hypo baths. *Cure:* Avoid strongly acid afterbaths. Agitate. Avoid developer with carbonate; use those which contain borax or sodium metaborate (Kodalk), neither of which produces a gas on contact with acid.

## RETICULATION

6. *Cause:* Sudden *drop* in temperature between developing and fixing, or the use of an exhausted fixing bath. *Cure:* Avoid both causes. Don't dry processed film in an air-conditioned room, where cold air currents may chill it.

## FOG

7. *Cause:* If margins are clear, fog was produced in camera; if margins are also fogged, fogging occurred while loading or developing and may have been caused by light leaks, unsafe safelight, high temperature, overdeveloping or old film. *Cure:* Obvious precautions.

To keep fog down, turn to the chemical with the unpronounceable name, *Benzotriazole* (Kodak Anti-Fog #1), which is the one to use when your film suffers from age or unfavorable storage conditions (as, for instance, long storage between exposure and processing). You can also use *potassium iodide;* 1 grain to each 20 ounces of developer.

## STUDY QUESTIONS

1. Note the brevity of causal explanation. Should the writer have included more information on causes?
2. The causes are clearly labelled. Where does the writer indicate the effect?
3. Is it appropriate here to give the effect before the cause? Why?
4. Does the causal information serve any purpose other than satisfying the reader's curiosity? Explain.
5. Does the writer give us immediate causes or remote causes? (Review Copi's "Causal Connections.")

# Analogy

The quickest way to understand *analogy* is to turn immediately to the following examples. Simply put, analogy involves making comparisons between two objects or processes or concepts. The comparisons are made either for the purpose of asserting that what is true for one object or process is also true for the other, or for the purpose of clarifying a complex mechanism or concept by comparing it with a simpler or more familiar one. Analogies, then, are generally used to explain the unfamiliar in terms of the familiar, and they may be divided into two categories: literal analogies and figurative analogies.

In a literal analogy we are comparing two rather similar objects or processes, and demonstrating that if they have certain characteristics in common they very well may share other important characteristics. For example, we could argue that Mars is very much like the earth. Like the earth, Mars has an atmosphere. Like the earth, Mars has oxygen present in its atmosphere. The earth rotates on its axis and so does Mars. There are seasons on earth and there are seasons on Mars. We have life on earth. What about Mars? By establishing such a point by point comparison we hope to demonstrate that the similarity between the two planets extends beyond what we know in fact. We are attempting, through analogy, to make the unknown known, the unfamiliar familiar.

A figurative analogy also attempts to make the unknown known, but its function is chiefly an illustrative one. A figurative analogy is essentially an extended metaphor, and, like most metaphors, it is generally used to cause a reader to perceive the unfamiliar in a new way by comparing the unfamiliar object with something the reader already knows well. To understand how the human body functions, for example, we often compare it to a simple machine that produces energy and requires fuel. Or we explain electricity by saying that the current "flows" through wires like water flows through pipes. These figurative analogies allow us to form a basic understanding of complex processes. Such analogies usually break down eventually, however, because the objects or processes being compared are not identical in all respects; they are only similar, and the figurative analogy is only serving an illustrative function.

The following examples demonstrate just how imaginative (and useful) figurative analogies can be. Richard Selzer's explanation of how the liver functions is obviously fanciful and employs several unusual metaphors in fabricating its basic analogy. The selections by Martin Gardner and Carl Sagan, both of whom are highly respected science writers, illustrate that certain very complex scientific principles can be made clear and understandable to the layman — when the proper analogies are found.

# The Cosmic Calendar

## Carl Sagan

The world is very old, and human beings are very young. Significant events in our personal lives are measured in years or less; our lifetimes in decades; our family genealogies in centuries; and all of recorded history in millennia. But we have been preceded by an awesome vista of time, extending for prodigious periods into the past, about which we know little — both because there are no written records and because we have real difficulty in grasping the immensity of the intervals involved.

Yet we are able to date events in the remote past. Geological stratification and radioactive dating provide information on archaeological, paleontological and geological events; and astrophysical theory provides data on the ages of planetary surfaces, stars, and the Milky Way Galaxy, as well as an estimate of the time that has elapsed since that extraordinary event called the Big Bang — an explosion that involved all of the matter and energy in the present universe. The Big Bang may be the beginning of the universe, or it may be a discontinuity in which information about the earlier history of the universe was destroyed. But it is certainly the earliest event about which we have any record.

| PRE-DECEMBER DATES | |
|---|---:|
| Big Bang | January 1 |
| Origin of the Milky Way Galaxy | May 1 |
| Origin of the solar system | September 9 |
| Formation of the Earth | September 14 |
| Origin of life on Earth | ~ September 25 |
| Formation of the oldest rocks known on Earth | October 2 |
| Date of oldest fossils (bacteria and blue-green algae) | October 9 |
| Invention of sex (by microorganisms) | ~ November 1 |
| Oldest fossil photosynthetic plants | November 12 |
| Eukaryotes (first cells with nuclei) flourish | November 15 |

~ = *approximately*

The most instructive way I know to express this cosmic chronology is to imagine the fifteen-billion-year lifetime of the universe (or at least its present incarnation since the Big Bang) compressed into the span of a single year. Then every billion years of Earth history would correspond to about twenty-four days of our

From Carl Sagan, *The Dragons of Eden* (New York: Random House, 1977), pp. 13–17.

## COSMIC CALENDAR
### DECEMBER

| SUNDAY | MONDAY | TUESDAY | WEDNESDAY | THURSDAY | FRIDAY | SATURDAY |
|---|---|---|---|---|---|---|
|  | **1** Significant oxygen atmosphere begins to develop on Earth. | **2** | **3** | **4** | **5** Extensive vulcanism and channel formation on Mars. | **6** |
| **7** | **8** | **9** | **10** | **11** | **12** | **13** |
| **14** | **15** | **16** First worms. | **17** Precambrian ends. Paleozoic Era and Cambrian Period begin. Invertebrates flourish. | **18** First oceanic plankton. Trilobites flourish. | **19** Ordovician Period. First fish, first vertebrates. | **20** Silurian Period. First vascular plants. Plants begin colonization of land. |
| **21** Devonian Period begins. First insects. Animals begin colonization of land. | **22** First amphibians. First winged insects. | **23** Carboniferous Period. First trees. First reptiles. | **24** Permian Period begins. First dinosaurs. | **25** Paleozoic Era ends. Mesozoic Era begins. | **26** Triassic Period. First mammals. | **27** Jurassic Period. First birds. |
| **28** Cretaceous Period. First flowers. Dinosaurs become extinct. | **29** Mesozoic Era ends. Cenozoic Era and Tertiary Period begin. First cetaceans. First primates. | **30** Early evolution of frontal lobes in the brains of primates. First hominids. Giant mammals flourish. | **31** End of the Pliocene Period. Quaternary (Pleistocene and Holocene) Period. First humans. |  |  |  |

196

**DECEMBER 31**

| | |
|---|---|
| Origin of *Proconsul* and *Ramapithecus,* probable ancestors of apes and men | ~ 1:30 P.M. |
| First humans | ~ 10:30 P.M. |
| Widespread use of stone tools | 11:00 P.M. |
| Domestication of fire by Peking man | 11:46 P.M. |
| Beginning of most recent glacial period | 11:56 P.M. |
| Seafarers settle Australia | 11:58 P.M. |
| Extensive cave painting in Europe | 11:59 P.M. |
| Invention of agriculture | 11:59:20 P.M. |
| Neolithic civilization; first cities | 11:59:35 P.M. |
| First dynasties in Sumer, Ebla and Egypt; development of astronomy | 11:59:50 P.M. |
| Invention of the alphabet; Akkadian Empire | 11:59:51 P.M. |
| Hammurabic legal codes in Babylon; Middle Kingdom in Egypt | 11:59:52 P.M. |
| Bronze metallurgy; Mycenaean culture; Trojan War; Olmec culture; invention of the compass | 11:59:53 P.M. |
| Iron metallurgy; First Assyrian Empire; Kingdom of Israel; founding of Carthage by Phoenicia | 11:59:54 P.M. |
| Asokan India; Ch'in Dynasty China; Periclean Athens; birth of Buddha | 11:59:55 P.M. |
| Euclidean geometry; Archimedean physics; Ptolemaic astronomy; Roman Empire; birth of Christ | 11:59:56 P.M. |
| Zero and decimals invented in Indian arithmetic; Rome falls; Moslem conquests | 11:59:57 P.M. |
| Mayan civilization; Sung Dynasty China; Byzantine empire; Mongol invasion; Crusades | 11:59:58 P.M. |
| Renaissance in Europe; voyages of discovery from Europe and from Ming Dynasty China; emergence of the experimental method in science | 11:59:59 P.M. |
| Widespread development of science and technology; emergence of a global culture; acquisition of the means for self-destruction of the human species; first steps in spacecraft planetary exploration and the search for extraterrestrial intelligence | Now: The first second of New Year's Day |

cosmic year, and one second of that year to 475 real revolutions of the Earth about the sun. I present the cosmic chronology in three forms: a list of some representative pre-December dates; a calendar for the month of December; and a closer look at the late evening of New Year's Eve. On this scale, the events of our history books — even books that make significant efforts to deprovincialize the present — are so compressed that it is necessary to give a second-by-second recounting of the last seconds of the cosmic year. Even then, we find events listed as contemporary that we have been taught to consider as widely separated in time. In the history of life, an equally rich tapestry must have been woven in other periods — for example, between 10:02 and 10:03 on the morning of April 6th or September 16th. But we have detailed records only for the very end of the cosmic year.

The chronology corresponds to the best evidence now available. But some of it is rather shaky. No one would be astounded if, for example, it turns out that plants colonized the land in the Ordovician rather than the Silurian Period; or that segmented worms appeared earlier in the Precambrian Period than indicated. Also, in the chronology of the last ten seconds of the cosmic year, it was obviously impossible for me to include all significant events; I hope I may be excused for not having explicitly mentioned advances in art, music and literature or the historically significant American, French, Russian and Chinese revolutions.

The construction of such tables and calendars is inevitably humbling. It is disconcerting to find that in such a cosmic year the Earth does not condense out of interstellar matter until early September; dinosaurs emerge on Christmas Eve; flowers arise on December 28th; and men and women originate at 10:30 P.M. on New Year's Eve. All of recorded history occupies the last ten seconds of December 31; and the time from the waning of the Middle Ages to the present occupies little more than one second. But because I have arranged it that way, the first cosmic year has just ended. And despite the insignificance of the instant we have so far occupied in cosmic time, it is clear that what happens on and near Earth at the beginning of the second cosmic year will depend very much on the scientific wisdom and the distinctly human sensitivity of mankind.

## STUDY QUESTIONS
1.  Sagan chooses to compress the history of the universe (from the Big Bang to the present) into one calendar year. Is the period of one year an effective choice? Would it have been as effective to choose a shorter time period — a week or a month — or a longer time period — a decade — for this analogy? Why or why not?
2.  In his analogy Sagan is measuring time quantitatively and compressing billions of years into one year. Would it also be possible to create an analogy that would make its point by *expanding* time quantitatively? What kind of situation might appropriately be described by such an analogy?
3.  *Writing Assignment:* Distance, weight, and volume are also measured quantitatively. Using the Sagan analogy as a model, create clear and reasonably accurate analogies for the weight of a butterfly or a railroad locomotive; the distance from Venus to Earth, or from Chicago to Melbourne, Australia; the volume of water in a nearby lake, or the volume of gasoline in a tanker truck. What do all these analogies of measurement have in common?

# Liver

## Richard Selzer

Envision, if you will, a house whose stones are living hexagonal tiles not unlike those forming the bathroom floors of first-class hotels. These are the hepatocytes, the cellular units of the liver. Under the microscope they have a singular uniformity, each as like unto its fellow as the antlers of a buck, and all fitted together with a lovely imprecision so as to form a maze of crooked hallways and oblong rooms. Coursing through this muralium of tissue are two arborizations of blood vessels, the one bringing food and toxins from the intestine, the other delivering oxygen from the heart and lungs. Winding in and among these networks is a system of canaliculi that puts to shame all the aqueductal glories of Greece and Rome. Through these sluice the rivers of bile, gathering strength and volume as the little ducts at the periphery meet others, going into ones of larger caliber, which in turn fuse, and so on until there are two large tubes emerging from the under-surface of the liver. Within this magic house are all the functions of the liver carried out. The food we eat is picked over, sorted out, and stored for future use in the cubicles of the granary. Starch is converted to glycogen, which is released in the form of energy as the need arises. Protein is broken down into its building blocks, the amino acids, later to be fashioned into more YOU, as old tissues die off and need to be replaced. Fats are stored until sent forth to provide warmth and comfort. Vitamins and anti-bodies are released into the bloodstream. Busy is the word for the liver. Deleterious substances ingested, inadvertently like DDT or intentionally like alcohol, are either changed into harmless components and excreted into the intestine, or stored in locked closets to be kept isolated from the rest of the body. Even old blood cells are pulverized and recycled. Such is the ole catfish liver snufflin' along at the bottom of the tank, sweepin', cleanin' up after the gouramis, his whiskery old face stirrin' up a cloud of rejectimenta, and takin' care of everything.

But there are limits. Along comes that thousandth literary lunch and — Pow! the dreaded wrecking ball of cirrhosis is unslung. The roofs and walls of the hallways, complaining under their burden of excess fat, groan and buckle. Inflammation sets in, and whole roomfuls of liver cells implode and die, and in their place comes the scarring that twists and distorts the channels, pulling them into impossible angulation. Avalanches block the flow of bile and heavy tangles of fiber impede the absorption and secretion. This happens not just in one spot but all over, until the gigantic architecture is a mass of sores and wounds, the old ones scarring over as new ones break down.

The obstructed bile, no longer able to flow down to the gut, backs up into the bloodstream to light up the skin and eyes with the sickly lamp of jaundice. The stool turns toothpaste white in commiseration, the urine dark as wine. The belly

From Richard Selzer, *Mortal Lessons* (New York: Simon and Schuster, 1974), pp. 15–23.

swells with gallons of fluid that weep from the surface of the liver, no less than the tears of a loyal servant so capriciously victimized. The carnage spreads. The entire body is discommoded. The blood fails to clot, the palms of the hands turn mysteriously red, and spidery blood vessels leap and crawl on the skin of the face and neck. Male breasts enlarge, and even the proud testicles turn soft and atrophy. In a short while impotence develops, an irreversible form of impotence which may well prod the invalid into more and more drinking.

Scared? Better have a drink. You look a little pale. In any case there is no need to be all that glum. Especially if you know something that I know. Remember Prometheus? That poor devil who was chained to a rock, and had his liver pecked out each day by a vulture? Well, he was a classical example of the regeneration of tissue, for every night his liver grew back to be ready for the dreaded diurnal feast. And so will yours grow back, regenerate, reappear, regain all of its old efficiency and know-how. All it requires is quitting the booze, now and then. The ever-grateful, forgiving liver will respond joyously with a multitude of mitoses and cell divisions that will replace the sick tissues with spanking new nodules and lobules of functioning cells. This rejuvenation is carried on with the speed and alacrity of a starfish growing a new ray from the stump of the old. New channels are opened up, old ones dredged out, walls are straightened and roofs shored up. Soon the big house is humming with activity, and all those terrible things I told you happen go away — all except that impotence thing. Well, you didn't expect to get away scot-free, did you?

And here's something to tuck away and think about whenever you want to feel good. Sixty percent of all cirrhotics who stop drinking will be alive and well five years later. How unlike the lofty brain which has no power of regeneration at all. Once a brain cell dies, you are forever one shy.

Good old liver!

## STUDY QUESTIONS

1. Selzer, who is a practicing surgeon, presents his knowledge about the liver in a very dramatic fashion. Is it too dramatic? Is this selection more entertaining than it is informative?
2. Selzer shifts his metaphors freely. He begins by comparing the liver to a house, a house through which the river of the bloodstream flows. At the end of the first paragraph, however, he describes the liver not as a house, but as a catfish "snufflin' along at the bottom of the tank." Is this shift disturbing or confusing? What other shifts in his analogy can you find? Do these distract the reader, or damage the analogy's effectiveness?

# Relativity

## Martin Gardner

Writers on relativity theory often explain it in the following way. Imagine a rubber sheet stretched out flat like a trampoline. A grapefruit placed on this sheet will make a depression. A marble placed near the grapefruit will roll toward it. The grapefruit is not "pulling" the marble. Rather, it has created a field (the depression) of such a structure that the marble, taking the path of least resistance, rolls toward the grapefruit. In a roughly (very roughly) similar way, spacetime is curved or warped by the presence of large masses like the sun. This warping is the gravitational field. A planet moving around the sun is not moving in an ellipse because the sun pulls on it, but because the field is such that the ellipse is the "straightest" possible path the planet can take in spacetime.

Such a path is called a geodesic. This is such an important word in relativity theory that it should be explained more fully. On a Euclidian plane, such as a flat sheet of paper, the straightest distance between two points is a straight line. It is also the shortest distance. On the surface of a globe, a geodesic between two points is the arc of a great circle. If a string is stretched as tautly as possible from point to point, it will mark out the geodesic. This, too, is both the "straightest" and the shortest distance connecting the two points.

In a four-dimensional *Euclidian* geometry, where all the dimensions are space dimensions, a geodesic also is the shortest and straightest line between two points. But in Einstein's *non-Euclidian* geometry of spacetime, it is not so simple. There are three space dimensions and one time dimension, united in a way that is specified by the equations of relativity. This structure is such that a geodesic, although still the straightest possible path in spacetime, is the *longest* instead of the shortest distance. This concept is impossible to explain without going into complicated mathematics, but it has this curious result: A body moving under the influence of gravity alone always finds the path along which it takes the longest proper time to travel; that is, the longest when measured by its own clock. Bertrand Russell has called this the "law of cosmic laziness." The apple falls straight down, the missile moves in a parabola, the earth moves in an ellipse because they are too lazy to take other routes.

It is this law of cosmic laziness that causes objects to move through space in ways sometimes attributed to inertia, sometimes to gravity. If you tie a string to an apple and swing it in circles, the string keeps the apple from moving in a straight line. We say that the apple's inertia pulls on the string. If the string breaks, the apple takes off in a straight line. Something like this happens when an apple falls off a tree. Before it falls, the branch prevents it from moving through space. The apple on the branch is at rest (relative to the earth) but speeding along its time

From Martin Gardner, *The Relativity Explosion* (New York: Vintage Books, 1976), pp. 96–99.

coordinate because it is constantly getting older. If there were no gravitational field, this travel along the time coordinate would be graphed as a straight line on a four-dimensional graph. But the earth's gravity is curving spacetime in the neighborhood of the apple. This forces the apple's world line to become a curve. When the apple breaks away from the branch, it continues to move through spacetime, but (being a lazy apple) it now "straightens" its path and takes a geodesic. We see this geodesic as the apple's fall and attribute the fall to gravity. If we like, however, we can say that the apple's inertia, after the apple is suddenly released from its curved path, carries it to the ground.

After the apple falls, suppose a boy comes along and kicks it with his bare foot. He shouts in pain because the kick hurts his toes. A Newtonian would say that the apple's inertia resisted his kick. An Einsteinian can say the same thing, but he can also say, if he prefers, that the boy's toes caused the entire cosmos (including the toes) to accelerate backward, setting up a gravitational field that pulled the apple with great force against his toes. It is all a matter of words. Mathematically the situation is described by one set of spacetime field equations, but it can be talked about informally (thanks to the principle of equivalence) in either of two sets of Newtonian phrases.

## STUDY QUESTIONS

1. The theory of relativity has the reputation of being an extremely difficult concept. Gardner's explanation, using such things as rubber sheets and an apple on a string, seems to indicate that relativity theory is something a child can understand. Is this misleading?

2. What guidelines might one who is creating an analogy use to test the effectiveness of his analogy? Is an analogy of any use if it involves oversimplification?

# Part Three

## Applications

# Section 1

## Informing

The remaining three sections of the text will be composed of long, usually complete samples of business and technical writing. The selections chosen will represent the three broad fundamental areas of report writing: writing that informs, writing that persuades and influences, and writing that directs, guides, and implements.

The first of these areas — writing that informs — is the type of writing you may encounter most frequently on the job. Informative writing, sometimes called expository writing, is writing that informs, clarifies, and explains — writing that expands the reader's fund of knowledge. What you have discovered in your research, your study, or your experimentation will be presented in expository writing to your superiors, the general public, or your peers in the scientific or business communities. Exposition is an integral part of most reports, and of course outside reports it is found in textbooks, in reference books, and in articles and books on business, science, and technology. In a world whose frontiers of knowledge are ever expanding, informative writing is an essential tool of technical communication.

In your informative writing, as in your writing that implements or persuades, you will find yourself combining many of the rhetorical forms you have examined in this book up to now. And in all three types of writing it will be necessary for you to apply problem-solving techniques, analyze your audience, and make the proper linguistic choices. At this point then, all of your writing skills must come together as you attempt to establish a communicative bridge between your mind and your reader's.

Informative writing requires that you become a teacher of sorts. You must explain a concept or a process to your reader and help that reader, as a teacher helps a student, to understand. If you have ever tried to be a teacher, either in a formal or informal way, you may have discovered how much teaching success depends on communicative abilities. Imagine for a moment that as a teacher you are attempting to help a student understand a concept, say the concept of multiplication. As a teacher, you will probably explain the concept carefully and to the best of your ability. But what do you do, when, after listening carefully, your student says he does not understand? You can, of course, repeat your explanation, word for word, and hope that the repetition clarifies the concept. It may, and then again it may not. Simply repeating the same thing over and over will not necessarily teach anyone anything. If repetition fails, you may soon find yourself looking for a different way of explaining the concept, a different set of words that you hope will evoke the proper response. In what sense does any teacher ever really teach? His teaching is limited to making the "proper" explanation, and that explanation is the one that consists of the right choice of words, the words that finally make sense to the student and explain the subject to him.

The best teachers and writers are not necessarily those who possess fantastic amounts of information; the best teachers and writers are those who are capable of choosing the proper language and shaping it into the explanation that is ideal for the occasion. The selections that appear in this chapter all do an excellent job of explaining and informing. Developing the skills and abilities that these writers possess would virtually guarantee anyone a successful career in whatever field he chose to enter.

# Anatomy and Function of the Eyeball

## John Eden

John Eden's explanation of the eyeball is taken from his work *The Eye Book*. This book attempts to answer the many questions that the average reader might have about eyes, glasses, and contact lenses. Eden is an M.D. and a practicing ophthalmologist.

The human eye is, of course, a dual organ — two eyes working together to transmit visual information to the brain. Although it is certainly possible to see with only one eye, it takes two normally functioning eyes to achieve normal vision.

Your eye is made up of numerous kinds of highly specialized cells, which perform different functions. It is equipped with muscular, fibrous connective, circulatory, and nervous systems of its own. Although they are similar to those systems that work throughout your body, they are designed to fill the special needs of the eyes.

From John Eden, *The Eye Book* (New York: Viking Press, 1978), pp. 2–9.

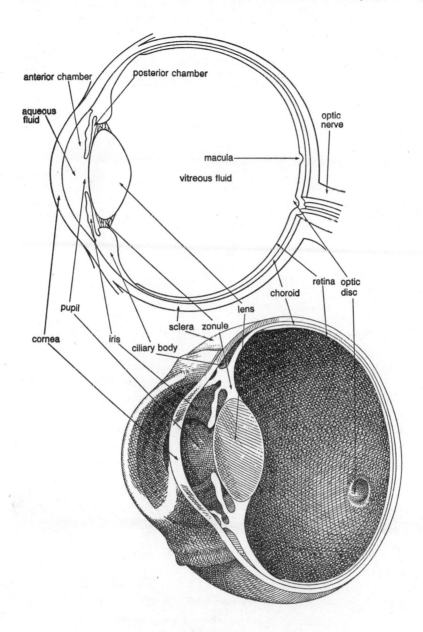

Figure 1.

The normal adult eyeball is an elliptical sphere, which means it is more egg-shaped than perfectly round. It has three distinct concentric tissue layers. The first serves to protect your eye's delicate internal structures, and it consists of the *sclera** — the opaque white of the eye — and the *cornea* — the transparent layer that lies in front of the pupil and iris.

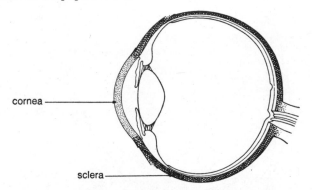

cornea

sclera

Figure 2. The protective layer of the eyeball.

The sclera covers about five-sixths of the surface of the eyeball. It is interrupted only by the cornea in front and the optic nerve, which enters the eyeball at the back. Although not much thicker than the page you are reading, the cornea and sclera are composed of extremely tough tissues. I will not say it is impossible to pierce them, but it takes a very sharp object traveling at high speed to do it.

A thin membrane called the *conjunctiva,* which is not technically a part of the eyeball, separates the exposed front and unexposed back portions of the eyeball. It covers the front part of the sclera and then laps over and continues forward onto the inner surface of the upper and lower eyelids. The conjunctiva thus closes off the back part of the eyeball, making it impossible for anything to get lost in your eye or travel back into your head.

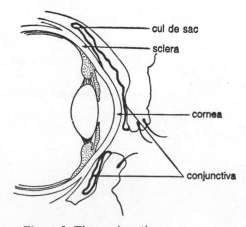

cul de sac

sclera

cornea

conjunctiva

Figure 3. The conjunctiva.

The second of the three layers is called the *uveal tract,* and its main functions are circulatory and muscular. The uveal tract is made up of the iris, the ciliary body, and the choroid.

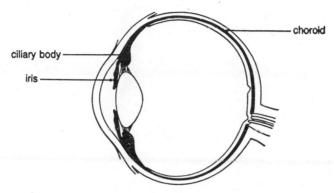

Figure 4. The uveal tract.

The *iris* is the round colored part of the eye that surrounds the pupil, and it is responsible for what we call the color of our eyes. The main function of the iris is to permit more or less light to enter your eye. The pupil itself is simply the hole surrounded by the iris and it is through this hole that light passes into your eye. The involuntary muscles of the iris respond primarily to the stimulus of light, constricting to make a smaller hole when light is bright and dilating to make a larger hole when light is dimmer. This action is like that of the iris diaphragm in a camera. But please don't take this analogy too literally. The human iris is not a mechanical device whose opening can be varied whenever you decide to do it. The action is involuntary. The muscles of your iris do not snap nearly shut when light is very bright and zoom open when light dims, but they constantly adjust and readjust to the level of light, remaining stationary only when the level of light stays the same. Your iris responds to all changes in light, no matter how subtle, and the adjustments are often extremely minute. The change from a darkened movie house to bright daylight is a dramatic one, but it is by no means the only sort of adjustment your iris makes.

The *ciliary body* lies between the iris and the choroid, and its function is also primarily muscular. It is connected to the lens by a ligament-like tissue called the *zonule,* from which the *lens* is suspended like a person in a hammock. The muscles of the ciliary body contract or relax to alter the shape of the lens, which allows your eye to focus for near vision, refocus to see at the distance, and then back again to see nearby objects clearly. When I speak of the *focusing muscles* of the eye, I am referring to the muscles of the ciliary body. When I speak of bringing an object into focus, I mean the bending of the lens by these muscles to change the shape of the surface through which light passes. The ciliary body also has a secretory function — it is from here that the *aqueous fluid* originates.

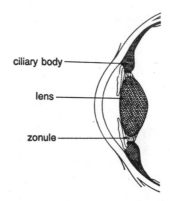

ciliary body

lens

zonule

Figure 5. The structures that allow your eye to focus for near vision.

Behind the ciliary body is the *choroid,* the main circulatory layer of the eye, through which blood is carried to nourish the various parts of the eye. This is not your eyeball's only blood supply; your retina, for example, has its own circulatory system. From the choroid, increasingly tiny arteries branch out to those portions of the eye that require blood for their metabolism, and then the veins return to the choroid blood laden with waste in the form of carbon dioxide. Some parts of your eye cannot be nourished in this way since the presence of blood vessels would interfere with their optical function, but nature has compensated for this by providing other ways of getting oxygen to those tissues and carrying away waste.

The innermost layer of the eye is the *retina,* an extremely thin sheet of specialized nerve tissue made up of ten distinct cell layers, each of which performs a specific part of the task of receiving visual images and transmitting them, via the optic nerve, to the brain. To go back to our camera analogy, the retina is like the film on which images are focused and recorded, but of course the retina is more complicated and can function more effectively than film. Your retina receives and passes along to your brain a very complex visual message, which includes size, shape, dimension, position in space, relative distance, and color.

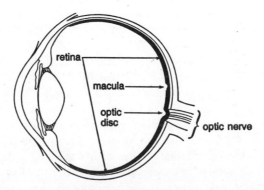

retina

macula

optic disc

optic nerve

Figure 6. The retina.

The key central area of the retina, located slightly to the outer side of the eyeball, is called the *macula.* This tiny area, which represents only a small part of the total retina, is its most vital part. It is responsible for your sharp central vision, and it is what permits normal 20/20 vision; the rest of the retina receives peripheral, or side vision, and delivers an image that is not so sharp as that coming from the macula.

Although blood vessels crisscross almost the entire retinal surface, the macula cannot be fed in this manner, since the highly sensitive receptor cells would be obscured by the blood vessels. Small capillaries feed into the edge of the macula, but the exchange of oxygen and carbon dioxide in the center takes place by absorption through cell walls. Nature's solution to the problem works quite well as long as nothing interferes with the delicate process, but it does make the macula more vulnerable to damage than the rest of the retina.

In addition to its receptor properties, the retina is able to adapt to light and dark. The iris performs the task of admitting or excluding light from the interior of your eye, but in addition to this, certain cells of the retina — the familiar *rods* and *cones* — undergo photochemical changes to enable you to see in various light levels. When you go from the daylight outdoors into a more dimly lit room, the rods in your retina are activated and the cones deactivated to adjust to the lower level of light; when you return to the sunlight, the cones are again activated and the rods function less so you can adjust to the brighter light. It takes a bit of time for your retina to adjust to the light change — an hour for complete *light* or *dark adaptation,* though you will see well in much less time — which is why when you go inside on a sunny day the room often seems quite dark for a while until your eyes adjust to the new light level. The cones are also responsible for your ability to perceive colors.

All the visual information collected and recorded by your eye is transmitted to the brain by the optic nerve, which enters the eye at the back of the retina. Because there is no retinal tissue at that point, this results in a *blind spot,* an area that cannot receive visual messages. Your eye doctor can locate and measure your blind spot by covering one eye at a time and performing a special test, but under normal conditions you do not notice your blind spot because the area it cannot see is seen by your other eye. However, even if you use only one eye, your blind spot is not a practical reality since it is so small.

## STUDY QUESTIONS

1. In the preface to his book Eden states that the work "is not intended to make you an expert." Is his explanation of the eyeball's anatomy and function consistent with this statement? Does it tell you more (or less) than it needs to?
2. Eden uses a camera analogy when discussing the eye (an analogy that is frequently used in such explanations). What are the weaknesses of this analogy? How does the reader discover them?
3. Eden establishes a distinctive relationship with his reader. He writes in the first

person and addresses the reader directly ("Your iris responds to all changes in light"). How would you characterize the tone that is achieved by this approach? Is it effective? In what situations would it not be effective? Why?

# The Farmer in the Lab

## Glenn R. Denlinger and Peter L. Sturla

Most large corporations publish some kind of newsletter, magazine, or journal for their many employees. These publications frequently describe new services, products, or production techniques for readers who work for the same company. Because of the diverse nature of many corporations, however, these readers — engineers, salesmen, secretaries, or executives — may be only slightly familiar with the products or services of other divisions. For example, the issue of *Sperry Technology* that the following article is taken from contains articles on avionics, plastic injection molding, computers, and this piece on Sperry New Holland farm equipment.

Beginning with the conception of every new piece or model of farm machinery, Sperry New Holland prescribes and drafts a set of maximum performance objectives. At the outset, functional and durability requirements for each machine are outlined and methods of achieving them are determined. To initiate these objectives, an experimental machine is designed, a prototype is built and it is subjected to tests in the field (Figure 1).

If early evaluation of these experimental machines is favorable, field tests are continued and a laboratory test program is initiated. Field tests are carried out in many different parts of the country under varying field and crop conditions. All aspects of a machine's performance (including failures) are monitored by field engineers and relayed to laboratory project engineers.

### Benefits of Laboratory Testing

Laboratory testing is not a cure-all for the problems which can occur on experimental machines, nor can it replace the need for functional design evaluation and endurance testing under actual field and crop conditions. However, with the farm industry's increasing demands for reliability, and management's demands for decreasing lead times between original concept and proposed marketing date, lab-

Glenn R. Denlinger and Peter L. Sturla, "The Farmer in the Lab," *Sperry Technology* (New York: Sperry Rand Corporation, 1974), pp. 36–38.

Figure 1. Sperry New Holland's mobile stress van runs alongside a Haybine mower-conditioner while monitoring it for field loading.

oratory testing has become a very important factor in the overall development of farm equipment.

Lab test acceleration is one of the techniques which can be used to meet industry demands. Accelerated testing programs can compress into minutes what would otherwise take hours of field time. For example, Sperry New Holland's Haybine® mower-conditioner has a rugged flotation head which absorbs the concussion-like impact of gulleys, ruts, bumps and other field contour phenomena. It takes a long time for these effects to produce cumulative fatigue on field units, but by increasing their frequency (not amplitude) in the laboratory, hours of tough field experience can be compressed into minutes. Barring situations in which heat build-up is a factor, fatigue tests can be accelerated by compressing as many as 100 field hours into one laboratory hour. It is this sort of extreme reliability and durability testing that is reassuring to the farmer and marketing management.

## Field Data Collection

Since the quality of any test program is directly proportional to the quality of data from which it is programed, realistic test programs depend upon the collection and analysis of representative amounts of actual field data. Complete data on field load conditions is relayed to Sperry New Holland's unique mobile laboratory through various transducers strategically mounted on the field machine (Figure 2).

The mobile laboratory personnel record field data both on FM tape and visual oscilloscope traces (Figure 3). By contrast with FM tape, the oscilloscope trace is not machine readable and its data is difficult to utilize. However, once the laboratory

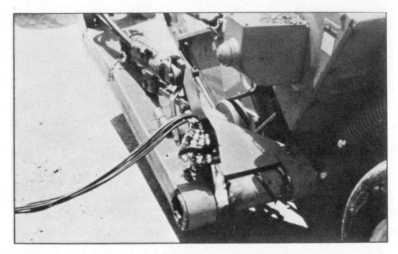

Figure 2. Physical parameters of field loads picked up by the header shoe and absorbed by the main frame header support are converted by transducers to electrical signals and transmitted to the stress van monitoring and recording equipment.

Figure 3. Stress van personnel supervise the recording of field data both for experimental machines and production models.

tests have been programed and units are under test, the visual field data trace becomes a valuable reference for verifying the accuracy of the laboratory simulation.

## Laboratory Simulation Techniques

After data from numerous field test runs has been reduced by computer devices (frequency counters, peak distribution analyzers, etc.) to a statistical or machine readable form, the data can be used to design lab test programs or load laboratory simulators. The specific methods used to simulate field conditions are not critical so long as physical parameters in the original data are faithfully transferred via the laboratory simulators to the unit under test.

Assurance that field conditions can be accurately simulated depends upon the common ability of loading devices to respond to electrical control signals. These signals might be generated by a wide variety of components, e.g., low frequency, random noise or ramp generators, punch-tape random selectors, magnetic tape and arbitrary wave programers. Many laboratory load sequences require a combination of these techniques in order to program the various average and range loads required in a complex test sequence.

Servo-hydraulic valves, through their linear response capability, respond in direct proportion to the electric signals generated by these components. The signals cause mechanical valve fluctuations which directly control hydraulic fluid movement.

Servo valve versatility thus enables program variation by frequency, amplitude, displacement, temperature and pressure. This proportional control, relayed through servo-hydraulic cylinders, motors, actuated platforms, power take-off shafts and other devices, mechanically induces loads on the test stand (Figure 4).

Figure 4. This electronically controlled hydraulic servo-valve converts electrical signals into physical displacement to simulate field conditions.

## Haybine Mower-Conditioner Under Test

The Haybine unit provides a good example of the thorough testing each prototype or production machine undergoes in the laboratory.

- On the test stand, an electric motor mounted on a hydraulically actuated dolly simulates the tractor's motive power. It can also be actuated to simulate the position of the tractor relative to the Haybine unit during field operation.

- The dolly is swung around as if the tractor were turning 90° relative to the Haybine unit. This simulates sharp right-angle turns such as would be made at the end of a field (Figure 5).

Figure 5.

- The header is hydraulically lifted simulating what the tractor operator would do if the Haybine unit were operated in rough terrain and had to be moved over a deep ditch or through unusually rough chuck holes.
- The header is then returned to the start position. Generators, hydraulic pumps and other equipment that simulate loads on rolls, reel and cutting knife are engaged. The data tracks, signal generators, cam banks, etc. are programed so that the average, peak and no-load conditions experienced in the field test runs are simulated by the loading devices.
- In addition to torque loading the various drive trains (Figure 6), structural loading is simulated by mounting the header shoe on hydraulically actuated platforms. These platforms move both sides of the header relative to the frame at the same velocity and amplitude that the header experiences during field operation.

In every instance, field data is fed to loading devices via the electrical control signal. The difference between an actual load condition and the control signal is indicated by the feedback signal magnitude. The laboratory test equipment automatically adjusts the load so that the error signal between the feedback loop and input is zero, which indicates an accurate relationship between the laboratory and the field data (Figure 7).

Figure 6. Rolls and reel are connected by universal assemblies to laboratory devices which simulate previously recorded field data.

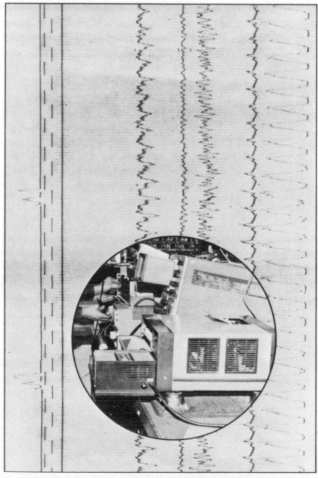

Figure 7. Data collected by the oscilloscope during field tests and while monitoring simulated lab tests, is visually reproduced for easy comparison.

## Computer Cost/Performance Analysis

As soon as the first production machines roll off the assembly line, they are put into service in the field. The product test department supervises this field operation to assure that the reliability designed into the prototype exists.

Performance of customers' machines also is reviewed by field service representatives who report all problems they encounter while servicing a particular machine. Through warranty forms, customers report parts which had to be replaced by their dealer or service representative. This cumulative data is stored in the computer to form a composite of each type of machine's operating characteristics. This composite can be readily analyzed for trends indicating the need for redesign or re-evaluation of particular components.

In the laboratory, this combined computerized information can be used to modify test set-ups so that each new generation of machines can be exposed to more exacting test parameters.

This data is also useful in determining alternate vendors who can supply reliable parts for use in Sperry New Holland farm equipment. Each vendor's parts can be subjected to the same simulated load conditions and evaluated on an equal basis with other vendors' products.

Sales figures for replacement parts for a given machine can also be analyzed from this computerized data. The sales ratio of a certain part to the number of units in the field becomes a good indicator of the reliability of that part.

## Conclusion

Prior to the advent of computers, FM magnetic tape, servo-hydraulic systems, and other complex electronic devices, it was impossible to reliably simulate field conditions in the laboratory. With the availability of these sophisticated devices, it is now possible to machine-read large quantities of field data, analyze it and apply it to solutions in a variety of management and marketing areas.

As far as modern farmer-managers are concerned, the days of the village blacksmith are truly gone. Having moved into the lab, the modern farmer can now rely totally upon the same engineering know-how as his counterpart in other allied industries depending on 20th century mechanical marvels.

### STUDY QUESTIONS
1. What evidence do you find that indicates this was written for a "special" audience, readers who work for the corporation that manufactures this equipment?
2. The authors fail to define many terms that the general reader might not understand, such as *peak distribution analyzer,* or *servo-hydraulic valve.* Why are these not defined? Is an understanding of these terms important if one is to understand the article?
3. How would you characterize the narrative voice these authors have chosen? Why have they chosen this manner of presenting the information?

# The Nature of Research in Mathematics [1]

## Raymond L. Wilder

How one goes about conducting research in the conventional sciences, such as biology, or in the social sciences, such as anthropology, seems clear enough. How one conducts research in pure (rather than applied) mathematics is not quite so clear to most people. In the following selection Raymond L. Wilder explains just what research in mathematics is, and how one goes about it.

No other field of science presents such a difficult problem in exposition as does mathematics, especially when one tries to explain the nature of its research to the layman — and here the term "layman" includes a large number of nonmathematicians who are scientists. As for nonscientists, most are unaware that research goes on in mathematics; after all, 2 + 2 = 4, and that's that — what is there to "research"?

Undoubtedly this situation is based on a popular misunderstanding of the nature of mathematics. Moreover, it is not just a modern misunderstanding, but has probably existed ever since the time of the Greeks, when a select few were working out the material later to be compiled in Euclid's *Elements* — the "Euclidean geometry." When Eudoxus, a contemporary of Plato, was formulating a theory of proportion, presumably to meet the criticisms directed at the foundations of mathematics by Zeno and others, the research atmosphere in mathematical circles must have been much like that to be found today. But it is doubtful if the average Greek of that period had any inkling of the crisis which had occurred in mathematics, or of the attempts under way to meet it by the invention of new concepts.

Today, when so many people are taught enough mathematics to include a sizable portion of Euclid's geometry, it may seem strange that there is so little comprehension of the nature of the field. Comprehension, however, was rarely a feature of the old "disciplinary" approach which prevailed in mathematics teaching until quite recently. The approach was generally dogmatic; the pupil *learned* that 2 + 2 = 4, and he *learned* that the sum of the angles of a triangle is two right angles. True, in the latter case his geometry book presented a proof of the fact; but unfortunately he *learned* this too, unless his teacher was perceptive enough and clever enough, to prevent it. With proper guidance into the "mysteries" of mathematics, he would have been led to ask, "Is the sum of the angles of a triangle always a certain magnitude, and if so, what is it?" Next he could have been helped to guess

[1] This is an expansion and adaptation of an article entitled "The Nature and Role of Research in Mathematics" which appeared in a volume of essays on the occasion of the University of Michigan's Sesquicentennial under the general title *Research: Definitions and Reflections*, D. E. Thackrey (ed.), 1967 (limited distribution).

Raymond L. Wilder, "The Nature of Research in Mathematics," in *The Spirit* and the *Uses of the Mathematical Sciences*, ed. By Saaty and Weyl (New York: McGraw-Hill, 1969), pp. 31–47.

the answer himself and then to ask "Why?" and when the proof was introduced, it would have been meaningful rather than a piece of mathematical Holy Writ. Fortunately in the so-called "new mathematics" now being taught in many schools, the question "Why?" as well as "What?" is being asked. To the bewildered parent who inquires, "Why is my child studying this 'set theory' which I never had in school? I got along all right without it" — the proper reply is that it is being introduced to give the child an understanding of why $2 + 2 = 4$ as well as to present the fact itself.

In no period of the past has there been so much research in mathematics as there is today. In a recent report by the National Academy of Sciences,[2] made to the Committee on Science and Astronautics of the House of Representatives, one of the most eminent of modern mathematicians stated that "mathematics is currently growing more explosively than any other science." What is the reason for this, and what constitutes "growth" in mathematics?

Let us return to $2 + 2 = 4$. The major difference between this mathematical statement and those which appear in current research articles is that it has become part of the common arithmetic that we use in our daily lives, because it was formulated during the earliest period of the evolution of mathematics (actually before there was any such term as "mathematics"). All ancient cultures — African, Asian, American Indian, etc. — found it necessary to devise elementary counting words or symbols such as "one" and "two." But only a few, such as the Babylonian and Mayan, extended these in order to enumerate large collections (the mathematician usually calls them *sets*) with a precision lacking in the commonly used indefinite term "many." The ability to communicate or record, say, the exact number of people in a large crowd was an achievement of a high order in the evolution of civilizations. Even this, however, did not imply the existence of what we call "arithmetic," with all its addition and multiplication rules that we learn as children.

It was only when human societies became more complex, with the development of large communities and their attendant problems concerning building plans, ownership and sale of property, taxes, etc., that mere counting was insufficient and arithmetic became a necessity. If one is going to build a brick wall, he wants to know, for example, how many bricks he will need; and with a well-developed arithmetic he can predict the number precisely. It was the "social" pressure resulting from such needs that forced the invention of arithmetic, one of the greatest of man's intellectual achievements. Moreover, arithmetic was not invented by one man or even by ten or twenty; it consists of a series of inventions participated in by a host of thinkers[3] from early pre-Babylonian down to modern times. Properly speaking, it was a cultural evolution. Although the use of zero, for instance, appeared relatively early (in both the Babylonian and Mayan cultures), only in modern times were *negative* numbers accepted as bona fide members of the arith-

---

[2] *Basic Research and National Goals,* March, 1965, p. 190.
[3] We cannot call them "mathematicians"; they were Babylonian temple priests, Arabian astronomers, and men of various scientific callings.

metic community. And they forced themselves in — in other words, they became a cultural necessity.

Thus modern arithmetic as we know and use it was the result of a long search — "research" — for a tool suitable to handle the complicated quantitative problems presented by growing societies. And this is one of the main reasons why mathematical research still goes on; indeed, the demands of a modern scientific community for new mathematical tools are greater than the mathematical community can supply. Much as the growing complexity of ancient societies forced the invention of simpler rules for doing arithmetic, so do the needs of modern science and industry instigate the search for appropriate new methods and concepts.

Just what is research in mathematics? The educated layman usually has some conception of the nature of research in such natural sciences as chemistry, physics, and biology. Each of these sciences is engaged in the search for reasonable explanations of certain types of natural phenomena. After a period of collecting data and making observations and measurements, the scientists begin constructing *theories* — conceptual structures which they hope will describe the forms and relationships which the phenomena appear to exhibit. If a theory "works" — that is, if subsequent experimentation and measurements conform to what the theory implies (or "predicts") — then it is called an "explanation" of the phenomena. With the maturity which modern physics, for example, has attained, the realization has grown that the theory may not be an exact description of reality — indeed, that an exact description may not even be possible. The theories are purely mental constructs — concepts — and the strange thing is that they *work;* as many have observed, the more abstract a science becomes, the better it conforms to the measurements that become ever more refined with the advances of technology. The classic example is the replacement, for certain aspects of reality, of Newtonian mechanics by Einsteinian mechanics.

Let us again consider the evolution of arithmetic. Was its development a result of research into the forms and relations of natural phenomena? It was this and something more, for arithmetic very early became involved in dealing with social phenomena. In counting the *number* of elements in a collection of sheep or a forest of trees, one was dealing only with natural phenomena; one was investigating the most elementary property of collections in nature — the numbers of elements in the collections. And as the collections grew more complex and numerous, concepts such as "base" and "place value"[4] were devised in order to make the counting feasible. However, arithmetic as we know it involves addition and subtraction, and these were developed as a result of the need for dealing with large collections encountered in social as well as physical situations. This social factor has grown as modern cultures have evolved.

Today the cultural demands on mathematics, in the form of problems posed by both the natural and the social sciences, predominate in the so-called "ex-

[4]The modern parent can ask his child what these terms mean if he wasn't told during his own childhood.

ternal" (i.e., outside mathematics) stimuli for mathematical research. Until recent times the chief source of such demands was the natural sciences, especially astronomy (which profoundly influenced the early development of arithmetic and geometry) and physics. That part of mathematics known as "analysis," which embodies such subjects as calculus and differential equations, was developed largely because of the need for mathematical expression of velocities, accelerations, moments of inertia, etc. Indeed, most of the mathematicians from the sixteenth to the nineteenth century were also natural scientists. For a mathematician of those times not to have at least a speaking acquaintance with the physical theories which were being worked out would have been almost inconceivable.

The striking feature of the mathematical concepts developed during that period, however, is their *generality*. Just as numbers can be used to count any collection, no matter whether the elements are animal, vegetable, or mineral,[5] so can the concepts of the calculus be applied to theories that have nothing to do with physical theories. For example, when the modern foundations of actuarial science, upon which the insurance business is based, were being laid, the calculus was already available and applicable; transition from a rate of falling bodies to rates of mortality takes little imagination. Similar generalization occurred in the case of statistical theory; moreover, both actuarial theory and statistics originated in problems related to social phenomena. (Ultimately statistics became a field of mathematics in its own right, with an ever expanding range of applications in both the social and the physical sciences.)

But the impression must not be given that research in mathematics is only the construction of concepts that will fit the patterns observed in the world of natural and social phenomena. Students of cultural evolution are familiar with the phenomenon of an embryonic set of related ideas growing until it attains the status of what can be termed a "cultural organism." This term is apt, since the organism becomes self-sufficient, and its developers achieve the status of a "profession" and form organizations devoted to the advancement of their calling. Examples are readily available — in the transition from primitive medicine men to the modern profession of medicine, from the early "great men" of the tribe to the profession of law, from the practice of alchemy to modern chemistry, and from the primitive shaman to theology. Today each of these professions embodies a complex of systematically developed concepts which have interrelations that are not entirely understood. What student of law, for instance, can outline precisely how the many legal principles and methods which he has learned have entered into the synthesis of a new legal principle that he has formulated? No matter how intensely he tries to re-

---

[5]This generality of numbers, which we take for granted, is in itself a remarkable achievement. Anthropologists have found many cultures in which the numerals — the symbols used in counting — vary according to the nature of things counted. Vestiges of such a situation exist today in the Japanese language, where different word forms for numbers from 1 to 10 are used in counting persons, dishes, and pencils, for instance. It is not known whether all number systems that were unaffected by diffusion from other cultures passed through such a stage in their evolution.

call them, he will certainly miss some. Or in music, what composer can specify the harmonies and principles which have combined to suggest to his mind the shape and form of the piece that he has created? Each of these individuals — the law student and the composer — has absorbed in his unique way a portion of the body of concepts constituting legal or musical theory, and this has influenced the synthesis which he has made.

The same is true for the mathematician; his theory has grown to such an extent that it has become self-sufficient, and it abounds in suggestions for new conceptual structures as well as in problems to be solved. A simple example can be found in the area of "number theory," which essentially is the study of the properties of the counting, or natural, numbers 1, 2, 3, and so on. This is an extremely interesting branch of mathematics because it is one of the oldest, if not *the* oldest. After the ancient Babylonians had managed to construct a sizable part of arithmetic, evidently the temple priests, who were the repository of most of the "mathematical" theory, began to notice some inherent properties of the natural numbers. For example, they saw that certain triples of numbers such as 3, 4, 5 have the property that the sum of the squares of two of them equals the square of the third; thus $3^2 + 4^2 = 5^2$ $(9 + 16 = 25)$. This observation may have originated from a knowledge that the sum of the squares of the lengths of the legs of a right triangle equals the square of the length of the hypotenuse (called the "Pythagorean theorem," but known by the Babylonians a thousand years before Pythagoras's time). Irrespective of why interest developed in these triples, however, they were evidently investigated for this interest alone. The same interest caused the Babylonians to study the relations between cubes of numbers and between numbers which are in arithmetic progression.

With the Pythagoreans (around the sixth century B.C.) this interest in relations between numbers became intensified. The heart of their philosophy was expressed by the statement, "Everything is number." A simple notion which they discovered is that of "perfect number," in which a number is "perfect" if the sum of its proper factors equals the number. Thus 6 is a perfect number, since its proper factors are 1, 2, and 3, and $1 + 2 + 3 = 6$. Similarly, 28 is perfect, since $1 + 2 + 4 + 7 + 14 = 28$. After 6 and 28, the next perfect numbers are 496 and 8,128; and the fifth perfect number is 33,550,336! The Pythagoreans divided numbers into odd and even, and one notices that all five of these perfect numbers are even; consequently it is natural to ask, are all perfect numbers even? We do not yet know the answer to this simple number-theoretic question.

Most people who studied geometry in high school are aware that the "bible" from which this geometry was derived was Euclid's *Elements,* compiled circa 300 B.C. Many do not know, however, that the *Elements* contained considerable number theory; for example, the proof which is still given for the fact that there is no greatest prime number[6] is Euclid's. Today, over 2,000 years later, research in

---

[6] A number is prime if its only factors are 1 and itself; thus 2 (the only even prime), 3, 5, 7, 11, etc., are prime numbers. For another discussion of these numbers, see Weissinger's article.

number theory still flourishes. One of its fascinations is that it contains so many simple, innocent-appearing, unsolved problems. As a consequence it has attracted many "amateurs"; indeed, some of the classical unsolved problems of number theory were proposed by amateurs. In 1742 C. Goldbach wrote to a famous mathematician (Euler) saying he had noticed that every even number, except 2, is the sum of two prime numbers. For example, $4 = 2 + 2$, $6 = 3 + 3$, $8 = 5 + 3$, $10 = 7 + 3$, $12 = 7 + 5$, and so on. Thus originated "Goldbach's conjecture," but it has never been proved or disproved. Another unsettled conjecture concerns "twin primes," which are primes that differ by 2, like 3 and 5, or 11 and 13. Are there twin primes that are arbitrarily large — or as the mathematician usually phrases the question, does an infinite number of twin primes exist? We do not yet know.

The development of number theory illustrates the manner in which a mathematical theory becomes an "organism" in the sense in which the term was used above. It has an inherent growth potential, independent of natural phenomena; this is established by the historical fact that number theory has been growing for some 3,000 years, and there are no signs that interest in it is waning. On the contrary, research in the field is as active as ever, especially since it was found in modern times that powerful tools from other branches of mathematics (in particular, the tool of analysis) can be brought to bear on its problems. To be a successful number theorist today, one must know a great deal more about mathematics than just the counting numbers 1, 2, 3, etc. And the average number theorist conducts research in number theory not because he thinks his discoveries may have "practical" applications, but because he is entranced by it. As we usually say, he "does mathematics for its own sake."

One of the most important modes of mathematical research is called "generalization." For example, the addition and multiplication employed in elementary algebra are generalizations of the same operations used in arithmetic; the particular rule $2 + 3 = 3 + 2$ is but a special case of the general formula $x + y = y + x$ for all numbers $x$ and $y$ (this is called the "commutative law"). In modern times, mathematicians began to notice that such operations were appearing in other parts of mathematics, such as geometry, and that they all obeyed certain common rules. In order to avoid duplication, it was natural to generalize further by assuming that one dealt with a group $G$ of objects, which could be numbers, algebraic entities, or geometric transformations, for example, but whose actual nature was not specified. Then one could assume that an "operation" was assigned to $G$, analogous to addition in arithmetic, whose nature was again not specified, but which was required to obey certain laws called "axioms." For example, one could require that if $x$ and $y$ are two objects of $G$ and the operation is designated by "∘", then $x \circ y = y \circ x$. However, for this to have meaning it is necessary to specify that the result of the operation is in $G$ (that is, $x \circ y$ is an object of $G$ just as $2 + 3$ is a number, namely 5) and that the sign "=" means "is the same object as." In this way, $x \circ y = y \circ x$ comes to mean that the objects $x \circ y$ and $y \circ x$ are the same. Also, just as in arithmetic there is a number, zero, whose addition to a number leaves that number unchanged, it may be assumed (as an axiom) that there exists in $G$ a special object $e$, such that $x \circ e = x$ for all $x$ in $G$. Continuing to generalize in this way, one sets up a concept called an "abelian group," a special case of the concept of "group" in which the rule "$x \circ y = y \circ x$" is *not* assumed.

Not only has the concept of group been a unifying influence in mathematics, seeming to draw together various branches, but it allows a proof to be made, once and for all, of a host of properties manifested by the various special cases of groups in arithmetic, algebra, geometry, etc. This is very important when a field is developing as many ramifications as mathematics. Today every mathematician can be expected to know group theory, much as every schoolchild is expected to know arithmetic. (Although he doesn't realize it, the child studying arithmetic is really studying certain aspects of group theory.) Inevitably group theory has found applications: in physics it was a valuable tool for quantum theory, and today it is being used in certain areas of chemistry. But the motive for the study of groups has always had its origin in mathematics, not in applied sources. It seems to be generally true that research directed exclusively toward applications leads to sterility; the mathematician must also be engaged in formulating mathematical theory for its own sake.

In the process of generalization, *abstraction* plays a prominent part. To arrive at the general concept of a group, the mathematician must abstract from the objects with which arithmetic, algebra, and geometry are concerned. He retains only those properties which have significance for all the objects, irrespective of their special roles as numbers, algebraic symbols, or geometric figures. And before he can apply mathematics to a physical or social situation, he must abstract from it to a greater or lesser degree, depending on whether the same or a similar situation has previously been treated in mathematics. Thus in applying numbers to counting, we abstract from the objects counted, but almost no effort is involved because we have already reduced the process to the status of a habit. If the situation is new, however, particularly if the type of mathematical tool to be employed is not immediately evident, a greater amount of abstraction is needed. And if it is a situation in which the kind of mathematics required has not even been developed yet, the highest degree of abstraction may be necessary.

A good example is the so-called "Königsberg bridge problem." In the old city of Königsberg, there were seven bridges whose locations are shown in Fig. 1 (a

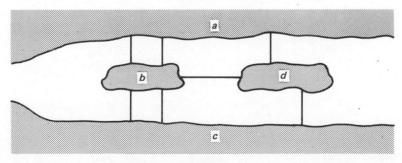

Figure 1.

schematic, not geographically accurate representation), where the shaded portions labeled *a, b, c,* and *d* represent land and the line segments represent the bridges. The question arose: Can one traverse all seven bridges without crossing any bridge more than once?

Divorcing the problem from its physical trappings, the mathematician Euler drew the diagram in Fig. 2. The seven line segments (straight and curved) connecting points *a, b, c,* and *d* represent the seven bridges, and points *a, b, c,* and *d* represent the land connections. Thus point *b* signifies the island designated *b* in Fig. 1, over which one can walk to any of the five bridges meeting it. The fact that the island has physical "extent," Euler perceived, is not relevant to the problem. Euler recognized that if it was impossible (and he probably surmised that this was the case) to start at any one of the points *a, b, c* or *d* and draw a continuous line over the whole figure, passing over each line segment only once, then the answer to the Königsberg bridge problem is negative.

To show this, Euler reasoned as follows: A continuous line, starting at, say, *a* in Fig. 2 and coming back to *a,* will have to cross two line segments having end points at *a.* Of course, it may pass through *a* several times, but if it begins and ends at *a,* it will cross an *even* number of the line segments which meet at *a.* And since at each of the four points *a, b, c,* and *d* an odd number of the line segments meet, then no continuous line crossing each segment exactly once is going to begin and end at any one of the four points. Using a similar argument, Euler also reasoned that if a hypothetical continuous line crossing each segment once neither begins nor ends at a given point, then again an even number of the line segments must meet at that point; and if it either begins (but does not end) or ends (but does not begin) at a point, then that point must be an end point of an *odd* number of the line segments. Consequently he concluded that a continuous line of the sort conjectured can begin and end at *different* points only if there are exactly two points at which an odd number of line segments meet.

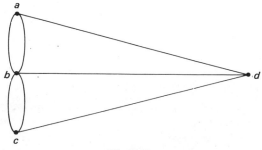

Figure 2.

The upshot of this is that if a continuous line can be drawn crossing each line segment exactly once, then either (1) an even number of line segments meet at every point, or (2) an even number meet at every point except two, at which an odd number meet. And since in Fig. 2 an odd number of line segments meet at every one of the four points *a, b, c,* and *d,* one must conclude that no continuous line can be drawn covering each segment exactly once. Returning to the original physical situation of the bridges of Königsberg, it follows that it is impossible to cross each of the seven bridges without traversing one of them more than once.

Euler's solution of this problem is interesting both for the manner in which he abstracted from the physical situation to obtain Fig. 2 and for the fact that he thereby obtained not only a solution to the original problem, but *a general answer to every conceivable problem of this sort.* For example, let us give the name *graph* to every figure like Fig. 2, which consists of a (finite) number of points *a,b,c,d,e,* . . . (called "vertices") and line segments (called "edges") connecting them. If the graph is such that one can draw a continuous line from any point in it to another, it is called a *connected graph.* The reasoning employed by Euler proved, as a *general theorem* about graphs, that a connected graph can be covered by a continuous line crossing each edge exactly once if, and only if, (1) at each vertex an even number of edges meet (in which case the continuous line ends at the vertex at which it began), or (2) at two and only two vertices an odd number of edges meet (in which case the continuous line begins at one of these vertices and ends at the other).

This is an excellent example of the method by which the creative mathematician frequently proceeds: By abstracting a pattern or "model" (e.g., Fig. 2) from a given physical or social situation, he establishes a theorem applying not only to the special situation, but to an infinite number of similar situations. Moreover, usually the process does not stop here. Following Euler's work, a number of distinguished eighteenth- and nineteenth-century mathematicians used graphs in order to abstract from particular physical situations and to settle problems concerning them. Noteworthy instances were G. Kirchhoff's studies of graphs and their application to electrical circuits and A. Cayley's work on graphs in which the graph represents a molecular structure, with vertices signifying atoms and edges denoting bonds between the atoms.

Today the theory of graphs is a thriving branch of mathematics; an increasing number of treatises set forth the general theory, and journal articles continually announce new results. As usually happens, many mathematicians who are engaged in such studies have temporarily lost sight of the connections between the graphs and the "real life" situations which they may represent. These researchers are motivated by problems that have arisen in the study of graphs as an object of interest in their own right — much as the worker in number theory may forget that the numbers with which he deals were originally invented for counting and measuring. However, this does not mean that the results obtained have no significance outside of mathematics. The history of mathematics is filled with instances of mathematical concepts which, although devised only for their contributions to the development of mathematical theory, were subsequently applied to other sciences. The case of group theory cited above is a good example, and so is the application of the calculus to develop the theory of mortality and its ramifications in the insurance industry. Calculus is a branch of mathematics which was originally created to handle problems concerning physical phenomena, and which underwent further development by generalization within mathematics and then was applied to social phenomena.

As a result of the great accumulation of mathematical theories and the consequent tendency to specialize, it is not surprising to find today that some mathematicians concentrate their efforts on mathematical theory, while others make a specialty of dealing with the problems arising from the interpretation of physical and social phenomena in mathematical terms. The former type of mathematician who "does mathematics for its own sake" is commonly called a "pure" mathematician, as opposed to the "applied" mathematician who is concerned with the mathematical problems encountered in studying the structures and patterns in social and physical phenomena. It should be emphasized that the distinction is not a clear one; the same person is often both a pure and an applied mathematician.[7] Moreover, most of the concepts of pure mathematics ultimately find application. In addition to the examples given above, there is the classical centuries-long attempt to show that Euclid's "parallel postulate"[8] was a consequence of the other postulates. At last, over a hundred years ago the Russian Lobachewski, the Hungarian Bolyai, and the German Gauss demonstrated independently that perfectly consistent "non-Euclidean" geometries could be devised in which the parallel axiom is false. This work was all "mathematics for its own sake," but today the non-Euclidean geometries are exploited in both physical and social theories.

Another interesting example is that of mathematical logic, which is usually considered as having started with the work of the English mathematician Boole. His 1847 treatise *The Mathematical Analysis of Logic* was an attempt to construct an algebra for logic which would make automatic logical conclusions starting from certain premises, much as the ordinary high school algebra devises formulae for conclusions concerning numerical data. The classic in the field is that of Russell and Whitehead, a massive three-volume work entitled *Principia Mathematica* 1910–1913), which attempted to derive all mathematics from a few logical axioms expressed in symbolic form. Their project is now known to be hopeless, but their work has formed a model for later research in mathematical logic. Although nothing could have been "purer" than this mathematical logic, it is now being applied in such fields as computer and automata theory and studies of the human nervous system, and it is almost certain to have other applications in the future.

Just why new mathematical concepts, created by the mathematician for their importance and interest to mathematics alone, should turn out later to have applications outside mathematics, is not thoroughly understood. As a living, growing body of concepts that transcends the capacity of any one individual to comprehend, and that has its roots in the necessity for coping with natural and social

---

[7]See the examples given by J. Weissinger in the preceding essay.

[8]This has several equivalent forms; one of the simplest is: "In a plane P, if L is a line and x is a point not on L, then there is one and only one line in P that contains x and does not meet L."

situations, mathematics apparently never loses its cultural ties, no matter how abstract it becomes.[9] This phenomenon is not peculiar to mathematics but is found in other sciences as well. The eminent physicist Werner Heisenberg has stated: "The interaction between technical developments and science is in the last resort based on the fact that both spring from the same sources. A neglect of pure science would be a symptom of the exhaustion of the forces which condition both progress and science."[10]

Just what role does research play in mathematics? As in the natural sciences, the primary role is to extend our knowledge, but there is an important subsidiary function that is probably not clearly apprehended by those who accuse universities of emphasizing research at the expense of teaching. It is a *sine qua non* that a good teacher must be interested in, not bored by, the subject he is teaching, and to do creative work in it is certainly one of the best ways to maintain his enthusiasm. For a teacher in a university, at least in scientific subjects, it appears that creative work is not only the best means for stimulating his interest, but also the most likely to produce good teaching. A creative scientist is accustomed to striking fearlessly into the unknown, accepting nothing on faith or prejudice, and this is the approach that should be taught to students. In mathematics especially, where teaching can degenerate into a dreary recital of formulae and of ways to derive one formula from another, the teacher's ability to stimulate and inspire is a most important consideration.

A corollary to the benefits that research provides for teaching is the enrichment that teaching can bring to research. In his exchanges with students the creative scientist frequently notices new problems and new concepts. Outstanding research scientists have often remarked that they value their teaching because they derive inspiration and new ideas from it.

The author knows one mathematician whose research is in a field where so far the applications have been rare, although its material seems to have increasing significance for other areas of mathematics. This man, however, has probably produced more mathematicians from his students than any other mathematics teacher in history. Even if his research should never find so-called "practical" applications, could one say that it has no value? To him it is an art, much the same as the artist's art is to the artist. Its "practical" significance, one can maintain, is that he is so enthusiastic about it that he attracts young minds to it whose mathematical powers might otherwise never have been discovered. Now they, together with their students and the students of their students, can be found not only in important universities but in the aircraft industry, the computing industry, government laboratories, and throughout research and development enterprises.

But one must not lose sight of the fact that the main function of research in mathematics is, like research in the other sciences, to extend the frontiers of

[9] A fuller discussion of the features inherent in mathematical reasoning which account for this uncanny relevance will be found in Eugene Wigner's article.

[10] *Naturwissenschaften*, vol. 40, p. 669, Oct. 5, 1934.

knowledge. And the structures and patterns that the mathematician discovers, both within the rapidly growing world of mathematics (if he is a pure mathematician) and in the problems presented by the other sciences (if he is an applied mathematician), do enormously increase man's knowledge. To be sure, it is not always a type of knowledge that explains or applies directly to a physical or social situation, but either it ultimately does, or it forms part of a general complex of concepts that do.

## STUDY QUESTIONS

1. The author uses several analogies in his explanation. Are they literal or figurative analogies?
2. Wilder includes a certain amount of world history in his discussion. Why does he do this? Is it necessary? Few of his historical facts are documented in any way. Why not?
3. Notice the many definitions Wilder uses. What do the number and nature of these definitions reveal about Wilder's audience?
4. To what extent is this "explanation" also a "justification"? Could this selection be labelled *persuasive* as well as *informative?*
5. *Writing Assignment:* Using this article as a model, write a similar report on your particular major or potential career. Entitle it, "The Nature of a Career in
   _____."

# Antennas

## John McVeigh

The following article by John McVeigh appeared in *Stereo Review,* a magazine for people who take their stereos very seriously. For such people (and you may be one of them), getting merely adequate performance from an FM tuner is not good enough, for excellence is the only standard they will accept. The quality and type of the FM antenna one uses, McVeigh points out, can be a crucial factor in the quality of sound one receives from his equipment.

### Why an Antenna?

Simply described, an antenna is an electrical conductor or group of conductors which extracts energy from passing radio waves and passes it on — via a *transmission line* — to a tuner's input terminals. The tuner amplifies these flea-power

John McVeigh, "Antennas," *Stereo Review,* September 1978, pp. 80–85.

radio-frequency signals and *demodulates* them (extracts the audio for subsequent amplification).

Today's FM tuners (or tuner sections) are quite sensitive, providing a listenable audio output from *very* low input levels. But just how "listenable" a tuner's output is with respect to noise and distortion depends, to a great extent, on how much input signal it has to work with. (A glance at the quieting and distortion curves in any Hirsch-Houck lab test of an FM tuner illustrates the point.) If you live in a high-signal-strength area, a relatively simple antenna will provide adequate signal levels. You might, however, need a more elaborate directional antenna to reduce the effects of multipath. And if you live in a weak-signal area, you might need help in the signal-strength department.

In situations between these two extremes there are many factors to be considered. *Each receiving location is unique,* with its own special characteristics (including those of the tuner employed) that determine which combination of antenna type, physical mounting, and auxiliary components (transmission line, possibly an antenna preamp, etc.) will be most effective.

## The Basic Antenna

The half-wave dipole can rightly be considered the fundamental antenna because most other types are derived from it. As its name implies, the conventional dipole comprises two conductors, each a quarter-wavelength at the frequency of interest. In free space, the dipole has a feedpoint impedance of 72 ohms at its *resonance* frequency (that is, at the frequency at which the dipole is a half-wavelength long). For example, a dipole 5.3 feet long resonates at 88 MHz (megahertz), the lower end of the FM broadcast band. A dipole 4.8 feet long resonates at 98 MHz, the center of the band, and one that is 4.3 feet long is tuned to 108 MHz, the top of the FM band.

Any dipole has a certain bandwidth — centered at its resonance frequency — over which it functions most efficiently. As it is operated at frequencies farther and farther away from resonance, it not only becomes less and less efficient, but a problem similar to multipath (see below) may occur.

However, there are ways to broaden the dipole's response. One is to use a *folded dipole* such as is shown in Figure 1. Here, a total of one wavelength of wire is used to construct the antenna. This dipole has increased surface area and therefore greater bandwidth. It also has a *feedpoint impedance* of about 300 ohms, offering a direct match to 300-ohm twinlead transmission line, and is widely used for FM reception.

Another means of broadening the dipole's frequency response is to increase the *surface area* of the conductors. It is for that reason, aside from structural strength, that metallic tubing is usually used in the construction of outdoor FM and TV antennas.

The dipole, be it conventional or folded, displays two fundamental antenna characteristics — *directivity* and *gain* — that are intimately related. Directivity refers

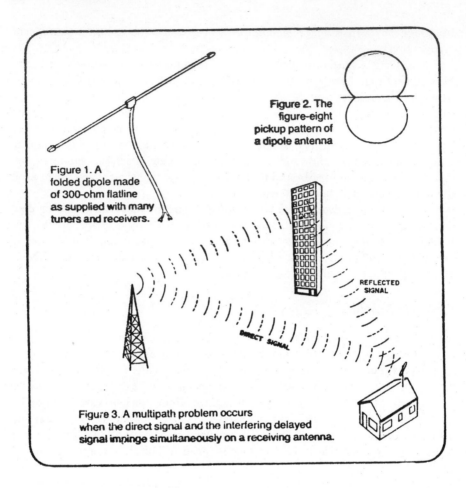

Figure 1. A folded dipole made of 300-ohm flatline as supplied with many tuners and receivers.

Figure 2. The figure-eight pickup pattern of a dipole antenna

REFLECTED SIGNAL

DIRECT SIGNAL

Figure 3. A multipath problem occurs when the direct signal and the interfering delayed signal impinge simultaneously on a receiving antenna.

to the fact that an antenna is more sensitive to signals from certain directions than from others. The characteristic response of a half-wave dipole is shown in Figure 2. Viewed in three dimensions, the pattern would be doughnut-shaped for both the single-wire and folded dipoles. It is obvious that the dipole is most sensitive to signals striking it broadside, and least sensitive to signals striking it on edge. It must therefore be oriented to favor those directions the signals you want to listen to are coming from. If mounted at least 5 feet (a half-wavelength at FM frequencies) above ground or any structure containing appreciable amounts of metal, a dipole's response becomes similar to the pattern of Figure 2.

This directivity can be a nuisance if you want to receive signals from several different directions, but more often it is an effective weapon in the fight against *multipath distortion*. This phenomenon, called "multipath" for short, results when two or more signals arrive at the antenna from the same transmitter (see Figure 3). One portion of the signal radiated by the transmitting antenna travels directly to the receiving antenna; another portion strikes a large *building* which *reflects* it toward the receiving antenna. And a third portion may be reflected toward the receiving antenna from another direction. Each signal follows a different path and takes a different amount of time to make the trip. Although this difference is only a matter of microseconds, it causes the signals to be out of phase with each other at the receiving antenna. When the signals combine randomly, they produce fading, "flutter," or distortion in the signal. If a tuner has good AM suppression and a good (low) capture ratio, it will resist the effects of multipath. However, the fading produced by multipath is often too much for the tuner to handle. That's when antenna directivity becomes important.

A dipole can be physically oriented to favor the direct signal and reject the reflections. But more directivity is needed in severe multipath situations, and a more complex antenna is required to achieve it.

Gain, the other basic characteristic of the dipole, is a measure of how much signal a particular antenna will deliver compared to the output of a reference antenna. Unfortunately, some manufacturers publish gain figures based on a theoretical "isotropic" (dBi) reference level, and others rate their products' gains using a half-wave dipole (dBd) reference. The dBi gain figure, if not the performance, will always be 2.1 dB "better" when *equivalent* antennas are being compared.

## Omnidirectional Antennas

If you want to listen to stations at various points on the compass, are not bothered with serious multipath problems, and don't need a lot of gain, an omnidirectional antenna is suitable. Two types are commonly used in FM applications – the "S" dipole and the "turnstile" antenna. The "S" dipole is a folded dipole composed of metallic tubing bent in the shape of the letter S (see Figure 4). This antenna has a

Figure 4. The "S" dipole has virtually omnidirectional response in the horizontal plane (see inset).

Figure 6. The yagi is intentionally directional, rejecting signals from the rear (R) and sides.

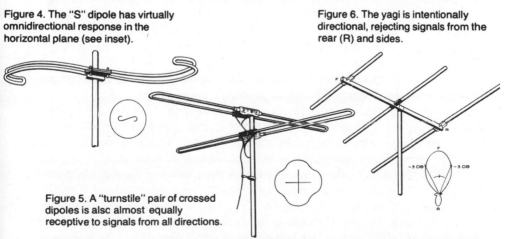

Figure 5. A "turnstile" pair of crossed dipoles is also almost equally receptive to signals from all directions.

polar response in the horizontal plane that is fairly circular or omnidirectional. However, the "S" dipole, like a straight one, is directional in the vertical plane. Its gain is about –1dB referenced to a "straight" dipole, or 1.1 dB over isotropic.

The turnstile antenna consists of two folded dipoles at right angles to each other. The turnstile and its horizontal polar response, shown in Figure 5, is roughly omnidirectional. At right angles to each dipole the turnstile has a 0-dB gain referenced to a single dipole. At intermediate angles, the polar responses of both dipoles comprising the turnstile combine to provide more received signal than either one alone would provide, but not as much as would be obtained from a single dipole oriented for optimum reception.

Several manufacturers offer antennas formed by stacking one turnstile on top of another. The two, spaced a few feet from each other, are connected together by a phasing harness to provide about 3 dB of gain over a dipole. Although these stacked turnstiles remain omnidirectional in the horizontal plane, they do discriminate in the vertical plane and may afford some relief from multipath.

## Directional Antennas

For really substantial amounts of gain and/or multipath suppression, a highly directional antenna must be used. Two types are commonly employed for FM reception — the *yagi* beam and the *log periodic.* Let's examine each in turn.

The yagi beam is composed of a dipole "driven" element and one or more "parasitic" elements coupled by electromagnetic fields. If a parasitic element slightly *longer* than the driven dipole is placed behind it, the element acts as a *reflector,* reinforcing the dipole's response to signals striking the dipole first. If a *shorter* parasitic element is placed in front of the dipole, it acts as a *director,* further focusing the dipole on signal sources to its front.

A three-element yagi and its horizontal polar pattern are shown in Figure 6. This antenna has about 7 dB of gain over a half-wave dipole and is much more directional. Adding more elements will enhance the antenna's directivity and increase its gain. Note that the polar pattern of the yagi beam contains one major lobe and several minor ones. As more elements are added the major lobe gets larger at the expense of the minor lobes.

Besides a yagi's gain, there are several other important specifications. Two are apparent from an examination of the antenna's polar pattern. The first is the *front-to-back ratio,* which describes in decibels how well the beam can discriminate between equal-strength signals coming from its front and back. Typical figures range from 10 to 30 dB. The higher the number, the more directional the antenna.

The second specification, the *half-power beamwidth,* is measured in degrees. It describes the width of the major lobe and is determined by locating the points where the antenna's response has fallen 3 dB (the "X" points in Figure 6). Lines are drawn from these points to the origin of the graph, and the angle thus formed is measured. The half-power beamwidths of typical yagis designed for FM service range from 50 to 70 degrees. This half-power or –3-dB beamwidth is analogous to the field of view of a telescope or binoculars. To obtain a higher degree of magnification, the field of view must usually be reduced. Accordingly, to obtain more gain, an antenna's "area of focus" or half-power beamwidth must be reduced.

The practical consequences of these three specifications are apparent. If the antenna has high front-to-back and front-to-side ratios, it will greatly favor signals from stations situated in the direction of its major lobe. This makes the yagi very effective against multipath.

An antenna with a narrow –3-dB beamwidth will have to be accurately aimed for best results. If all the stations you want to listen to have transmitting antennas at one common site, the antenna can be oriented so that the major lobe points towards that site. However, if the stations you want to receive have scattered broad cast-antenna locations, you'll need to aim the yagi in different directions at different times. This is best accomplished by using an *antenna rotator* on the yagi's supporting mast (more on this later).

The yagi might appear to be exceptionally well suited for FM broadcast reception, but it does have some drawbacks. The most serious is that it is inherently a relatively narrow-band device. Of course, some of its elements can be folded to broaden its frequency coverage. There is much more than this involved, however, and special design techniques must be employed to produce a yagi that offers a good performance across the entire FM band.

A directional beam antenna that enjoys many of the advantages of the yagi but does not have bandwidth limitations is the *log-periodic dipole array,* or simply the *log periodic.* As shown in Figure 7, the log periodic is made up, in effect, of a series of dipoles. At the upper end of the FM band those dipoles that serve as

Figure 7. The log periodic is also directional. The −3-dB points are explained in the text.

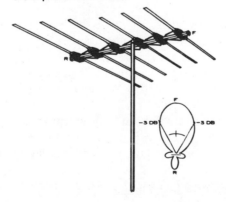

"driven" (active) elements for the lower frequencies become reflectors and some of the directors behave like driven elements. At the lower end of the FM band the roles change, and what were driven elements become directors, what were reflectors become driven elements. This smooth transition is a basic characteristic of the log periodic and accounts for its extremely wide bandwidth. In fact, log-periodic arrays can function over as much as a 4:1 frequency range (say, from 50 to 200

MHz) with relatively constant gain, feed-point impedance, front-to-back ratio, -3-dB beamwidth, etc. A well-designed log periodic can exhibit 8 dB or more of gain over a dipole and a front-to-back ratio comparable to that of a yagi. The log periodic is usually connected directly to 300-ohm twinlead.

The directionality that is a boon in some situations can be a liability in others, requiring the use of a rotator. Also, yagis and log periodics designed for high gain and directivity tend to be physically large. The boom (the horizontal beam that supports the active elements) can approach 12 feet in length and the elements 6 feet. Besides being unwieldy, large directional arrays require secure mounts and heavy-duty rotators, and they are vulnerable to damage by gusty winds because of their large surface areas.

How large an antenna you need depends on many factors. And, as is the case with other audio components, the specifications provide the basic — if not complete — information you need for making an informed choice. The key antenna specs required, if not supplied by advertisements or electronics catalogs, can usually be obtained by writing to the manufacturer or consulting his local distributors.

## Television Antennas

Because the FM band lies between TV channels 6 and 7, it might be supposed that a wideband TV antenna, such as a log periodic, could be used for FM as well. In some cases this is true, in others it is not.

To provide consistent performance on television channels 2 through 13, many manufacturers produce antennas that have *two* separate active sections. One covers channels 2 through 6; the other channels 7 through 13. These vhf antennas are frequently not good performers on the FM band — the "low-vhf" section's performance falls off abruptly above 88 MHz, the lower end of the FM-broadcast band. As a matter of fact, some TV antennas are deliberately made to perform poorly on FM because strong FM signals can interfere with TV reception. In some instances, however, manufacturers give the purchaser a choice by installing FM trap or control elements that are left on if FM rejection is desired or snapped off if the antenna is to be used to receive FM signals.

Finally, there are some "all-band" TV antennas that give a good account of themselves on the FM band as is. If you have a TV antenna that does not have the FM-reject feature permanently built in, try connecting its lead-in to the antenna terminals of your FM tuner. You might find it performs suitably, in which case all you'll need is a simple two-set coupler or "signal splitter" to feed the television set and FM receiver simultaneously.

## Indoor Antennas

For best performance, an antenna should be mounted as high as possible and have a clear line of sight to the transmitting antenna. These conditions are hardly ever met indoors. However, there are many reasons for *not* mounting an antenna outdoors, ranging from inconvenience to absolute impossibility (for example, when a landlord prohibits them). What can an apartment-dwelling FM listener do when his lease prohibits outdoor antennas — or he simply doesn't want the bother or expense.

There are a few options open to him. If the apartment building has a master TV antenna system, it can frequently be used to drive the FM tuner. An inexpensive accessory (the signal splitter mentioned above) can be installed to provide signals to both the TV and FM tuner. If you subscribe to cable TV, try connecting your tuner to the cable system. Consult the cable company for how-to-do-it details, costs, and exactly what channels in the FM band are available. (The full 88- to 108-MHz band may not be.) More and more cable systems are offering high-quality FM service, either as part of the package or as an option. Sometimes the only possibility open to the apartment dweller is an indoor antenna — which is certainly worth a try if there is no alternative.

Most manufacturers pack a simple 300-ohm flatline folded dipole with their tuners. The dipole can be taped or tacked to a wall — or, better yet, to a simple wooden "T"-frame that can be rotated to position the antenna for best reception. You may find that the folded dipole provides adequate results. On the other hand, the dipole may not be able to provide enough signal to quiet the tuner (eliminate hiss) adequately or to prevent multipath-interference distortion. Or you might find that the signal fluctuates as you move around the room.

If the folded dipole proves unsatisfactory, try an indoor TV/FM or FM-only antenna. Many such antennas are being marketed, ranging from relatively simple "rabbit ears" to rather elaborate models with built-in phasing switches to attenuate multipath or TV ghosting (they are related phenomena). Some even have internal preamps to boost received signal levels.

Some years ago (June 1973), *Stereo Review* featured a comprehensive article on FM indoor antennas. Tests were conducted on a number of models. Interestingly, few of them, except those with built-in preamplifiers, performed appreciably better than the reference folded dipole (some performed much worse), though adjusting the physical orientation of these antennas could, to varying degrees, result in the rejection of unwanted signal reflections. The key findings of the tests were:

- If you have a choice between a good, directional outdoor antenna and an indoor antenna, use the outdoor one.
- If you are limited to an indoor antenna and don't mind orienting it carefully and adjusting element lengths, choose the least expensive, least elaborate, rabbit-ears type.
- Indoor antennas are quite sensitive to positioning. Keep them away from metal surfaces. The presence of a human body can affect an indoor antenna and make adjustments tricky.

Since that article appeared, two relatively sophisticated "Beam Box" indoor antennas have been introduced by B.I.C. A Beam Box has inside it a crossed pair of quarter-wavelength dipoles, with switching to choose among the elements. In addition, there are 75- and 300-ohm outputs and a tunable signal filter that operates in switchable wide- and narrow-band modes. In the narrow-band mode it can attenuate by 12 dB unwanted signals 4 MHz above or below the frequency to which it is tuned. This antenna is theoretically as directional as a dipole (and much more convenient) and hence can attenuate multipath. However, the gain of the Beam Box is rated at –12 dB (in the wide-band mode) or –5 dB (narrow-band) referenced to a dipole. Hence it may not be suitable if you do not live in a relatively high-signal-strength area. But if an outdoor antenna is an impossibility, it's worthwhile to consider a Beam Box or some other, less elaborate indoor antenna.

## Transmission Line

The technical name for the cable that carries the received signals from the antenna to the input of the tuner is the *transmission line.* A good one will have low loss and a constant impedance. It will also be easy to handle and resist the destructive effects of weather and pollution. Three basic types of transmission line are commonly used in FM reception: *coaxial cable* (also called "coax"), *twinlead,* and *shielded twinlead,* which is a hybrid of the first two.

Twinlead, which has a characteristic impedance of 300 ohms, is available in several varieties. What might be called *standard* twinlead is composed of two parallel, stranded copper wires spaced about ½-inch apart and embedded in an insulating plastic. This is the least-expensive transmission line available for FM reception. Although standard twinlead has very low losses (only about 1.25 dB per 100 feet at 100 MHz), it is not really intended for outdoor use. Signal loss increases dramatically when the cable is wet, dirty, or aged by sunlight.

Twinlead is a *balanced* transmission line, providing a good match for a balanced (physically symmetrical) antenna such as the folded dipole. Like the balanced microphone lines used in recording studios, twinlead resists noise pickup, but only if its electrical balance is maintained. If one conductor is closer to, say, a metal antenna mast or window frame than the other, the twinlead will become unbalanced and susceptible to electrical noise and interfering signals. This tendency of twinlead to become unbalanced can be countered by twisting the transmission line during installation so that it has about one turn per foot. This will also help prevent the flat twinlead from whipping about in the wind.

The next step up in quality is flat, foam-filled twinlead — a polyethylene outer layer and an inner layer of plastic foam surrounding the two conductors. Compared to standard flat twinlead, foam twinlead is less affected by moisture and aging and has lower losses — but costs about one-and-one-half times to twice as much. In any case, even foam-filled twinlead should be replaced after a few years of outdoor service, especially in urban and coastal areas.

In *tubular* twinlead, the two conductors are bonded to a round or oval plastic tube whose center is filled with either air or polyethylene foam. Besides having low losses (about 0.75 dB per 100 feet dry) and being less prone to wind flutter, tubular twinlead is also more resistant to the effects of moisture, dirt, and pollutants. *Foam*-filled tubular twinlead is best in this respect, but even this type of line should be replaced after a few years' exposure to the elements. Twisting the line, securing it with standoffs, and keeping it away from metal objects is also recommended.

Unlike twinlead, coaxial cable, because of its physical construction, is not affected by proximity to metal surfaces. An inner solid copper-wire conductor is surrounded by either a solid or foam polyethylene insulating layer. A copper braid is woven over the insulation (or aluminum foil is wrapped around it), and the entire assembly is protected from the environment by an outer insulating jacket of vinyl. Besides serving as one of the signal conductors, the braid or foil behaves like a shield. Properly designed coax can provide almost complete isolation from noise and other extraneous signals. It can be run close to or inside metal masts, pipes, drain spouts, etc., without disturbing its electrical characteristics. Coax is also highly weather-resistant (assuming the cable is of good quality) and can even be buried in the ground. It has a normal outdoor lifetime of about ten years.

Coaxial cable is an unbalanced type of transmission line, and the type used in FM installations, RG-59/U, has a characteristic impedance of 75 ohms. These two properties necessitate the use of a *balun* (an abbreviation for "balanced-to-unbalanced" transformer) between the 300-ohm antenna and the 75-ohm coaxial transmission line. At 100 MHz, the RG-59/U has a signal loss of about 3.5 dB per 100 feet. Although coax has greater loss than *dry* twinlead, it is certainly more efficient than *wet* twinlead. Coax is also physically easier to handle and is not prone to whipping about in the wind. Good-quality coax, however, is more expensive than twinlead — it runs about 13 cents per foot.

The shielding properties of coaxial cable have been highly touted and are beneficial in many installations. However, the outer coaxial conductor, be it braid or foil or both, can act as an efficient shield only if it covers all or most of the surface area of the insulating dielectric. For that reason, do *not* use coax that has a loosely woven braid. (If you can see an appreciable amount of the foam insulation through "holes" in the weave after carefully removing a portion of the outer vinyl jacket, the braid will be an ineffective shield.) It would be wiser to buy a high-quality, tightly woven braid-shielded cable intended for transmitting use by hams, or a high-quality foil-shielded cable produced by a reputable manufacturer such as Belden or Alpha Wire.

Shielded twinlead enjoys many of the advantages of both coax and twinlead. Its feedpoint impedance and balance allow it to be connected to 300-ohm antennas and FM receivers without a balun. If it is grounded, the shield acts as a barrier to noise and other undesired signals and also prevents the inner conductors from being disturbed by nearby metal masses. Like coax, shielded twinlead can be run close to or even inside metal masts and pipes. Its electrical characteristics, including signal loss (which is comparable to that of RG-59/U), do not vary with weather conditions, and its price and lifetime are about the same as for coax.

## Choosing and Installing an FM Antenna

Among the factors you must consider when choosing an antenna are: the distance to the transmitting antenna, the power it radiates, and the terrain between it and your antenna. On the tuner side, you need to know its sensitivity, capture ratio, and overload characteristics. The further you are from a transmitter, the more antenna gain you will need. In multipath-affected areas, high directivity will be required. First evaluate your listening location and habits, and then compare your needs with the features of each antenna type we have discussed.

If you are physically able, you can install the chosen antenna yourself. All that's required is some mechanical ability and a standard assortment of tools. Otherwise, plan on paying your local TV service technician or antenna retailer $35 to $100 for an installation, depending on the complexity of the job.

It should be clear that it's best to use either coax or shielded twinlead transmission line. They are not only electrically superior but also more economical in the long run. The choice between shielded twinlead and coax is largely a matter of personal preference and the type of antenna connector on the rear apron of your tuner. Whatever type of line is chosen should be supported at frequent intervals with *standoff insulators* to minimize strain on the connections.

The antenna should be mounted as high and as much in the clear as possible. At a minimum, the antenna should be mounted at least 5 feet above ground or any large structure. Television and FM receiving antennas are usually installed on 5- or 10-foot masts, but in some installations telescoping masts or towers are used. If the terrain is relatively flat and there are no large intervening objects, a roof-mounted 10-foot mast is a good antenna support in an urban or near-suburban installation. For greater height, a 5-foot mast can be placed over a 10-foot mast if proper hardware is used. Two antennas (one, say, for FM and one for television reception, or two for FM) can be mounted on the same mast provided they are spaced at least 5 feet from each other.

In some locations greater antenna height is required, and this can be provided by a telescoping mast. Radio Shack, Lafayette, and Jerrold all offer galvanized-steel masts which telescope to 20, 30, 40, or 50 feet, depending on the model. These masts should be guyed with either nylon rope or aluminum- or plastic-clad, galvanized-steel wire. Collars with holes drilled for the guys are mounted on the masts at the factory.

There are several types of mounts for antenna masts available, including chimney, vent pipe, wall, eave, and roof mounts. Chimney mounts employ two straps that are wrapped around the chimney and tightened to provide the required support. Wall, eave, and roof mounts are the most secure of all because they are bolted to the supporting structure. They can be attached to masonry after holes have been drilled and lead expansion fasteners installed, or they can be attached to wood beams with large wood screws. Tripod roof mounts and "snap-in" wall mounts are very convenient to use because they will temporarily support the antenna assembly before the mast-retaining hardware has been tightened — you won't need another person to hold the mast while you tighten the hardware.

It's best to stay away from chimney strap-on and vent-pipe clamp-on mounts because they're not as secure as those that are bolted to the supporting structure. The chimney is a convenient place to mount an antenna, and wall mounts can be attached to chimney brick, but the antenna will corrode rapidly if mounted so that exhaust gases pass over it and ash builds up on it. If you are using a telescoping mast, a bolt-on wall mount, tripod, or pivoted-flange roof mount is a *must*.

When a *great* deal of antenna height is needed, such as in fringe or deep-fringe reception areas or in a valley, an antenna tower should be used. A tower can also be installed if your roof cannot support (or you have aesthetic objections to) a telescoping mast. Towers are available in sectional, crank-up, or tilt-over configurations. Besides these categories, a tower can be classified as either guyed or free-standing.

A crank-up or tilt-over tower (the former telescopes) allows you to install, remove, inspect, or repair the antenna without disassembling all or part of the tower. Also, a crank-up or tilt-over tower can be lowered temporarily to reduce the risk of damage to both tower and antenna if high winds or an ice storm are expected. A sectional tower, on the other hand, is simply a collection of (usually) 8- or 10-foot sections bolted together to form one rigid unit. Sectional towers are less expensive than the other types, and guyed towers less expensive than free-standing ones. Of course, before purchasing any tower you should investigate local zoning ordinances to make sure they are permitted where you live!

Lightning protection for your home will be afforded by connecting the tower to a good earth ground. Grounding the tower or large antenna mast should be done *in addition to* and *not* instead of installing lightning arrestors on the transmission and rotator control lines. *And, for safety's sake, NEVER attempt to install an antenna and mast in a location where it can in any conceivable way come in contact with overhead power lines.* Hundreds of people are killed or severely injured every year because they fail to observe this simple precaution.

As you can see, the "antenna question" is one that is not simply answered. But once you understand the ground (or grounding) rules, the task of selecting a "sky hook" suitable for your particular circumstances is not really all that difficult — and the payoff is cleaner, clearer, and altogether more satisfying FM reception.

## STUDY QUESTIONS

1. The section entitled "Why an Antenna?" begins the article, yet this section is very brief? Is it too brief? Why is not more space devoted to this seemingly crucial question?

2. McVeigh includes some very specific instructions and guidelines for improving FM reception. Was the nature of the publication a factor in this? Might he do this regardless of where the article was published?

3. This article is written in a fairly formal style. Does that seem to fit the audience of this particular publication? Why or why not?

# Solar Cells and Your Future Commuter Car

## Ted Lucas

Ted Lucas received a degree in physics from M.I.T. and now works as a technical writer in Los Angeles. His book *How to Build a Solar Heater* is basically a "how-to-do-it" text that gives tips for constructing solar panels and heaters. But in almost every book that tells "how to" there is also a chapter that explains "why." The following excerpt is from just such a chapter that explains why, an explanatory chapter on solar cells.

As the price of gasoline keeps climbing, it's inevitable that there should be increasing interest in other kinds of cars than those with internal combustion engines. Electricity is a logical substitute, especially if this electric power can be derived from the sun's energy. Moreover, it is not necessary to build a solar power plant to do this.

An elegant way to produce electricity from the sun is the direct approach using photovoltaic devices, or solar cells. These are a group of materials, all of which are *semiconductors* — classified between the good conductors of electricity like copper, aluminum and other metals, and the insulators, or poor conductors of electric power, like glass and most ceramics. The specific characteristic of solar cells that makes them useful in converting the sun's energy is that these materials, when exposed to solar radiation, produce electrical energy.

In solar cells, heat is transformed into electric power entirely by electronic methods. This involves the excitation and massive flow of electrons in the semiconductor without any visible physical or chemical change or movement. There are no moving parts in such a photo-generator — no gears, wheels, pistons, or heat exchangers handling fluids or gases. Favorite materials used for solar cells are silicon wafers containing minute impurities; thin films of cadmium sulfide plated on copper sulfide and other metallic compounds; and gallium arsenide. There are many other possible photovoltaic substances, but these represent the most popular in terms of research and development efforts in various parts of the world.

### A Brief History of Solar Cells

The first applications of solar cells began less than twenty-five years ago. In 1953 it was thought that the conversion efficiency of such materials was very poor indeed, about 0.6%. Quite recently, however, scientists of Communications Satellite Corporation have announced silicon cells with an efficiency of about 20%; and commercial devices with efficiencies of 12% to 14% have been available for some time.

From Ted Lucas, *How to Build a Solar Heater* (New York: Ward Ritchie Press, 1975), pp. 191–195.

Since solar cells have been used on panels to power satellites and space vehicles, or to supply electricity for navigation and communications equipment on oil wells far at sea or at other remote locations, the price of these photovoltaic devices has come down dramatically. Dr. Valentin A. Baum, noted Soviet solar scientist, pointed out at the United Nations energy conference in 1961 that silicon "solar-batteries" had been reduced in price from $500 per watt in 1959 to $275 per watt in 1960 and $175 a year later.

## Declining Cost of Solar Cells

At present, the price of these silicon solar cells ranges from $15 to $30 per watt, depending on whether you price merely the device itself or include it as part of a system installed on a roof to deliver electric power to operate appliances in a building. Even this price is far too high to permit solar cells to compete effectively against other solar energy collectors such as the various kinds of panels described in earlier chapters. But scientists and engineers working in this field have a target: they expect to be able to supply photovoltaic power at $.50 per watt by 1980, in limited quantities of devices, and at a somewhat lower price in huge quantities by 1985.

Certainly, the whole history of semiconductor development by the electronic industry makes these targets appear not only feasible but modest; they may well be achieved somewhat sooner. There is a massive effort under way now, directed by the Energy Research and Development Administration (ERDA) with several National Aeronautics and Space Administration (NASA) facilities, including the Jet Propulsion Laboratory (JPL) of the California Institute of Technology, Pasadena; Goddard Space Flight Center, Beltsville, Maryland; the National Science Foundation, Washington, D.C.; and a large team at the Sandia Laboratories of Western Electric Company, Albuquerque; as well as many other industrial research organizations.

## How Solar Cells Work

A typical silicon solar cell has two layers, as shown on the left in Figure 1. Depending upon the tiny amount of impurities added to the silicon, one layer is called a P-type because it conducts positive charges called holes, while the other layer is called an N-type because it conducts negative charges called electrons. Between the two layers is an intermediate surface called a P-N junction and formed by means of another diffused impurity in the silicon cell.

When an energy particle from sunlight, called a photon, strikes near the P-N junction, it produces both an electron and a hole (absence of electron, or a positive charge). As photons continue to strike the junction, electrons move toward the N-type layer and holes move toward the P-type. These charges result in creating a voltage across the cell. If you have enough silicon cells connected together, you can provide adequate electricity to operate a d-c light bulb directly from sunlight.

A cadmium sulfide (CdS) cell, diagrammed on the right in Figure 1, consists of an N-type cadmium sulfide film on the backing of P-type copper sulfide ($Cu_2S$). Between them is a P-N junction. Operation of this cell is similar to the silicon type.

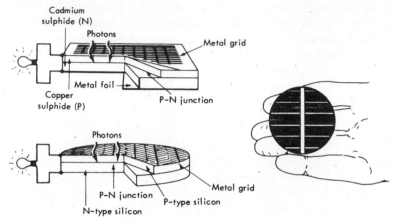

Figure 1. Components of two types of solar cells.

Among the scientific teams working on devices of this kind is one at Sandia under the direction of Dr. D. G. Schueler, supervisor of the Solid State Electronics Division. Their three main objectives are: development of cadmium sulfide and copper sulfide thin film solar cells; polycrystalline silicon film cells to achieve lower cost and improved efficiency; integration of solar cells into a solar system like the installation at Solar One House, University of Delaware and on the Mitre Corporation building in Washington, D.C.

### Improvements in Silicon Solar Cells

One of the leaders in developing a commercial process for making cheaper and better polycrystalline silicon solar cells is Mobil Tyco Solar Energy Corporation, Waltham, Massachusetts. This is a joint venture between Tyco, a relatively small but sophisticated instrument maker, and Mobil, the giant oil company which has bank-rolled this new program to the extent of thirty million dollars.

As pointed out by Dr. Abraham I. Mlavsky, executive vice president of Mobil Tyco, the big problem is not the initial cost of the raw material. Silicon is the second most plentiful element in the earth's crust. Materials needed to make silicon solar cells, including aluminum and boron, are also plentiful. Cost of the raw materials to make one panel of solar cells, good for delivering 1 kw or 1,000 watts of electric power on a satellite in outer space, is only $150. But the cost of processing this material runs the price up enormously — at least $50,000 per kw.

A reason for this is that the conventional method for making single-crystal silicon is to slice an ingot of pure Si into thin slices. The labor cost is very large. And the cutting or grinding process destroys more than ¾ of the ingot.

So the Tyco scientists began about nine years ago to develop a method for growing Si crystals. Actually, the first material grown by their new technique was sapphire for microcircuits. They called their process "edge-defined, film-fed growth (EFG)" and have nearly one hundred patents in existence or pending.

Here's how EFG works to produce long, continuous ribbons of silicon in a method that may reduce the cost of silicon solar cells. A die is lowered into molten silicon. The hot liquid rises through its center by capillary action. The silicon flows to the top of the die and no further. Shape of the die makes the silicon crystal into a ribbon. Dr. Mlavsky of Mobil Tyco has produced ribbons up to 30 feet long and expects to reach 100 feet by the end of 1975. A typical ribbon is 1 inch wide and should be about 0.004 inches thick. This thickness is sufficient to absorb all the solar radiation that a useful photovoltaic converter needs.

A parallel effort is to make sure that these silicon ribbons have a conversion efficiency of 10% or better. This is now being achieved, but the goal is to reach something like 16% efficiency and, correspondingly, reduce the area of solar cells needed for any specific installation.

Still another thrust of the Mobil Tyco research is to grow many long ribbons at once from a single crucible of molten silicon. This is going to take time, but the experience of Tyco with sapphire makes the goal look reasonable. Sapphire tubes for arc lamps are now mass-produced by Corning Glass under license from Tyco with eighteen sapphire tubes at a time. These sapphire arc tubes for efficient sodium vapor lamps are made at lower cost than ceramic arc tubes and are hence replacing them. The raw material is a semi-precious gem and more expensive than ceramic. But the manufacturing process is more efficient.

## Cost Estimates for Future Production

Mobil Tyco has estimated the cost of producing twenty silicon ribbons simultaneously with each ribbon being 2 inches wide and whatever length — perhaps 100 feet — seems economical. With the cooperation of major producers of the raw silicon, such as Dow Chemical, in reducing the cost of the silicon to perhaps $10 per pound, it seems as if manufacturing cost might come in at $15 per pound. This would mean a cost of $125 per kilowatt for silicon ribbons.

Another advantage of the ribbons is that this EFG material can be converted into solar cells, with appropriate packaging and leads for wiring, by automatic machinery. This again will result in savings.

As Dr. Mlavsky says: "Despite the enormous cost of solar cells developed for space applications, the overall process of converting sand (silicon dioxide) to solar cells is much less complex than, for example, the production of an automobile from iron ore, various other minerals, petrochemicals, etc. And the automobile sells for $2–$3 per pound at retail!"

So within a few years, after an investment estimated to be about one hundred million dollars in development and improved production machinery, we should have solar cells at a cost of $.50 or less per watt. Then this technique of generating electric power will be highly competitive with other methods. And our supplier of energy, the sun, is considered inexhaustible for at least another five billion years.

**STUDY QUESTIONS**

1. What part of this selection would you label narrative? Is this narrative portion necessary?
2. How would you characterize the audience Lucas is trying to inform here? How would you characterize the tone of this chapter? Do you think his tone is appropriate for the audience he is trying to reach?
3. Is his explanation of how a solar cell operates sufficiently detailed? What knowledge does he assume his reader possesses?
4. Lucas is very specific when he names companies and agencies at work on solar cells, listing the complete names of these companies and agencies and including the cities where they are located. Why does he give this information in such detail?

# Germs

## Lewis Thomas

Lewis Thomas's book *The Lives of a Cell* was a best seller and is considered a classic in the field of scientific writing. In several chapters of his book Thomas encourages his readers to re-examine their basic premises about life on this planet. His chapter on "Germs" attempts to set the record straight on these misunderstood microorganisms.

Watching television, you'd think we lived at bay, in total jeopardy, surrounded on all sides by human-seeking germs, shielded against infection and death only by a chemical technology that enables us to keep killing them off. We are instructed to spray disinfectants everywhere, into the air of our bedrooms and kitchens and with special energy into bathrooms, since it is our very own germs that seem the worst kind. We explode clouds of aerosol, mixed for good luck with deodorants, into our noses, mouths, underarms, privileged crannies — even into the intimate insides of our telephones. We apply potent antibiotics to minor scratches and seal them with plastic. Plastic is the new protector; we wrap the already plastic tumblers of hotels in more plastic, and seal the toilet seats like state secrets after irradiating them with ultraviolet light. We live in a world where the microbes are always trying to get at us, to tear us cell from cell, and we only stay alive and whole through diligence and fear.

We still think of human disease as the work of an organized, modernized kind

From Lewis Thomas, *The Lives of a Cell* (New York: The Viking Press, 1974), pp. 75–80.

of demonology, in which the bacteria are the most visible and centrally placed of our adversaries. We assume that they must somehow relish what they do. They come after us for profit, and there are so many of them that disease seems inevitable, a natural part of the human condition; if we succeed in eliminating one kind of disease there will always be a new one at hand, waiting to take its place.

These are paranoid delusions on a societal scale, explainable in part by our need for enemies, and in part by our memory of what things used to be like. Until a few decades ago, bacteria were a genuine household threat, and although most of us survived them, we were always aware of the nearness of death. We moved, with our families, in and out of death. We had lobar pneumonia, meningococcal meningitis, streptococcal infections, diphtheria, endocarditis, enteric fevers, various septicemias, syphilis, and, always, everywhere, tuberculosis. Most of these have now left most of us, thanks to antibiotics, plumbing, civilization, and money, but we remember.

In real life, however, even in our worst circumstances we have always been a relatively minor interest of the vast microbial world. Pathogenicity is not the rule. Indeed, it occurs so infrequently and involves such a relatively small number of species, considering the huge population of bacteria on the earth, that it has a freakish aspect. Disease usually results from inconclusive negotiations for symbiosis, an overstepping of the line by one side or the other, a biologic misinterpretation of borders.

Some bacteria are only harmful to us when they make exotoxins, and they only do this when they are, in a sense, diseased themselves. The toxins of diphtheria bacilli and streptococci are produced when the organisms have been infected by bacteriophage; it is the virus that provides the code for toxin. Uninfected bacteria are uninformed. When we catch diphtheria it is a virus infection, but not of us. Our involvement is not that of an adversary in a straightforward game, but more like blundering into someone else's accident.

I can think of a few microorganisms, possibly the tubercle bacillus, the syphilis spirochete, the malarial parasite, and a few others, that have a selective advantage in their ability to infect human beings, but there is nothing to be gained, in an evolutionary sense, by the capacity to cause illness or death. Pathogenicity may be something of a disadvantage for most microbes, carrying lethal risks more frightening to them than to us. The man who catches a meningococcus is in considerably less danger for his life, even without chemotherapy, than meningococci with the bad luck to catch a man. Most meningococci have the sense to stay out on the surface, in the rhinopharynx. During epidemics this is where they are to be found in the majority of the host population, and it generally goes well. It is only in the unaccountable minority, the "cases," that the line is crossed, and then there is the devil to pay on both sides, but most of all for the meningococci.

Staphylococci live all over us, and seem to have adapted to conditions in our skin that are uncongenial to most other bacteria. When you count them up, and us, it is remarkable how little trouble we have with the relation. Only a few of us are plagued by boils, and we can blame a large part of the destruction of tissues on the zeal of our own leukocytes. Hemolytic streptococci are among our closest intimates, even to the extent of sharing antigens with the membranes of our muscle cells; it is our reaction to their presence, in the form of rheumatic fever, that gets us into trouble. We can carry brucella for long periods in the cells of our reticuloendothelial system without any awareness of their existence; then cyclically, for reasons not understood but probably related to immunologic reactions on our part, we sense them, and the reaction of sensing is the clinical disease.

Most bacteria are totally preoccupied with browsing, altering the configurations of organic molecules so that they become usable for the energy needs of other forms of life. They are, by and large, indispensable to each other, living in interdependent communities in the soil or sea. Some have become symbionts in more specialized, local relations, living as working parts in the tissues of higher organisms. The root nodules of legumes would have neither form nor function without the masses of rhizobial bacteria swarming into root hairs, incorporating themselves with such intimacy that only an electron microscope can detect which membranes are bacterial and which plant. Insects have colonies of bacteria, the mycetocytes, living in them like little glands, doing heaven knows what but being essential. The microfloras of animal intestinal tracts are part of the nutritional system. And then, of course, there are the mitochondria and chloroplasts, permanent residents in everything.

The microorganisms that seem to have it in for us in the worst way — the ones that really appear to wish us ill — turn out on close examination to be rather more like bystanders, strays, strangers in from the cold. They will invade and replicate if given the chance, and some of them will get into our deepest tissues and set forth in the blood, but it is our response to their presence that makes the disease. Our arsenals for fighting off bacteria are so powerful, and involve so many different defense mechanisms, that we are in more danger from them than from the invaders. We live in the midst of explosive devices; we are mined.

It is the information carried by the bacteria that we cannot abide.

The gram-negative bacteria are the best examples of this. They display lipopolysaccharide endotoxin in their walls, and these macromolecules are read by our tissues as the very worst of bad news. When we sense lipopolysaccharide, we are likely to turn on every defense at our disposal; we will bomb, defoliate, blockade, seal off, and destroy all the tissues in the area. Leukocytes become more actively phagocytic, release lysosomal enzymes, turn sticky, and aggregate together in dense masses, occluding capillaries and shutting off the blood supply. Complement is switched on at the right point in its sequence to release chemotactic signals, calling in leukocytes from everywhere. Vessels become hyperreactive to epinephrine so that physiologic concentrations suddenly possess necrotizing properties. Pyrogen is released from leukocytes, adding fever to hemorrhage, necrosis, and shock. It is a shambles.

All of this seems unnecessary, panic-driven. There is nothing intrinsically poisonous about endotoxin, but it must look awful, or feel awful, when sensed by cells. Cells believe that it signifies the presence of gram-negative bacteria, and they will stop at nothing to avoid this threat.

I used to think that only the most highly developed, civilized animals could be fooled in this way, but it is not so. The horseshoe crab is a primitive fossil of a beast, ancient and uncitified, but he is just as vulnerable to disorganization by endotoxin as a rabbit or a man. Bang has shown that an injection of a very small dose into the body cavity will cause the aggregation of hemocytes in ponderous, immovable masses that block the vascular channels, and a gelatinous clot brings the circulation to a standstill. It is now known that a limulus clotting system, perhaps ancestral to ours, is centrally involved in the reaction. Extracts of the hemocytes can be made to jell by adding extremely small amounts of endotoxin. The self-disintegration of the whole animal that follows a systemic injection can be interpreted as a well-intentioned but lethal error. The mechanism is itself quite a good one, when used with precision and restraint, admirably designed for coping with intrusion by a single bacterium: the hemocyte would be attracted to the site, extrude the coagulable protein, the microorganism would be entrapped and immobilized, and the thing would be finished. It is when confronted by the overwhelming signal of free molecules of endotoxin, evoking memories of vibrios in great numbers, that the limulus flies into panic, launches all his defenses at once, and destroys himself.

It is, basically, a response to propaganda, something like the panic-producing pheromones that slave-taking ants release to disorganize the colonies of their prey.

I think it likely that many of our diseases work in this way. Sometimes, the mechanisms used for overkill are immunologic, but often, as in the limulus model, they are more primitive kinds of memory. We tear ourselves to pieces because of symbols, and we are more vulnerable to this than to any host of predators. We are, in effect, at the mercy of our own Pentagons, most of the time.

## STUDY QUESTIONS

1. Compare this essay with others you have read that are addressed to the general reader. Is this one easier or more difficult to understand than the other essays? Why?

2. Thomas uses several examples in his attempt to prove that germs are misunderstood. Which of these examples are the most effective? Why is this so?

3. For an article that presents very technical information "Germs" remains highly interesting. Why?

4. Go through the essay carefully and note the different analogies Thomas selects. Do these analogies seem to be similar in any way? Is there anything "natural" or "appropriate" about them?

5. *Writing Assignment:* This essay attempts to correct a popular misunderstanding about germs. Choose a topic that you feel is also misunderstood and write an article that attempts to correct this misunderstanding.

# Poison Ivy, Poison Oak, and Poison Sumac

## Donald M. Crooks and Dayton L. Klingman

The United States government prepares thousands of informative pamphlets for its citizens. The storehouse of knowledge that lies at one's fingertips in government pamphlets is something that not everyone is aware of, yet these publications, covering an amazing variety of topics, are usually free or sold for a very small fee. Here, for example, is what you can learn about poisonous plants from a government pamphlet.

Many people are accidentally poisoned each year from contact with plants that they did not know were harmful. If they had known how to recognize these poisonous plants, they could have escaped the painful experience of severe skin inflammation and water blisters. Few persons have sufficient immunity to protect them from poisonous plants. Many people do not recognize these plants although they occur in almost every part of the United States in one or more of their various forms.

Poisoning is largely preventable. A knowledge sufficient to identify plants in their various forms is easily gained by anyone who will study pictures and general descriptions, then train himself by diligent practice in observing the plants in his locality. Children should be taught to recognize the plants and to become poison ivy conscious.

### Poison Ivy and Poison Oak

Poison ivy and poison oak are neither ivy nor oak species. Rather, they belong to the cashew family and are known by a number of local names, and several different kinds of plants are called by these names. Plants vary greatly throughout the United States. They grow in the form of: (1) woody vines attached to trees or objects for support, (2) trailing shrubs mostly on the ground, or (3) erect woody shrubs entirely without support. They may flourish in the deep woods, where soil moisture is plentiful, or in very dry soil on the most exposed hillsides. Plants are most frequently abundant along old fence rows and edges of paths and roadway. They ramble over rock walls and climb posts or trees to considerable height. Often they grow with other shrubs or vines in such ways as to escape notice.

Leaf forms among plants or even on the same plant are as variable as the habit of growth; however, the leaves almost always consist of three leaflets. The old saying, "Leaflets three, let it be," is a reminder of this consistent leaf character, but may lead to undue suspicion of some harmless plant. Only one three-part leaf leads off from each node on the twig. Leaves never occur in pairs along the stem.

Donald M. Crooks and Dayton L. Klingman, "Poison Ivy, Poison Oak, and Poison Sumac," Farmers' Bulletin No. 1972, U.S. Department of Agriculture.

Flowers and fruit are always in clusters on slender stems that originate in the axils, or angles, between the leaves and woody twigs. Berrylike fruits usually have a white, waxy appearance and ordinarily are not hairy, but may be so in some forms. The plants do not always flower and bear fruit. The white or cream-colored clusters of fruit, when they occur, are significant identifying characters, especially after the leaves have fallen.

For convenience, these plants are discussed under three divisions — common poison ivy, poison oak, and Pacific poison oak.

## Common Poison Ivy[1]

The plant is known by various local names — poison ivy, three-leaved ivy, poison creeper, climbing sumac, poison oak, markweed, picry, and mercury. Common poison ivy may be considered as a vine in its most typical growth habit.

Vines often grow for many years, becoming several inches in diameter and quite woody. Slender vines may run along the ground, grow with shrubbery, or take support from a tree. A plant growing along the edge of a lawn and into the shrubbery may be inconspicuous as compared with a vine climbing on a tree. The vine develops roots readily when in contact with the ground or with any object that will support it. When vines grow on trees, these aerial roots attach the vine securely. A rank growth of these roots often causes the vines on trees to have the general appearance of a fuzzy rope.

The vines and roots apparently do not cause injury to the tree except where growth may cover the supporting plant and exclude sunlight. The vining nature of the plant makes it well adapted to climbing over stone walls or on brick and stone houses.

Poison ivy may be mixed in with ornamental shrubbery and vines. This often results in its cultivation as an ornamental vine by people who do not recognize the plant. An ivy plant growing on a house may be prized by an unsuspecting owner. The vine is attractive and sometimes turns a brilliant color in the fall. Do not tolerate its use as an ornamental. This use can result in cases of accidental poisoning, and these plants may serve as propagating stock for more poison ivy in the vicinity.

Poison ivy, mixed in with other vines, may be difficult to detect, unless you are trained in recognizing the plant. Virginia creeper and some forms of Boston ivy often are confused with it. You can recognize Virginia creeper by its five leaflets radiating from one point of attachment. Boston ivy with three leaflets is sometimes difficult to detect. Study a large number of Boston ivy leaves and you will usually find some that have only one deeply lobed blade or leaflet. Poison ivy has the three leaflets. A number of other plants are easily confused with poison ivy. Learn to know poison ivy on sight, through practiced observation, then make sure by looking at all parts of the suspected plant.

[1] *Rhus radicans*

Common poison ivy in full sunlight grows more as a shrub than as a vine along fence rows or in open fields. In some localities, the common form is a low-growing shrub that is 6 to 30 inches tall. Both forms usually have rather extensive horizontal systems of rootstocks or stems at or just below the ground level. Under some conditions, the vining form later becomes a shrub. Plants of this type may start as a vine supported on a fence and later extend upright stems that are shrub-like. In some localities, the growth form over a wide range is consistently either vine or shrub type. In other areas, common poison ivy apparently may produce either vines or shrubs.

Leaves of common poison ivy are extremely variable, but the three leaflets are a constant character. The great range of variation in the shape or lobing of the leaflets is impossible to describe. The five leaves shown in figure 1 give a fair range of patterns. Other forms may be found. One plant may have a large variety of leaf forms, or it may have all leaves of about the same general character. The most common type of leaf having leaflets with even margins is shown in figure 1, *A*. Other forms in figure 1 are not quite so widespread, but may be the usual type throughout some areas.

Most vines or shrubs of poison ivy produce some rather inconspicuous flowers (fig. 1, *A*) that are always in quite distinct clusters arising on the side of the stem immediately above a leaf (fig. 2, *A*). Frequently, the flowers do not develop or are abortive and no fruit is produced. Poison ivy fruits are white and waxy in appearance and have rather distinct lines marking the outer surface, looking like the segments in a peeled orange (fig. 2, *B*).

In some forms of poison ivy, the fruit is covered with fine hair, giving it a downy appearance; however, in the more common form fruits are entirely smooth. The fruit is especially helpful in identifying plants in late fall, winter, and early spring when the leaves are not present.

## Poison Oak[2]

Poison oak is more distinctive than some other types. Some people call it oakleaf ivy while others call it oakleaf poison ivy.

Poison oak usually does not climb as a vine, but occurs as a lowgrowing shrub. Stems generally grow upright. The shrubs have rather slender branches, often covered with a fine pubescence that gives the plant a kind of downy appearance. Leaflets occur in threes, as in other ivy, but are lobed, somewhat as the leaves of some kinds of oak. The middle leaflet usually is lobed alike on both margins and resembles a small oak leaf, while the two lateral leaflets are often irregularly lobed (fig. 3). The lighter color on the under side of one of the leaves is caused by the pubescence, or fine hairs, on the surface. The range in size of leaves varies considerably, even on the same plant.

---

[2] *Rhus toxicodendron*

Figure 1.

Figure 2.

## Pacific Poison Oak[3]

Pacific poison oak of the Pacific Coast States, usually known as poison oak, occasionally is referred to as poison ivy or yeara. This species is in no way related to the oak but is related to poison ivy.

The most common growth habit of western poison oak is as a rank upright shrub that has many small woody stems rising from the ground. It frequently grows in great abundance along roadsides and in uncultivated fields or an abandoned land.

Pacific poison oak sometimes attaches itself to upright objects for support and takes more or less the form of a vine. The tendency is for individual branches to continue an upright growth rather than to become entirely dependent on other objects for support. In some woodland areas, 70 to 80 percent of the trees support vines extending 25 to 30 feet in height.

In open pasture fields, Pacific poison oak usually grows in spreading clumps from a few feet to several feet tall. Extensive growth greatly reduces the area for grazing. It is a serious menace to most people who frequent such areas or tend cattle that come in contact with the plants while grazing.

Low-growing plants, especially those exposed to full sunlight, often are quite

[3] *Rhus diversiloba*

Figure 3.

woody and show no tendency for vining. These plants are common in pasture areas or along roadsides. Livestock in grazing do not invade the poison ivy shrub. As a rule, these plants spread both by root-stock and seed.

As in other poison ivy, leaves consist of three leaflets with much irregularity in the manner of lobing, especially of the two lateral leaflets. Sometimes lobes occur on both sides of a leaflet, giving it somewhat the semblance of an oak leaf. The middle, or terminal, leaflet is more likely to be lobed on both sides, and resembles an oak leaf more than the other two (fig. 4, *A*). Some plants may have leaflets with an even margin and no lobing whatsoever (fig. 4, *B*). The surface of the leaves is usually glossy and uneven, giving them a thick leathery appearance.

Figure 4.

Flowers are borne in clusters on slender stems diverging from the axis of the leaf. Individual flowers are greenish white and about one-fourth inch across. The cluster of flowers matures into greenish or creamy white berrylike fruits about mid-October. These are about the size of small currants and much like other poison ivy fruits. Many plants bear no fruit, although others produce it in abundance (fig. 5). Fruits sometimes have a somewhat flattened appearance. They remain on plants throughout fall and winter and help identify poison oak after leaves have fallen.

Figure 5.

## Poison Sumac

Poison sumac[4] grows as a coarse woody shrub or small tree and never in the vine-like form of its poison ivy relatives. This plant is known also as swamp sumac, poison elder, poison ash, poison dogwood, and thunderwood. It does not have variable forms, such as occur in poison oak or poison ivy. This shrub is usually associated with swamps and bogs. It grows most commonly along the margin of an area of wet acid soil.

Mature plants range in height from 5 or 6 feet to small trees that may reach 25 feet. Poison sumac shrubs usually do not have a symmetrical upright treelike appearance. Usually, they lean and have branched stems with about the same diameter from ground level to middle height.

Isolated plants occasionally are found outside swampy regions. These plants apparently start from seed distributed by birds. Plants in dry soil are seldom more than a few feet tall, but may poison unsuspecting individuals because single isolated plants are not readily recognized outside their usual swamp habitat.

[4] *Rhus vernix*

Leaves of poison sumac consist of 7 to 13 leaflets, arranged in pairs with a single leaflet at the end of the midrib (fig. 6).

Figure 6.

The leaflets are elongated oval without marginal teeth or serrations. They are 3 to 4 inches long, 1 to 2 inches wide, and have a smooth velvetlike texture. In early spring, their color is bright orange. Later, they become dark green and glossy on the upper surface, and pale green on the lower, and have scarlet midribs. In the early fall, leaves turn to a brilliant red-orange or russet shade.

The small yellowish-green flowers are borne in clusters on slender stems arising from the axis of leaves along the smaller branches. Flowers mature into ivory-white or green-colored fruits resembling those of poison oak or poison ivy, but usually are less compact and hang in loose clusters that may be 10 to 12 inches in length (fig. 7).

Because of the same general appearance of several common species of sumac and poison sumac, there is often considerable confusion as to which one is poisonous. Throughout most of the range where poison sumac grows, three non-poisonous species are the only ones likely to confuse. These are the smooth sumac[5], staghorn sumac[6], and dwarf sumac[7], which have red fruits that together form a distinctive terminal seed head. These are easily distinguished from the slender handing clusters of white fruit of the poison sumac. Sometimes more than one species of harmless sumac grow together.

## Introduced Poisonous Sumac and Related Species

The small Japanese lacquer-tree (*Rhus verniciflua*), uncommon in the United States, is related to native poison sumac. Native to Japan and China, it may be a source of

[5]*Rhus glabra*
[6]*Rhus typhina*
[7]*Rhus copallina*

Figure 7.

Japanese black lacquer. Poisoning has followed contact with lacquered articles. Never plant this tree.

A native shrub or small tree (*Metopium toxiferum*) called poisonwood, doctor gum, Metopium, Florida poison tree, or coral sumac is commonly found in the pinelands and hummocks of extreme southern Florida, the Keys, and the West Indies. It is much like, and closely related to, poison sumac. The shrub or small tree has the same general appearance as poison sumac. However, the leaves have only three to seven, more-rounded leaflets. Fruits are borne in clusters in the same manner as those of poison sumac, but they are orange colored and each fruit is two to three times as large. All parts of the plant are poisonous and cause the same kind of skin irritation as poison ivy or poison sumac.

When seed heads or flower heads occur on plants, it is easy to distinguish poisonous from harmless plants; however, in many clumps of either kind, flowers or fruit may not develop. The leaves have some rather distinct characteristics. Leaves of the three harmless species, often mistaken for poison sumac, are shown in figure 8.

Leaves of the smooth sumac (fig. 8, *A*) and of staghorn sumac (fig. 8, *B*) have many leaflets, which are slender and lance shaped and have a toothed margin. These

Figure 8.

species usually have more than 13 leaflets. Leaves of dwarf sumac (fig. 8, *C*) and poison sumac (fig. 8, *D*) have fewer leaflets; these are more oval shaped, and have smooth or even margins. The dwarf sumac is readily distinguished from poison sumac by a winged midrib. Poison sumac never has the wing margin on the midrib.

## Poisoning

Many people know through experience that they are susceptible to poisoning by poison ivy, poison oak, or poison sumac. Others, however, either have escaped contamination or have a certain degree of immunity. The extent of immunity appears to be only relative. In repeated contact with the plants, persons who have shown a degree of immunity may develop poisoning.

The skin irritant of poison ivy, poison oak, and poison sumac is a nonvolatile phenolic substance called urushiol, found in all parts of the plant. The danger of poisoning is greatest in spring and summer and least in late fall or winter.

Poisoning usually is caused by contact with some part of the bruised plant, as actual contact with the poison is necessary to produce dermatitis. A very small amount of the poisonous substance can produce severe inflammation of the skin. The poison is easily transferred from one object to another.

Clothing may become contaminated and is often a source of prolonged infection. Dogs and cats frequently contact the plants and carry the poison to children or other unsuspecting persons. The poison may remain on the fur of animals for a considerable period after they have walked or run through poison ivy plants.

Smoke from burning plants carries the toxin, and can cause severe cases of poisoning.

Children who have eaten the fruit have been poisoned although the fruit when fully ripe is reported as nonpoisonous. A local belief that eating a few leaves of the plant will develop immunity in the individual is unfounded. Never taste or eat any part of the plant.

Cattle, horses, sheep, hogs, and other livestock apparently do not get the skin irritation caused by these plants, although they graze on the foliage occasionally. Bees collect nectar from the flowers, but no ill effects from use of the honey have been reported.

The time between contamination of the skin and first symptoms varies greatly with individuals and probably with conditions. The first symptoms of itching or burning sensation may develop in a few hours or even after 7 days or more. The delay in development of symptoms is often confusing when an attempt is made to determine the time or location when contamination occurred. The itching sensation and subsequent inflammation that usually develops into water blisters under the skin may continue for several days from a single contamination. Persistence of symptoms over a long period is most likely caused by new contacts with plants, or by contact with previously contaminated clothing or animals.

Severe infection may produce more serious symptoms, which result in much pain through abscesses, enlarged glands, fever, or other complications.

If it is necessary to work among poisonous plants, some measure of prevention can be gained by wearing protective clothing. It is necessary, however, to remember that the active poison can be easily transferred. Some protection also may be obtained by using protective creams or lotions. They prevent the poison from contact with the skin, or make it easily removable by washing with soap and water, or neutralize it to a certain degree.

All measures to get rid of the poison must be taken within a few minutes after contact. A 10-percent water solution of potassium permanganate, obtainable in any drug store, usually is effective if applied within 5 to 10 minutes after exposure.

Many ointments and lotions are sold for prevention of poisoning by chemical or mechanical means. Their use should always be followed by repeated washings with soap and water to remove the contaminant.

Contaminated clothing and tools often are difficult to handle without causing further poisoning. Automobile door handles or steering wheels may, after trips to the woods, cause prolonged cases of poisoning among persons who have not been near the plants. Decontaminate such articles by thorough washing in several changes of strong soap and water. Do not wear contaminated clothing until it is thoroughly washed. Do not wash it with other clothes. Take care to rinse thoroughly any implements used in washing. Dry cleaning processes will probably remove any contaminant; but there is always danger that clothing sent to commercial cleaners may poison unsuspecting employees.

Dogs and cats can be decontaminated by washing; take care, however, to avoid poisoning while washing the animal.

There seems to be no absolute, quick cure for all individuals, even though many studies have been made to find effective remedies. Remedies may be helpful in removing the poison or rendering it inactive, and for giving some relief from the irritation. Mild poisoning usually subsides within a few days, but if the inflammation is severe or extensive, consult a physician.[8]

## Control by Mechanical Means

Poison ivy and poison oak can be grubbed out by hand quite readily early in spring and late in fall, only if a few plants are involved. Roots are most easily removed when soil is thoroughly wet. Grubbing when soil is dry and hard is almost futile because the roots break off in the ground, leaving large pieces that later sprout vigorously. Grubbing is effective if well done.

Poison ivy vines climbing on trees should be severed at the base, and as much of the vine as possible should be pulled away from the tree. Often the roots of the tree and weed are so intertwined that grubbing is impossible without injury to the tree. Bury or destroy roots and stems removed in grubbing, because the dry material is almost as poisonous as the fresh.

*Smoke from burning poison ivy plants or contaminated articles may carry the poison in a dispersed form. Take extreme caution to avoid inhalation or contact of smoke with the skin or clothing.*

Old plants of poison ivy produce an abundance of seeds, and these are freely disseminated, especially by birds. A poison ivy seedling 2 months old usually has a root that one mowing will not kill. Seedling plants at the end of the first year have well-established underground runners that only grubbing or herbicides will kill. Seedlings are a threat as long as old poison ivy is in the neighborhood.

Plowing is of little value in combating poison ivy and poison oak.

Mowing with a scythe or sickle is not an efficient means of controlling poison ivy and poison oak. It has little effect on the roots unless frequently repeated.

Weed burners are also inefficient in controlling poison ivy and poison oak.

## Control by Herbicides

Poison ivy and poison oak can be destroyed with herbicides without endangering the operator. One usually may stand at a distance from the plants and apply the herbicide without touching them. Most herbicides are applied as a spray solution by sprayers equipped with nozzles on extensions 2 feet or more in length. The greatest danger of poisoning occurs in careless handling of gloves, shoes, and clothing after the work is finished.

The most satisfactory herbicides for control of poison ivy, poison oak, and poison sumac are: (1) *amitrole* (3–amino-s-triazole); (2) *silvex* [2–(2.4,5–trichlorophenoxy) propionic acid] ; (3) *ammonium sulfamate;* and (4) *2,4-D* [2,4–(dichloro-

---

[8]See Public Health Service Publication No. 1723, Poison Ivy, Oak, and Sumac; for sale by the Superintendent of Documents, U.S. Government Printing Office, Washington, D.C. 20402. Price 5 cents.

phenoxy) acetic acid]. These herbicides are sold under their common names and under various trade names.

Any field or garden sprayer, or even a sprinkling can, can be used for applying the spray liquid, but a common compressed-air sprayer holding 2 to 3 gallons is convenient and does not waste the spray.

Use moderate pressure giving relatively large spray droplets, rather than high pressure giving a driving mist, because the object is to wet the leaves of the poison ivy and poison oak and avoid wetting the leaves of desirable plants. High pressures cause formation of many fine droplets that may drift to desirable plants.

Follow the manufacturer's recommendations shown on the container label in preparing the spray solution. Cover all foliage, stems, shoots, and bark of poison plants with herbicide spray. Although best results normally are obtained soon after maximum foliage development in the spring, applications may be made up to 3 weeks before fall frost is normally expected under good growing conditions in the humid areas.

Many herbicides used on poison ivy and poison oak will injure most broad-leaved plants. Apply them with caution if the surrounding vegetation is valuable. During the early part of the growing season, the leaves of poisonous plants usually tend to stand conspicuously apart from those of adjacent plants, and they can be treated separately if sprayed with care. Later the leaves become intermingled, and injury to adjacent species is unavoidable. Chemicals other than oil are not injurious to the thick bark of an old tree, and poison ivy clinging to the trunk safely can be sprayed with them. However, cutting the vine at the base of the tree and spraying regrowth may be more practical.

Apply sprays when there is little or no air movement. Early morning or late afternoon, when the air is cool and moist, usually is a favorable time.

No method of herbicidal eradiction can be depended on to kill all plants in a stand of poison ivy and poison oak with one application. Retreatments made as soon as the new leaves are fully expanded are almost always necessary to destroy plants missed the first time, to treat new growth, and to destroy seedlings. Plants believed dead sometimes revive after many months. An area under treatment must be watched closely for at least a year and retreated where necessary.

Dead foliage and stems remaining after the plants have been killed with herbicides are slightly poisonous. Cut off dead stems and bury or burn them, taking care to keep out of the smoke.

## Precautions

Herbicides used improperly may cause injury to man and animals. Use them only when needed and handle them with care. Follow the directions and heed all precautions on the labels.

Keep herbicides in closed, well-labeled containers in a dry place. Store them where they will not contaminate food or feed, and where children and animals cannot reach them.

When handling a herbicide, wear clean, dry clothing.

Avoid repeated or prolonged contact of herbicides with your skin.

Wear protective clothing and equipment if specified on the container label. Avoid prolonged inhalation of herbicide mists.

Avoid spilling herbicide concentrate on your skin, and keep it out of your eyes, nose, and mouth. If you spill any on your skin, wash it off immediately with soap and water. If you spill it on your clothing, remove clothing immediately and wash contaminated skin. Launder the clothing before wearing it again.

After handling a herbicide, do not eat, drink, or smoke until you have washed your hands and face. Wash any exposed skin immediately after applying a herbicide.

Avoid drift of herbicide to nearby crops.

To protect water resources, fish, and wildlife, do not contaminate lakes, streams, or ponds with herbicide. Do not clean spraying equipment or dump excess spray material near such water.

It is difficult to remove all traces of herbicides from equipment. For this reason, do not use the same equipment for applying herbicides that you use for insecticides and fungicides.

Dispose of empty herbicide containers at a sanitary land-fill dump, or crush and bury them at least 18 inches deep in a level, isolated place where they will not contaminate water supplies. If you have trash-collection service, wrap small containers in heavy layers of newspapers and place them in the trash can.

## Contents

**STUDY QUESTIONS**

1. What segment of the population might be most interested in this document? Does the pamphlet seem designed to best appeal to this group?
2. Would it be better, in your opinion, if this information were presented in several smaller pamphlets? What might these pamphlets be and what might they contain?
3. Notice the overall organization of this pamphlet in its table of contents. Is the section on "Poisoning" located in the best place? Where would you suggest this information be presented? What other changes would you make in the ordering?

# Ocean Thermal Energy Conversion

## William G. Pollard

The Institute for Energy Analysis is a division of Oak Ridge Associated Universities, a group of forty-six colleges and universities that conduct research programs for government and private industry. The work of the Institute for Energy Analysis involves the study of national and international energy issues. This report on thermal conversion of ocean thermal energy was prepared by the Institute under contract with the U.S. Department of Energy.

## Abstract

Ocean thermal energy conversion (OTEC) systems are briefly described, as well as some of the engineering problems encountered in their development. Such systems utilize stable thermal gradients in tropical oceans for thermal input to a closed cycle for electric generation. Thermal-to-electric conversion efficiencies on the order of 2 percent are contemplated. Large areas of heat exchangers in marine service are required, and primary objectives of government R&D are concerned with biofouling and corrosion of heat exchanger surfaces and materials. Transmission of power from the floating OTEC station at sea to load centers on land is by submarine cable or liquid hydrogen for longer distances. This paper provides a general description and assessment of OTEC systems to judge their future utilization in electric utility systems. A bibliography of detailed studies and engineering designs is provided for those desiring more specific assessments of particular aspects.

## Ocean Thermal Energy Conversion

The oceans of the earth constitute an immense reservoir for the storage of solar energy. Insolation in the surface of the oceans leads to an equilibrium between thermal input and output which maintains stable thermal gradients. In tropical oceans at latitudes between ±20°, the temperature difference between warm surface water and cold water at depths of 1,500–3,000 feet varies from approximately 32°F to 40°F. This difference is quite stable and remains essentially constant throughout the year. At higher latitudes, the temperature differences are smaller and the temperature of the surface water varies by increasing amounts between summer and winter as the latitude increases.

An ocean thermal energy conversion (OTEC) system is one which seeks to tap this immense solar-maintained heat reservoir for the generation of electricity using the deep cold water as the sink for a Rankine cycle driving a turbine and generator.

William G. Pollard, "Ocean Thermal Energy Conversion," Institute for Energy Analysis, Oak Ridge Associated Universities, 1978.

Because of its low efficiency, such a system requires the pumping of large quantities of water and is correspondingly massive compared with fuel-fired systems or other systems benefiting from larger temperature differences.

A design of such a system prepared by Lockheed in collaboration with the Bechtel Corporation and T. Y. Lin International consists of an anchored spar of reinforced concrete construction with a displacement of 300,000 tons on which are mounted eight ammonia-drive 30 MW(e) turbogenerators. Cold water is drawn through a telescoping concrete pipe 1,000 feet long, with a diameter tapered from 126 feet to 102 feet, to a condenser consisting of 120,000 titanium tubes 2 inches in diameter and 5.2 feet long. The cold water passes through the tubes and the ammonia outside. Water flow through the condenser is 120,000 gal/sec which is approximately the flow through Boulder Dam. The evaporator heat exchanger, also of titanium, is the same size and is fed by warm surface water. A third of the power generated is used for pumping, and the net output of the system is 160 MW(e). The design calls for 263,000 tons of concrete and 29,000 tons of steel for the supporting structure; 36,000 tons of concrete for the cold water pipe; 12,000 tons of steel for the mooring line; and 8,700 tons of cold rolled steel and 8,600 tons of titanium for the heat exchangers and power system. A subsequent modification of this design extends the cold water pipe to 2,500 feet in length, optimizes the relative flows of seawater to the evaporators and condensers, and substitutes aluminum coil panel heat exchangers for titanium. The net power output for this modified design is increased to 250 MW(e) at a more favorable cost. A number of other fairly detailed designs of OTEC systems have been carried out.

The most difficult problem in the development of OTEC systems has to do with the heat exchangers. Heat exchangers of a Cu-Ni alloy have been used for a long time in marine service for conventional power plants with condenser cooling from the sea. This material is, however, incompatible with ammonia but could be used with propane and some other possible working fluids. For long-term stability in a marine environment, titanium appears to be best, but a single OTEC unit would exhaust the entire annual U.S. production. Aluminum would be much cheaper, but corrosion in a marine environment could be serious with it, although extensive research is devoted to the development of corrosion-resistant aluminum alloys. Numerous small leaks between seawater and ammonia over a very large area could be a severe problem.

The goal of most designs is a 2–3 percent thermal-to-electric conversion efficiency, but any slippage in heat exchanger performance would degrade the entire thermal cycle and could easily result in no electric power generation. The problem of biofouling of marine hardware by slime growth has been extensively studied. OTEC experiments carried out so far indicate that only one-thousandth of an inch of slime would reduce the plant's performance by 60 percent. There appears to be no practical way to prevent slime formation so it would have to be cleaned off. The Lockheed design includes a means for doing so that uses air pressure to force rubber balls through the titanium tubes. In addition, if an OTEC plant were shut down for repairs, barnacles would attach to the surfaces and continue growing after operations were resumed. The cost of removing such an incrustation from a whole OTEC system could be insupportable.

At 2 percent efficiency, approximately 30,000 Btu per hour must be exchanged for each kilowatt hour generated (170,000 Btu into the evaporator, 170,000 back out in the condenser). Typical heat exchangers in marine service will transfer between 300 and 450 Btu per hour per ft$^2$ per degree Fahrenheit. The temperature drop across the exchangers is about 5°F so the heat exchanger area for each kW(e) gross capacity at 2 percent efficiency is between 150 and 225 ft$^2$. The Lockheed design calls for 6,600,000 ft$^2$ or 150 acres of titanium heat exchangers. It is possible that vigorous research and development (R&D) may improve heat-transfer coefficients, but corrosion and slime accumulation would of course reduce them. In any event, the areas involved are very large and their maintenance and repair could prove in practice to be quite costly.

Some of the most promising locations for OTEC systems from the standpoint of available thermal gradients are the South Atlantic equatorial current area, the Gulf of Panama, Micronesia, and the northwest coast of Australia south of Java. From the standpoint of demand, this source of electricity will be of greatest interest for locations along the coast of Brazil, coastal areas in equatorial Africa, Melanesia, Indonesia, Australia, and Micronesia. For the United States and Mexico, OTEC systems have been considered for Puerto Rico, Hawaii, the Atlantic coast of Southern Florida, and for coastal sites along the Gulf of Mexico, although the latitude of all these areas makes them hurricane prone and would involve seasonal variations in output.

Other problems in the OTEC development program are the assembly and security of the large cold water pipe up to half a mile in length, damage to or even loss of whole systems in higher latitudes from severe ocean storms, and the delivery of power from its site of generation to demand centers on land. This last problem is perhaps the most serious for the commercial feasibility of such systems. If the OTEC system can be operated close enough to shore for transmission by submarine cable, this problem is not severe. But matching such sites with load centers that need to be supplied limits this application to a minor contribution to total electric supply. For a general contribution to electric power demand, a long-range transmission of the energy generated is essential. This could be accomplished by on-board electrolysis followed by hydrogen liquefaction with transport of the liquid hydrogen in cryogenic tankers. At generation facilities on land, the hydrogen could be used in fuel cells followed by DC-AC inversion. Since about one-third of the energy content of hydrogen is required to liquefy it, the on-board efficiency of this process with 85–90 percent efficient electrolytic cells is around 60 percent. At the generation sites on land, the fuel cells have an efficiency of 50 percent and DC-AC inverters, 80–90 percent. Thus the overall efficiency of converting on-board power to land-site power by this means is about 25 percent not counting losses. This means that the cost per kW(e) capacity on land is about four times that of the net on-board capacity exclusive of the cost of the electrolysis and fuel cells, hydrogen liquefaction and transport, and DC-AC inversion.

The cost of power from OTEC systems cannot be determined until all of these problems are solved and contracts are let for the construction of such systems. The Lockheed estimate for the capital cost of the first unit under shipyard construction conditions was $3,610 per kW(e), and for an expanded multiunit construction schedule, $2,621 per kW(e) exclusive of the cost of power transmission to shore. If such transmission is by hydrogen, the total capital cost is increased to between $11,000 and $15,000 per net on-land electrical kilowatt. At this stage of development, such an estimate is subject to wide uncertainties and the capital cost of actual operational systems could be several times greater. Operation and maintenance costs can only be determined on the basis of some operating experience with such unconventional systems. Although such systems do seem to be a technically possible means of generating electricity, their commercial feasibility appears very dubious. Compared with the intermittency of direct solar generation, they have the great advantage of availability whenever needed.

Because of favorable sites for such systems, the Florida Power and Light Company has studied OTEC perhaps more than any other utility and has found the arguments for its eventual commercial feasibility unconvincing. The Electric Power Research Institute is also not interested and does not include any funding for OTEC in its solar program. The Department of Energy is, however, pursuing the concept vigorously with the goal of eventually constructing a demonstration plant.

The effect of a large number of OTEC systems on the thermal equilibrium of the ocean in the area in which they are located is a matter of possible concern. A $2°F$ reduction in surface temperature would reduce the long-wave heat reradiation to space by a little over 1 percent and result in a 6 percent decrease in the rate of evaporation. At 2 percent efficiency, 10 billion watts of OTEC power would result in the transfer of heat from the surface to the deep ocean at the rate of 500 billion watts (thermal). Ultimately the ocean must reach a new steady state in which this additional heat input to the deep water is continuously removed. The extent to which this new equilibrium might shift established ocean currents or alter established thermal gradients is difficult to evaluate. On the other hand, it is equally difficult to visualize a world in which the demand for OTEC power would be sufficiently great to generate the immense capital investment needed for the deployment of large numbers of such systems.

**BIBLIOGRAPHY**

Avery, W. H. et al. (1976). *Maritime and Construction Aspects of Ocean Thermal Energy Conversion (OTEC) Plant Ships: Detailed Report,* Applied Physics Laboratory, NTIS MA–RD–940–T76074, Johns Hopkins University, Laurel, Maryland.

Florida Solar Energy Center (1977). *Ocean Thermal Energy Conversion (OTEC)*

*Resource and Environmental Assessment Workshop, Proceedings,* RD-77-2, Cape Canaveral, Florida, October.

Lockheed Missiles and Space Company, Inc. (1975). *Ocean Thermal Energy Conversion (OTEC) Power Plant Technical and Economic Feasibility,* 2 volumes, LMSC-DO56566, Sunnyvale, California.

"Ocean Thermal Energy: The Biggest Gamble in Solar Power," SCIENCE, *198,* October 14, 1977, pp. 178-180.

McGowan, J. G. (1976). "Review Paper, Ocean Thermal Energy Conversion — A Significant Solar Resource," SOLAR ENERGY, *18,* pp. 81-92.

Perry, A. M. et al. (1977). *Net Energy Analysis of Five Energy Systems,* Chapter 2, "OTEC," Institute for Energy Analysis, Oak Ridge Associated Universities, ORAU/IEA(R)-77-12, Oak Ridge, Tennessee.

TRW (1975). *Ocean Thermal Energy Conversion: Final Report,* 5 volumes, Redondo Beach, California.

United Engineers and Constructors, Inc. (1975). *A Conceptual Feasibility and Cost Study for a 100 MW(e) Sea Solar Power Plant,* Philadelphia, Pennsylvania.

U.S. Energy Research and Development Administration, Division of Solar Energy (1976). *Ocean Thermal Energy Conversion (OTEC), Program Summary,* ERDA 76-142, Washington, D.C., October.

Winer, B. M. and J. Nicol (1977). "Electrical Energy Transmission from OTEC Power Plants," in *Proceedings: Fourth OTEC Conference,* University of New Orleans, New Orleans, Louisiana.

## STUDY QUESTIONS

1. Essentially this report is a summary of information that can be found in the works listed in the bibliography. How would you describe the style in which the report is written? How does the style reflect that this is a summary?
2. Outline this report. Is outlining it particularly difficult? Why or why not?
3. Examine the scope and type of information found in "Ocean Thermal Energy Conversion." How would this report be used? Might it be used to persuade or implement as well as inform?

# Section 2

## Persuading

When you prepare proposals, feasibility studies, and recommendations, you will be attempting to convince your reader of the correctness of your point of view. You will be trying to persuade someone to take action, follow your plan, buy your product or service, or simply believe as you do. Traditionally, however, *persuasion* is a term that is associated with an appeal to emotion. Strictly emotional appeals are suspect in scientific and technical writing because most scientists and technicians deal with factual data all day long; obviously these people feel that issues are rightly resolved by *facts* and *logic* rather than by *opinion* and *emotion*. In the business world we also expect people to be impressed largely with factual data — production costs, development schedules, rates of return — rather than emotional appeals. It is only in certain segments of the business world, such as marketing and advertising, where we find writers devoting their energies to strictly emotional appeals. Even here, however, these emotional appeals are directed only to a specific audience — the purchasing public. In-house proposals and recommendations still must be based on facts rather than catchy slogans or jingles.

The kind of persuasive writing normally practiced in business, science, and industry, then, is somewhat different from the persuasive writing that may be practiced in other fields, for the persuasive writing you will engage in will stick to the facts, resist emotional appeals, and never distort the truth (even a little) in order to influence opinion. Technically, then, what we are calling *persuasive* writing here, is sometimes called *argumentative* writing, for *argumentative* writing (according to this distinction) resists emotional appeals and sticks to the facts. To call the writing we find in a proposal *argumentative* is also slightly misleading, however, for that label suggests the type of writing that is intended to resolve a debate or confrontation. The kind of influential writing you will likely produce does not fit either of the traditional categories neatly. A sound proposal sticks to the facts, but it does not literally argue; a solid recommendation is intended to persuade, but it does not use an emotional appeal to achieve that persuasion.

If one must stick to the facts, you may say, in what sense can a writer persuade? That is, if an emotional appeal is not allowed, what can a writer do to influence the reader, beyond simply presenting the data and letting the facts speak for themselves? The major technique that can be employed involves the emphasis of data, presumably the data you feel is most important to your case. Emphasis is achieved in several ways: 1.) through repetition, which is obviously a dangerous practice in any kind of writing; 2.) through *proportion,* by elaborating at length on the things you feel are important, and giving a brief treatment to those things you do not wish to call to your reader's attention; 3.) through *position,* which involves manipulating the order in which you present your materials. Of these methods, both emphasis by proportion and emphasis by position are sound strategies in technical communication, but emphasis by position is far preferable, because such a strategy is obviously more objective; by arranging evidence carefully you can be persuasive without distorting the truth.

When you are writing a proposal or a recommendation, there will probably be several main points that you feel are most important to your position. These points in your presentation will presumably be supported with solid technical data. And, if you are honest, there may be some aspects of your proposal that are not really as strong as you would like them to be; there may in fact be some negative aspects that you would prefer not mentioning, for these factors seem to contradict the conclusions you want your reader to draw. This is where arranging your evidence becomes important. Studies have shown that readers, after finishing a report or an article, tend to remember that material they read last, and tend to forget the material they read first. Such a discovery is not at all surprising; we simply remember best those things that are closest to us in time. Here then we have a simple rhetorical strategy. To leave a lasting impression upon your reader and influence him, place your strongest evidence *last* in your presentation of data; place your weakest points *before* your stronger ones. If there is any contradictory information, that should be placed *first* of all, where your reader (you hope) will eventually forget it.

Do not be afraid to include this negative data. Even in a persuasive report you are obligated to be objective. Omitting such material will not be objective, and will of course be distorting the truth. Including this evidence can actually assist you in influencing your reader favorably, for it demonstrates that you are being open-minded and fair, and have examined all aspects of the subject. Presenting and acknowledging such negative data also demonstrates the overall strength of your proposal, for it asserts that you believe strongly in a particular position in the face of certain evidence which must, therefore, by implication truly be essentially unimportant and inconsequential. Besides, a person who seems to be stating flatly "I am always right; believe me and trust me," is not a person we feel inclined to believe or trust, for few positions lack weaknesses. Briefly acknowledging your proposal's weaknesses, therefore, can, paradoxically, make your attempt at persuasion more plausible and influential, rather than less so.

# Asbestos, Chloroform, Peanuts, and the Deerfly

## Phillip J. Wingate

Phillip J. Wingate is a retired vice-president of the Dupont Corporation, a company whose name is synonymous with chemicals and "better living through chemistry." Naturally enough, Dupont is concerned with public opinion in an age when environmentalists and consumer advocates have made *chemical* a dirty word. This essay is taken from *Dupont Magazine,* a publication directed primarily to employees of the company.

During the past decade or two, a growing mass of scientific information has emerged identifying new hazards related to the manufacture and use of chemicals. For example, it has been found that vinyl chloride, long thought to be a harmless chemical, can cause cancer many years after exposure occurs. Similarly, the tragic effects that resulted from the use, by pregnant women, of a sedative known as Thalidomide have been documented.

New scientific insights and knowledge were needed in proving that some chemicals, once thought to pose no·threat to human health or the environment, were not always the harmless materials they seemed to be. Such revelations by the scientific community are invaluable and must be encouraged and supported. But at the same time, it must be recognized that these revelations have generated much unwarranted public alarm. Some people call for a total and immediate ban of "all these chemical assassins." Take us back, they say, to a zero-risk world where no chemicals threaten us. Such demands ignore the fact that a zero-risk world, free of chemicals, never did exist and never can.

The world around us is a seething mass of chemicals and always has been. People consist of chemicals, and so do the oceans and continents around us and the air above us. Trees and other green plants on earth discharge over four hundred billion pounds of complex organic chemicals into the air each year and most, in the wrong amounts and concentrations, are hazardous to people. How can they be banned?

Similarly, the freshest fruits and vegetables, grown using only natural fertilizers, are filled with an astonishing array of chemicals — ketones, esters, lactones,

Phillip J. Wingate, "Asbestos, Chloroform, Peanuts, and the Deerfly," *Dupont Magazine,* September/October 1977, Inside Cover.

acids, amines, amino acids, alcohols, mercaptans, terpenes, metal ions and chelates of these ions. And again, most of them, in the wrong concentrations and amounts, are hazardous to people. Should they all be banned? If so, how?

Finally, there is the matter of peanuts. These lowly delicacies usually contain, clinging to them, a few molecules of two potent carcinogens — aflatoxin B and aflatoxin G. These two carcinogens are produced by molds and if the peanuts are harvested and dried properly, the amount of the two aflatoxins left on the peanuts is usually vanishingly small but, nevertheless, higher than would be permitted in a zero-risk world.

To ignore the harm which may lurk within either natural or synthetic chemicals would be irresponsible, but quantities and concentrations simply must be a part of whatever precautions are proposed.

Consider the case of asbestos. It is a natural product — an inorganic chemical that will not burn and does not irritate the skin of people working with it. A hundred years ago asbestos was thought to be "safe" in every sense of the word. Now we know that exposure to asbestos dust can lead to two forms of lung cancer — bronchiogenic carcinoma and mesothelioma — ten or twenty years after the exposure occurs.

Now that this hazard is known — and it is tragic that it wasn't known twenty-five years ago — defensive measures can be taken. Precautions in production and handling will reduce the hazard. But no matter how complete the precautions, the hazards of asbestos cannot be reduced to zero because asbestos exists in nature and cannot be made to disappear.

Chloroform is another interesting case in the anthology of chemical hazards. It is a man-made product and was once considered to be so safe that it was used as an anesthetic and as a solvent in medicines and cough syrups. Now we know that chloroform, in substantial amounts, can damage the liver and perhaps do other harm. Clearly this is a good thing to know, but some people have overreacted. Eager to achieve zero risk, they would prohibit one part per billion of chloroform in a meat wrapper. Nothing in the long history of chloroform justifies such alarm.

The cases of asbestos, chloroform, and peanuts underscore the need for a blend of scientific knowledge and common sense. Science must be encouraged to explore and define, as quantitatively as possible, all chemical hazards which may exist. Then society must apply the lubricant of common sense or the wheels of progress will grind to a halt.

Most Americans have no problem with science and tend to hold scientists in awe. But sadly, even the lofty pronouncements of scientists should be studied with the same wary eye which must be applied to the rest of us.

The legend of the deerfly, as told by the author John B. White in "A Fly Too Fast To Die," shows the need of wariness even when dealing with scientists. In 1926, Dr. Charles Townsend, an entomologist with a Ph.D. degree from a respected university, wrote an article for *Scientific Monthly.* In it he discussed the possibility of flying around the world in a single day. At the time, 250 mph was fast for an airplane. But Dr. Townsend wrote that a globe-encircling flight in one day could be done if airplanes could be designed to fly as fast as the deerfly which, he said, buzzed around deer at speeds up to 400 yards per second.

The piece of information caught the eye of scientists and soon the deerfly — or Cephenomyia as the entomologists called him — was widely reported in scientific literature as the fastest living thing on earth, because 400 yards per second figures out to be 818 miles per hour. Perhaps people were fascinated by this speed because it is faster than the speed of sound — so the deer couldn't hear the deerfly coming, only going away — or because Dr. Townsend said the female deerfly could do only 600 miles per hour since "the male has to catch her." The deerfly went zooming through the journals of science at supersonic speeds for more than ten years.

Then Dr. Irving Langmuir, 1932 Nobel prize winner in chemistry, ran across the trail of Cephenomyia and made some calculations which convinced him that the story was a phony. Langmuir calculated that to attain 818 mph the deerfly would have to generate half a horsepower and consume 1.5 times his own weight in fuel per second. What's more, at 818 mph the air pressure on the deerfly would mash him flat as a pancake.

When Langmuir published his conclusions, there were some red faces among the entomologists. Some of them then made independent checks on the speed of the deerfly and concluded that he really could do only about 25 to 30 mph. But this mighty swat by Langmuir did not kill the deerfly; it only knocked him down for a while. Soon he was once more zooming through the journals of science at a great rate. *General Entomology*, published in 1959, credited him with supersonic speed and *Webster's Third International Dictionary*, published in 1961, clocked old Cephenomyia at "more than 800 mph." Clearly, when scientists make a mistake, it is not always corrected promptly.

The moral: Science is invaluable in helping to lift the veil of ignorance, still the world's greatest hazard. But before any one scientist is authorized to hoist his flag and lead humanity in a stampede toward a zero-risk world, other scientists should be given time to scrutinize his findings. He just might be wrong — and one fly too fast to die is enough.

## STUDY QUESTIONS

1. Wingate begins his essay by discussing the hazards that have been associated with "the manufacture and use of chemicals." In the course of his discussion, however, he identifies what he regards as a greater hazard, what he calls "the world's greatest hazard," in fact. What is that hazard?

2. Does Wingate argue fairly? Does he acknowledge the opposition's arguments?

3. How would you characterize Wingate's persuasive strategy? What information does he present? What facts does he omit? What questions does he focus on closely? What questions does he not discuss?

4. It might be said that this article has an overt and a covert message. What is its overt message? What is its covert message, the message that is not boldly stated but is always implied?

# Parkinson's Law or the Rising Pyramid

## C. Northcote Parkinson

Whenever there is work to be done, even in an automated society, people must be hired to carry out some phase of that work. Completing a job with a minimum of waste and a maximum of efficiency is a need all companies face, so the problems associated with hiring staff are perennial ones for business and industry. Doing work efficiently is essentially the issue that C. N. Parkinson explores in this article. The questions he raises about the nature of work have implications for us all.

Work expands so as to fill the time available for its completion. General recognition of this fact is shown in the proverbial phrase "It is the busiest man who has time to spare." Thus, an elderly lady of leisure can spend the entire day in writing and dispatching a postcard to her niece at Bognor Regis. An hour will be spent in finding the postcard, another in hunting for spectacles, half an hour in a search for the address, an hour and a quarter in composition, and twenty minutes in deciding whether or not to take an umbrealla when going to the mailbox in the next street. The total effort that would occupy a busy man for three minutes all told may in this fashion leave another person prostrate after a day of doubt, anxiety, and toil.

Granted that work (and especially paperwork) is thus elastic in its demands on time, it is manifest that there need be little or no relationship between the work to be done and the size of the staff to which it may be assigned. A lack of real activity does not, of necessity, result in leisure. A lack of occupation is not necessarily revealed by a manifest idleness. The thing to be done swells in importance and complexity in a direct ratio with the time to be spent. This fact is widely recognized, but less attention has been paid to its wider implications, more especially in the field of public administration. Politicians and taxpayers have assumed (with occasional phases of doubt) that a rising total in the number of civil servants must reflect a growing volume of work to be done. Cynics, in questioning this belief, have imagined that the multiplication of officials must have left some of them idle or all of them able to work for shorter hours. But this is a matter in which faith and doubt seem equally misplaced. The fact is that the number of the officials and the quantity of the work are not related to each other at all. The rise in the total of those employed is governed by Parkinson's Law and would be much the same whether the volume of the work were to increase, diminish, or even disappear. The importance of Parkinson's Law lies in the fact that it is a law of growth based upon an analysis of the factors by which that growth is controlled.

The validity of this recently discovered law must rest mainly on statistical proofs, which will follow. Of more interest to the general reader is the explanation of the factors underlying the general tendency to which this law gives definition. Omitting technicalities (which are numerous) we may distinguish at the outset

C. N. Parkinson, *Parkinson's Law, and Other Studies in Administration* (Boston: Houghton Mifflin, 1975), pp. 2–13.

two motive forces. They can be represented for the present purpose by two almost axiomatic statements, thus: (1) "An official wants to multiply subordinates, not rivals" and (2) "Officials make work for each other."

To comprehend Factor 1, we must picture a civil servant, called A, who finds himself overworked. Whether this overwork is real or imaginary is immaterial, but we should observe, in passing, that A's sensation (or illusion) might easily result from his own decreasing energy: a normal symptom of middle age. For this real or imagined overwork there are, broadly speaking, three possible remedies. He may resign; he may ask to halve the work with a colleague called B; he may demand the assistance of two subordinates, to be called C and D. There is probably no instance in history, however, of A choosing any but the third alternative. By resignation he would lose his pension rights. By having B appointed, on his own level in the hierarchy, he would merely bring in a rival for promotion to W's vacancy when W (at long last) retires. So A would rather have C and D, junior men, below him. They will add to his consequence and, by dividing the work into two categories, as between C and D, he will have the merit of being the only man who comprehends them both. It is essential to realize at this point that C and D are, as it were, inseparable. To appoint C alone would have been impossible. Why? Because C, if by himself, would divide the work with A and so assume almost the equal status that has been refused in the first instance to B; a status the more emphasized if C is A's only possible successor. Subordinates must thus number two or more, each being thus kept in order by fear of the other's promotion. When C complains in turn of being overworked (as he certainly will) A will, with the concurrence of C, advise the appointment of two assistants to help C. But he can then avert internal friction only by advising the appointment of two more assistants to help D, whose position is much the same. With this recruitment of E, F, G, and H the promotion of A is now practically certain.

Seven officials are now doing what one did before. This is where Factor 2 comes into operation. For these seven make so much work for each other that all are fully occupied and A is actually working harder than ever. An incoming document may well come before each of them in turn. Official E decides that it falls within the province of F, who places a draft reply before C, who amends it drastically before consulting D, who asks G to deal with it. But G goes on leave at this point, handing the file over to H, who drafts a minute that is signed by D and returned to C, who revises his draft accordingly and lays the new version before A.

What does A do? He would have every excuse for signing the thing unread, for he has many other matters on his mind. Knowing now that he is to succeed W next year, he has to decide whether C or D should succeed to his own office. He had to agree to G's going on leave even if not yet strictly entitled to it. He is worried whether H should not have gone instead, for reasons of health. He has looked pale recently — partly but not solely because of his domestic troubles. Then there is the business of F's special increment of salary for the period of the conference and E's application for transfer to the Ministry of Pensions. A has heard that D is in love with a married typist and that G and F are no longer on speaking terms — no one seems to know why. So A might be tempted to sign C's draft and have done with it.

But A is a conscientious man. Beset as he is with problems created by his colleagues for themselves and for him — created by the mere fact of these officials' existence — he is not the man to shirk his duty. He reads through the draft with care, deletes the fussy paragraphs added by C and H, and restores the thing back to the form preferred in the first instance by the able (if quarrelsome) F. He corrects the English — none of these young men can write grammatically — and finally produces the same reply he would have written if officials C to H had never been born. Far more people have taken far longer to produce the same result. No one has been idle. All have done their best. And it is late in the evening before A finally quits his office and begins the return journey to Ealing. The last of the office lights are being turned off in the gathering dusk that marks the end of another day's administrative toil. Among the last to leave, A reflects with bowed shoulders and a wry smile that late hours, like gray hairs, are among the penalties of success.

From this description of the factors at work the student of political science will recognize that administrators are more or less bound to multiply. Nothing has yet been said, however, about the period of time likely to elapse between the date of A's appointment and the date from which we can calculate the pensionable service of H. Vast masses of statistical evidence have been collected and it is from a study of this data that Parkinson's Law has been deduced. Space will not allow of detailed analysis but the reader will be interested to know that research began in the British Navy Estimates. These were chosen because the Admiralty's responsibilities are more easily measurable than those of, say, the Board of Trade. The question is merely one of numbers and tonnage. Here are some typical figures. The strength of the Navy in 1914 could be shown as 146,000 officers and men, 3249 dockyard officials and clerks, and 57,000 dockyard workmen. By 1928 there were only 100,000 officers and men and only 62,439 workmen, but the dockyard officials and clerks by then numbered 4558. As for warships, the strength in 1928 was a mere fraction of what it had been in 1914 — fewer than 20 capital ships in commission as compared with 62. Over the same period the Admiralty officials had increased in number from 2000 to 3569, providing (as was remarked) "a magnificent navy on land." These figures are more clearly set forth in tabular form.

**Admiralty Statistics**

| Year | Capital ships in commission | Officers and men in R.N. | Dockyard workers | Dockyard officials and clerks | Admiralty officials |
|---|---|---|---|---|---|
| 1914 | 62 | 146,000 | 57,000 | 3249 | 2000 |
| 1928 | 20 | 100,000 | 62,439 | 4558 | 3569 |
| *Increase or Decrease* | −67.74% | −31.5% | +9.54% | +40.28% | +78.45% |

The criticism voiced at the time centered on the ratio between the numbers of those available for fighting and those available only for administration. But that comparison is not to the present purpose. What we have to note is that the 2000 officials of 1914 had become the 3569 of 1928; and that this growth was unrelated to any possible increase in their work. The Navy during that period had diminished, in point of fact, by a third in men and two-thirds in ships. Nor, from 1922 onward, was its strength even expected to increase; for its total of ships (unlike its total of officials) was limited by the Washington Naval Agreement of that year. Here we have then a 78 per cent increase over a period of fourteen years; an average of 5.6 per cent increase a year on the earlier total. In fact, as we shall see, the rate of increase was not as regular as that. All we have to consider, at this stage, is the percentage rise over a given period.

Can this rise in the total number of civil servants be accounted for except on the assumption that such a total must always rise by a law governing its growth? It might be urged at this point that the period under discussion was one of rapid development in naval technique. The use of the flying machine was no longer confined to the eccentric. Electrical devices were being multiplied and elaborated. Submarines were tolerated if not approved. Engineer officers were beginning to be regarded as almost human. In so revolutionary an age we might expect that storekeepers would have more elaborate inventories to compile. We might not wonder to see more draughtsmen on the payroll, more designers, more technicians and scientists. But these, the dockyard officials, increased only by 40 per cent in number when the men of Whitehall increased their total by nearly 80 per cent. For every new foreman or electrical engineer at Portsmouth there had to be two more clerks at Charing Cross. From this we might be tempted to conclude, provisionally, that the rate of increase in administrative staff is likely to be double that of the technical staff at a time when the actually useful strength (in this case, of seamen) is being reduced by 31.5 per cent. It has been proved statistically, however, that this last percentage is irrelevant. The officials would have multiplied at the same rate had there been no actual seamen at all.

It would be interesting to follow the further progress by which the 8118 Admiralty staff of 1935 came to number 33,788 by 1954. But the staff of the Colonial Office affords a better field of study during a period of imperial decline. Admiralty statistics are complicated by factors (like the Fleet Air Arm) that make comparison difficult as between one year and the next. The Colonial Office growth is more significant in that it is more purely administrative. Here the relevant statistics are as follows:

| 1935 | 1939 | 1943 | 1947 | 1954 |
|------|------|------|------|------|
| 372  | 450  | 817  | 1139 | 1661 |

Before showing what the rate of increase is, we must observe that the extent of this department's responsibilities was far from constant during these twenty years. The colonial territories were not much altered in area or population between 1935 and 1939. They were considerably diminished by 1943, certain areas being in enemy hands. They were increased again in 1947, but have since then shrunk steadily from year to year as successive colonies achieve self-government. It would be rational to suppose that these changes in the scope of Empire would be reflected in the size of its central administration. But a glance at the figures is enough to convince us that the staff totals represent nothing but so many stages in an inevitable increase. And this increase, although related to that observed in other departments, has nothing to do with the size — or even the existence — of the Empire. What are the percentages of increase? We must ignore, for this purpose, the rapid increase in staff which accompanied the diminution of responsibility during World War II. We should note rather, the peacetime rates of increase: over 5.24 per cent between 1935 and 1939, and 6.55 per cent between 1947 and 1954. This gives an average increase of 5.89 per cent each year, a percentage markedly similar to that already found in the Admiralty staff increase between 1914 and 1928.

Further and detailed statistical analysis of departmental staffs would be inappropriate in such a work as this. It is hoped, however, to reach a tentative conclusion regarding the time likely to elapse between a given official's first appointment and the later appointment of his two or more assistants.

Dealing with the problem of pure staff accumulation, all our researches so far completed point to an average increase of 5.75 per cent per year. This fact established, it now becomes possible to state Parkinson's Law in mathematical form: In any public administrative department not actually at war, the staff increase may be expected to follow this formula —

$$x = \frac{2k^m + l}{n}$$

$k$ is the number of staff seeking promotion through the appointment of subordinates; $l$ represents the difference between the ages of appointment and retirement; $m$ is the number of man-hours devoted to answering minutes within the department; and $n$ is the number of effective units being administered. $x$ will be the number of new staff required each year. Mathematicians will realize, of course, that to find the percentage increase they must multiply $x$ by 100 and divide by the total of the previous year, thus:

$$\frac{100 \, (2k^m + l)}{yn} \, \%$$

where $y$ represents the total original staff. This figure will invariably prove to be between 5.17 per cent and 6.56 per cent, irrespective of any variation in the amount of work (if any) to be done.

The discovery of this formula and of the general principles upon which it is based has, of course, no political value. No attempt has been made to inquire whether departments *ought* to grow in size. Those who hold that this growth is essential to gain full employment are fully entitled to their opinion. Those who doubt the stability of an economy based upon reading each other's minutes are equally entitled to theirs. It would probably be premature to attempt at this stage any inquiry into the quantitative ratio that should exist between the administrators and the administered. Granted, however, that a maximum ratio exists, it should soon be possible to ascertain by formula how many years will elapse before that ratio, in any given community, will be reached. The forecasting of such a result will again have no political value. Nor can it be sufficiently emphasized that Parkinson's Law is a purely scientific discovery, inapplicable except in theory to the politics of the day. It is not the business of the botanist to eradicate the weeds. Enough for him if he can tell us just how fast they grow.

## STUDY QUESTIONS

1. Examine your own work experiences. Does Parkinson's law seem to apply to any of them?
2. Whenever a "law" like Parkinson's is created, critics disagree by contending that the generalization is far too hasty, that it does not apply to all cases. In your opinion does Parkinson defend his "law" well? Are you convinced?
3. What kinds of evidence does Parkinson cite? Does he argue fairly or has he stacked the cards in his favor?

# Vivitar Series 1

## Vivitar Corporation

The Vivitar Corporation markets quality photographic equipment of all kinds — cameras, lenses, electronic flashes, and accessories. One of their products, the Vivitar Series 1 lens described here, is a kind of "super lens," a lens that combines features of several separate lenses. This technical advertisement appeared in *Popular Photography.*

Vivitar Corporation, 1977.

# This is the lens that is changing the way photographers think about macro lenses. The Vivitar Series 1™ 90mm f2.5 macro lens-life-size to infinity with exceptional edge-to-edge sharpness, contrast and true flat-field image.

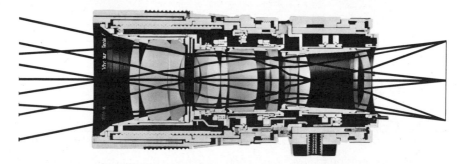

Traditionally, optical designers have had to make serious compromises when designing macro lenses of longer than normal focal lengths, sacrificing resolution and contrast as the focus approaches infinity in order to optimize performance in the near life-size (1:1) focusing range. The Vivitar Series 1 90mm f2.5 macro lens provides almost perfect edge-to-edge sharpness and a true flat-field image with excellent contrast, particularly in the difficult 1:2 to 1:1 reproduction ratio range. The floating group concept, incorporated in the design of the lens, assures optimum performance throughout the entire focusing range. In tests for resolution and contrast from infinity to life-size (1:1) reproduction, the Series 1 90mm f2.5 macro lens received some of the highest overall axial to corner ratings obtained for 35mm SLR macro lenses.

In designing the Series 1 90mm lens, Vivitar engineers opted for a unique 8 element/7 group configuration in the primary lens. Aberrations have been reduced to an absolute minimum and stabilized throughout the entire focusing range from a reproduction ratio of 1:2 to infinity. These same stringent performance demands necessitated the use of optical glass of very high index of refraction and several unusually thick elements. Borrowing the concept of a null lens from astronomical optics, Series 1 designers developed a 3 element macro corrector-lens adapter that provides a true flat-field image, high resolution and excellent contrast in the 1:2 to 1:1 reproduction range. Its design function is to correct close focus aberrations for use in macro photography.

This lens provides an unusual degree of versatility to the 35mm photographer. It is an ideal focal length for portraiture and most general purpose photographic applications. When coupled with its macro adapter, the lens allows for life-size photography at a greater working distance than would a shorter focal length macro lens, thus increasing the photographer's options in illuminating macro subjects uniformly and reducing the chance of disturbing live macro subjects.

As with all lenses in the Vivitar Series 1 group, the mechanical configuration has been given the same degree of attention as the optics. The lens engravings provide maximum information and legibility, and lens barrel styling is purely functional with all controls placed in the most appropriate positions for precise, comfortable operation.

**Specifications:**
Focal length: 90 mm. Aperture range: (2.5 to f22. Construction: Main lens-8 elements, 7 groups. Macro adapter-3 elements, 3 groups. Angle of Acceptance: 27°. Weight: Main lens-644gm (23 oz.), w/adapter-936gm (33 oz.). Length: 90mm (3.5 in.), w/adapter-138 mm (5.4 in.). Diameter: 70mm (2.8 in.). Accessory size: 58mm. Lens Coating: VMC Vivitar multicoating. Closest focusing distance from film plane: Main lens-39.3cm (15.5 in.), w/adapter-35.5cm (14 in.). Maximum Reproduction Ratio: Main lens-1:2, w/adapter-1:1.

Available in mounts to fit: **Nikon, Canon, Minolta, Olympus, Pentax, Vivitar** and universal thread mount cameras. The 90mm f2.5 lens is now available in combination with the macro adapter and case or as a separate lens.

Vivitar Corporation, 1630 Stewart Street, Santa Monica, CA 90406.
In Canada: Vivitar Canada Ltd./Ltée

# Vivitar® Series 1

## STUDY QUESTIONS

1. Photography has become such a popular hobby that a lens such as this one, which at one time might have been purchased only by a professional photographer, is now sold to the amateur photographer and hobbyist. How does the language and style of this advertisement reflect the type of consumer who will purchase this product?
2. What terms are not defined in this advertisement that you would like to have defined? Why are they not defined?

**3.** Where does this advertisement seem to be "selling" and not simply describing? Would you characterize the writing you find here as being basically persuasive, or informative? Why?

# Advent /1

## Advent Corporation

The Advent Corporation has for several years produced high fidelity loudspeaker systems that are both very popular and widely imitated. The following selection is taken from a promotional brochure, prepared by Advent, introducing a new model.

For the past several years, the most popular and imitated speaker in the United States has been the Advent Loudspeaker, which, including its newly redesigned format, is approaching the 700,000 mark in sales. Right behind it has been the Smaller Advent Loudspeaker, a speaker we carefully designed to have the same frequency range and much the same overall performance for less money in a smaller cabinet.

Advent Corporation, 1978.

The Advent/1 is a new speaker system that replaces the Smaller Advent. It is a redefinition of just how close we can come to the performance of our flagship speaker system in a smaller, less expensive speaker.

The Advent/1 has 2½ dB less response at 30 Hz than its bigger brother. That is the only difference worth quantifying. Its overall sound is as close to the New Advent's as any speaker can come to another. Its power-handling capabilities are the same, and its efficiency is high enough to allow it to be well driven by low-power amplifiers and receivers.

We feel that the performance-per-dollar (and per-cubic-foot) of the Advent/1 is unsurpassed by anything we or anyone else can offer in a speaker.

## Our Approach to Speaker Design

With almost twenty years' worth of experience in the design and manufacture of fine speakers (speakers good enough to have persisted in people's sound systems, seldom appearing on the used-speaker market), we have learned these basic facts:

- It is not difficult or esoteric to design an excellent speaker. Any of several design techniques will do the job, and no "breakthroughs" or new materials are needed to produce a speaker that does full justice to music.

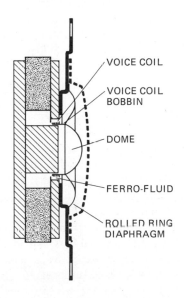

The trick in speaker design — as in *all* design — is to do just enough to make something right for its purpose, without cutting any corners. The customer shouldn't have to pay for needlessly elaborate engineering or complex "solutions" to non-existent problems.

- For home listening, two-way speaker design is the best, least problematical way to cover the audible frequency range. It is the closest practical alternative to the theoretical ideal of a *single* speaker, avoiding interference and crossover-effect problems common in three-and-more-way design. More complex systems aren't necessary for full-range coverage, and they don't have subtler characteristics than two-way designs.

■ Speakers can and should be designed to sound good not just under theoretically "ideal" or laboratory conditions, or with the best program material, but under the wide range of conditions that exist in real listening situations in people's homes — on a very wide range of musical material recorded in very different ways. Speakers designed to fulfill a single theoretical aim, such as "equal energy response," simply don't sound good much of the time under real conditions of use. What we have found best is to combine very wide-range, relatively smooth response (no sudden peaks or dips) with a frequency balance — octave to octave across the audible range — that sounds convincing with the widest variety of recording techniques.

## The Basics of the Advent/1.

The Advent/1 is a two-way, two-speaker system that combines an acoustic-suspension low-frequency driver with a small, high-dispersion, direct-radiator tweeter. The crossover is an LCR type that keeps low frequencies out of the tweeter and allows the woofer to roll off smoothly on its own without an abrupt cut-off. (Crossovers that employ an abrupt electrical cut-off to compensate for irregularities at the upper end of a woofer's range tend to have phase irregularities that produce a "hole in the middle" and other audible problems.)

The acoustic-suspension woofer replaces the stiffness of mechanical cone suspensions with the controlled acoustic stiffness of the air sealed inside a deliberately small enclosure. This "air spring" is much more uniform in its action than the best mechanical suspension, and it allows the woofer to move long distances to push and pull air for deep bass response without distortion.

The acoustic-suspension design also has the advantage of predictable, reliable behavior, with no side effects. There are designs that permit greater sound output for a given amplifier power but all of them have significant drawbacks. Horn enclosures have to be *very* big (and very expensive) for deep bass response, and sound reflected back into the mouth of the horn from a listening room can (and usually does) cause severe peaks and dips in mid-bass and low-bass response. Standard vented enclosures have also been big and expensive, and have tended toward boomy mid-bass which is why they have virtually disappeared from the marketplace. And the latest generation of smaller vented enclosures (along the lines originally suggested by A. N. Thiele) are problematical for true low-bass response because their cones are not "loaded" (controlled) below their bass cut-off point. This means that subsonic disturbances such as record-warp pulses, rumble, and feedback cause cone "bobble" that seriously interferes with response in the audible mid-range.

In the Advent/1, we have balanced the four interrelated factors of acoustic-suspension design — the mass of the moving system, magnet strength, enclosure volume, and cone size — to produce extended bass in a small enclosure without a great sacrifice of efficiency. We have kept the ability (found in the large New Advent Loudspeaker) to reach into the very lowest octave of bass response. But by giving up 2½ dB of output at the very bottom of the range, we have kept efficiency to the level needed to allow the Advent/1 to be driven well by a low-power amplifier or receiver.

To accomplish that, the Advent/1 uses the same woofer employed in its bigger brother. This also helps achieve power handling fully equal to that of the bigger speaker.

We believe that the 2½ dB trade-off in bottom bass response will be inaudible with the great majority of program material, giving up none of the solidity needed for a convincing bass foundation for music.

The high-frequency driver is also identical to that of the New Advent. And its unique advantages are also worth spelling out.

In a two-way system, a good tweeter has to be light and small enough to make the rapid movements needed for accurate high-end response. It also has to handle power effectively down into the mid-range region. And it has to radiate high-frequency energy efficiently over a wide listening area. (This last requirement of good dispersion isn't just to make highs audible at various seating locations in a room, but to keep highs in proper overall balance with the more easily spread mid-range frequencies, which reflect back from room surfaces. If the proportion of highs to mid-range in the overall sound-spread falls off, the excess mid-range can produce sound that is dull or lifeless or "nasal.")

The design of the tweeter used in the Advent/1 allows it to combine small size, efficiency, power-handling and excellent dispersion. The unique rolled-ring diaphragm (see our illustration) serves as its own free-moving suspension so that it can move more effectively down into the mid-range region without resistance and without reflections of energy back along the diaphragm. The center dome area is driven by the voice coil at its periphery, so that (as in conventional small-dome designs) no part of its diaphragm is far from the driving surface — which prevents flexing and "breakup." But the design also allows the voice coil to radiate energy outward into the rolled ring. The inner half of the ring becomes a provider of extra radiating area for excellent efficiency, and the outer half serves as a suspension that allows natural roll-off with a minimum of reflections. The radiating inner half of the ring is no further from the voice coil than is the center dome section.

The tweeter's voice coil is suspended in ferro-fluid, an expensive but well-worthwhile innovation that we think will be used increasingly in speaker design. This "magnetic fluid" suspension, plus the use of anodized aluminum for the voice-coil bobbin, allows very effective heat dissipation to let the tweeter radiate very high levels without thermal damage. And the ferro-fluid also provides added mechanical damping action that simplifies the design of the crossover network between woofer and tweeter and makes it easier for an amplifier to drive the overall system.

The design of the Advent/1's tweeter, originally conceived for our most expensive speaker, overcomes the limitations we see in dome tweeters and conventionally suspended tweeter cones. We think it is an unbeatable design for superb high-frequency performance at the lowest cost to the buyer.

## Octave-To-Octave Balance.

Although there are no visible parts to point to, a vital contributor to the Advent/1's sound quality is the octave-to-octave tonal balance designed into the system (primarily through its crossover network) to make it sound convincing with the widest range of music and recordings. This is a crucial part of designing a speaker for real-world enjoyment rather than for theoretical factors or easily marketed "showroom sound" of one kind or another. A speaker has to be lived with, day after day, with many different kinds of music and recording or broadcast techniques.

The musical balance we painstakingly choose for Advent speakers — and carefully maintain in speaker to speaker off the production line — not only makes them sound "right" initially to so many people, but also helps account for the scarcity of Advents on the used-speaker market. They go on sounding right over years of listening.

## Just How Good Is The Advent/1?

We think the Advent/1 is so close to the abstract "best" that can be done in speaker design that most of us involved with speakers at Advent could use it indefinitely with no sense of anything missing in its performance.

We can see reasons for some people to value the added 2½ dB of bottom bass offered by the Advent/1's bigger brother — if they listen consistently to the rare material, such as pipe organ music, that contains significant energy under 40 Hz. But since the Advent/1 goes low enough to do justice to the 43 Hz low E on a rock-bass guitar (the lowest frequency most people will ever want to reproduce), and since its high end is the same as the best we have to offer, we think the Advent/1's sound is right for virtually any purpose.

Inexpensive enough (and efficient enough) to be a part of a very reasonably priced stereo system, and small enough to be used in virtually any living room, the Advent/1 is an ideal product for most people who want a maximum of music in their lives.

### Specifications

| | |
|---|---|
| Crossover Frequency | 1500 Hz |
| Impedance | 8 ohms nominal |
| System Resonance | 52 Hz |
| Suggested Minimum Amplifier Power | 15 watts per channel |

| Cabinet Dimensions | 559 X 337 X 235 mm deep<br>22 X 13¼ X 9¼" deep |
|---|---|
| Weight | 13.2 kg<br>29 lbs. |
| Cabinet Construction | With wood finish, the Advent/1's cabinet is constructed of non-resonant particle board finished in genuine walnut veneer.<br><br>With vinyl finish, the Advent/1's cabinet is constructed of non-resonant particle board finished in walnut-grained vinyl. |

## Quality at Advent

The performance of any high fidelity product depends at least as much on care of manufacture as on design. We go to exceptional lengths at Advent to make sure that our loudspeakers leave the factory sounding the way they should.

Every driver of every Advent speaker system is tested before being installed in a cabinet. Then the response on every finished speaker system is tested across the frequency range. Small deviations from the desired response bring rejection of the system.

The drivers used in the Advent/1 are manufactured by Advent, not bought from outside suppliers (as most speakers from most manufacturers are). This not only helps us to do what we want to do in the design of Advent speakers (in ways that often wouldn't be possible with standard "stock" parts), but also helps maintain the exceptional uniformity in sound quality of speaker to speaker off the production line.

# ADVENT

Advent Corporation
195 Albany Street
Cambridge
Massachusetts 02139
USA

## STUDY QUESTIONS

1. What features of the loudspeaker system does Advent seem to think are the most important? What rhetorical strategy do they use to emphasize the importance of these features? Would most readers also agree that these are the most important features?
2. This brochure, although it is fairly technical in its presentation, is prepared to

help sell speakers. Do you find any places where the text seems to avoid "sticking to the facts"?

3. What aspects of this product are omitted from the brochure? Why do you suppose these omissions occur?

# High-Speed Drill Noise and Hearing

## Bonnie Forman - Franco, Allan L. Abramson, and Theodore Stein

If you were a dentist you might be concerned about the fairly common occupational health hazards dentists encounter. One of these problems is the alleged loss of hearing some claim occurs when a dentist is exposed to the sound of a high-speed drill on a daily basis. The following report examines the issue and contradicts the beliefs that most dentists hold about potential hearing loss.

Increasing attention has been focused on the relationship between exposure to noise from the high-speed drill and noise-induced loss of hearing within the dental profession.

In 64 dental students who had a limited history of drill usage and exposure, Hopp[1] found no significant shifts in auditory thresholds. Skurr and Bulteau[2] also supported this conclusion in their study of 56 third-year dental students; however, they suggested a noise-induced loss of hearing could be expected in later years.

Forty practicing dentists, who were exposed to the air-turbine drill for a median time of 3.7 years, were studied by Taylor and others.[3] They found that a 5 to 7 dB drop in hearing thresholds occurred at 4,000 and 6,000 Hz. At the Minnesota State Dental Convention in 1968, Ward and Holmberg[4] reported that dental drills had little effect on the hearing thresholds of 142 dentists.

There is no universal agreement on the association of the dental drill and a resulting loss of hearing. In this study, the effects of the high-speed dental drill on the hearing acuity of dentists were examined.

### Materials and Methods

Seventy-two dentists (70 men and two women) from eight dental specialties were randomly selected from the department of dentistry at a large suburban

Bonnie Forman-Franco, Allan L. Abramson, and Theodore Stein, "High-Speed Drill Noise and Hearing," *Journal of the American Dental Association*, September 1978, pp. 479–482.

medical center. Their specialties included oral surgery, pedodontics, prosthodontics, endodontics, periodontics, orthodontics, general dentistry, and dental anesthesiology.

Two dentists were eliminated from the study; one dentist had had a sudden loss of hearing in one ear a year before the study, and the other had a severe to profound sensorineural loss of hearing, possibly hereditary. Seventy dentists remained in the study population.

Each participant was given a detailed questionnaire to determine dental specialty, years in practice, manufacturer of present drill, frequency of drill use, total hours each week engaged in practice, and previous history of exposure to noise.

Routine audiometric testing was performed on each person with the use of a Maico 18 single-channel audiometer conforming to American National Standards Institute (ANSI) standards and an IAC 402 test booth. Testing included air-conduction thresholds at 500, 1,000, 1,500, 2,000, 3,000, 4,000, 6,000, and 8,000 Hz. Tests of bone-conduction thresholds were performed when indicated. Monitored live-voice, speech-reception thresholds (SRT) and speech-discrimination tests (35dB re:SRT) using the CID W-22 word lists were administered to all participants. An American Electromedics model 83 Impedance Audiometer was used for tympanometry and acoustic reflex threshold levels. Acoustic reflex thresholds were tested at 500, 1,000, 2,000, and 4,000 Hz for each ear.

The audiometric findings for the dentists were adjusted for age and compared with findings from a normal population. The latter statistics were taken from the National Hearing Health Statistics[5] and adjusted to ANSI standards. A Friedman two-way analysis of variance[6] was used to compare both populations. Correlations of two or more (partial) variables were determined by the Kendall Rank correlations.[6]

## Results

The largest percentage (56%) of the study population consisted of general practitioners. The mean age ranged from a minimum of 37 years for the oral surgeons to a maximum of 50.5 years for the orthodontists (Table). Oral surgeons had the least mean time in practice (nine years), and the orthodontists had the greatest (22.6 years). Twenty-nine dentists had more than 20 years of exposure to drills.

The mean hearing thresholds of the right and left ears are presented in Figure 1. When the hearing levels of the general dentists were adjusted for age and compared with a similar normal population,[5] no statistical differences ($P > .5$) were found. In the 45- to 64-year-old group, hearing was better in the higher frequencies (3,000, 4,000, and 6,000 Hz) for the general dentists than for the control group.

A direct relationship existed when hearing threshold levels were compared (correlation coefficient, $r > 0.3811$, $P < .05$) to years in practice. However, with increasing age came increasing years in practice. Partial rank correlations were per-

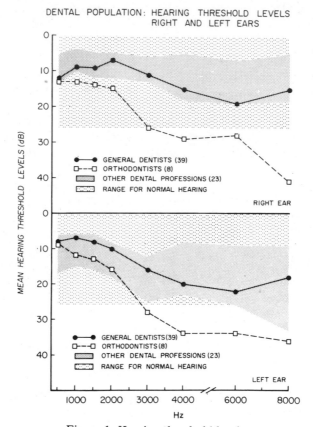

DENTAL POPULATION: HEARING THRESHOLD LEVELS
RIGHT AND LEFT EARS

GENERAL DENTISTS (39)
ORTHODONTISTS (8)
OTHER DENTAL PROFESSIONS (23)
RANGE FOR NORMAL HEARING

RIGHT EAR

GENERAL DENTISTS (39)
ORTHODONTISTS (8)
OTHER DENTAL PROFESSIONS (23)
RANGE FOR NORMAL HEARING

LEFT EAR

Figure 1. Hearing threshold levels.

formed to determine the effects of aging and years in practice individually and jointly. The correlation coefficient for age alone was $r = 0.0949$ ($P > .20$) and for years in practice, $r = 0.0920$ ($P > .20$). A high correlation coefficient existed ($r = 0.8615$, $P < .0001$); thus, the relationship of age to years in practice was statistically found to be inseparable.

When compared with the remaining dental specialties, the orthodontists had poorer mean hearing thresholds in both ears for 3,000 Hz and above. When adjusted for age and compared with a normal population, only one of the eight orthodontists exceeded the 95% confidence limits for the high frequencies.

Tympanometric findings indicated normal bilateral middle ear function (type A) for all participants. Reflex levels (dB) were determined on the basis of responses at each frequency: 2,000 Hz, right ear 94.3%, left ear 97.1%; 4,000 Hz, right ear 92.9%, left ear 94.3%. The mean reflex threshold levels at both of these frequencies are within normal limits (Figure 2).

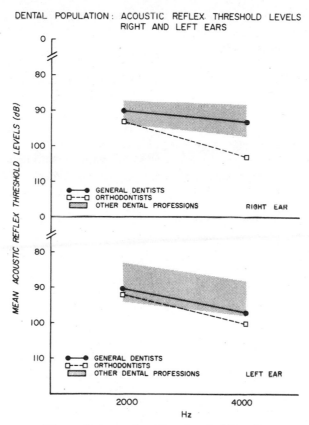

Figure 2. Acoustic reflex threshold levels.

## Discussion

Many dentists believe that the high-speed drills are a hazard to their hearing. The factors that must be considered in acoustic trauma are frequency of vibration, intensity, length of exposure, and intervals between exposures.

Different opinions have been expressed about the effects of the noise that is generated by the high-speed drill on the dentists' hearing acuity. Some investigators[1,2,4] found no evidence that the use of the drill is detrimental to hearing, yet others found a significant loss of hearing.[3,7]

Individuals with presbycusis, or loss of hearing resulting from aging, initially show a decrease in hearing threshold at 8,000 Hz. As aging continues, a gradual loss of hearing is seen in the higher frequencies in a descending pattern (8,000, 6,000, 4,000, 3,000, and 2,000 Hz). A noise-induced loss of hearing occurs in the frequencies between 3,000 and 6,000 Hz, with an improvement in threshold levels at 8,000 Hz.

Most of the energy from the air-turbine, high-speed dental drill is concentrated in the higher frequencies (above 4,000 Hz[4,7]). When a ball bearing type of drill was used on level with the ear, intensities slightly above 85 dB sound pressure level (SPL) at 4,000 and 8,000 Hz were found.[3] Air-bearing drills have a level 10 dB lower; when either type of handpiece is subjected to a cutting torque, an additional drop of 8 to 12 dB occurs.[8]

The damage risk criterion for continuous eight-hour exposure is 90 dB SPL.[9] Depending on the individual and the type of dental practice, variations of total exposure time for eight hours ranged between 12 and 45 minutes.[10,11] Thus, on the basis of length of exposure and noise intensity, air-turbine drills in proper working order should be within tolerable limits regarding damage to hearing.

**Table ■ Dental Specialties Investigated.**

| Dental Specialty | No. | Mean Age | Mean Years in Practice |
|---|---|---|---|
| General dentistry | 39 | 42.8 | 16.1 |
| Pedodontics | 6 | 48.7 | 21.0 |
| Orthodontics | 8 | 50.5 | 22.6 |
| Oral surgery | 5 | 37.0 | 9.0 |
| Prosthodontics | 4 | 38.5 | 12.8 |
| Endodontics | 3 | 39.0 | 13.7 |
| Periodontics | 4 | 37.5 | 9.5 |
| Anesthesiology | 1 | 35.0 | 10.0 |

The general practitioners composed the largest portion of our study population. In addition to using the drill for the longest period of daily exposure, they also had the greatest incidence of previous exposure to noise. Hearing at frequencies less than 3,000 Hz, which are commonly referred to as the speech range, was not affected by the drilling, although a slight loss of hearing, which was not statistically significant, was found at 6,000 Hz. Other investigators have reported a loss of hearing at 4,000 and 6,000 Hz.[3,4,7]

Hearing threshold levels provide a means of predicting an individual's sensitivity for speech. The normal limits for hearing within the 500 to 2,000 Hz range is between –10 dB and 26 dB.[12] The effect of decibel levels on word discrimination is often related. Fifty percent of word intelligibility is found between 1,000 and 2,450 Hz,[13] therefore, few, if any, discrimination difficulties would be expected in our dental population. In our study, mean hearing threshold levels were within normal limits and the slight loss noted at 6,000 Hz was not enough to interfere with speech discrimination.

The remaining specialties, periodontics, endodontics, oral surgery, pedodontics, prosthodontics, and anesthesiology, had hearing threshold levels in close agreement with those of the general practitioners. The orthodontic population (eight) superficially appears to have the most significant loss of hearing at high frequencies, but when adjusting this group for age and comparing them with the control population (95% confidence level), only one specialist exceeded normal limits.

When comparing the hearing levels as adjusted for age of the general population[5] to the hearing levels of the general practitioners, no statistical differences were found. A correlation appears to exist between years in dental practice and progressive loss of hearing. However, this mimics the relationship of advancing age and loss of hearing in the normal population and suggests that when a loss of hearing occurred, it was primarily an effect of aging.

Dr. Leon Eisenbud, department of dentistry, Long Island Jewish-Hillside Medical Center, New Hyde Park, NY, provided the dental population necessary for this study.

## ENDNOTES

[1] Hopp, E. S. Acoustic trauma in high-speed dental drills. Laryngoscope 72:821 June 1962.

[2] Skurr, B. A., and Bulteau, V. G. Dentists' hearing: the effect of high speed drill. Aust Dent J 15:259 Aug 1970.

[3] Taylor, W.; Pearson, J.; and Mair, A. The hearing threshold levels of dental practitioners exposed to air turbine drill noise. Br Dent J 118:206 March 2, 1965.

[4] Ward, W. D., and Holmberg, C. J. Effects of high-speed drill noise and gunfire on dentists' hearing. JADA 79:1383 Dec 1969.

[5] National Health Survey, Vital and Health Statistics Series 11, no. 11. Hearing levels of adults by age and sex. United States 1960–1962. Oct 1965, p 1–34.

[6] Sokal, R. R., and Rohlf, J. J. Biometry. San Francisco, W. H. Freeman and Co., 1969.

[7] Weatherton, M. A.; Melton, R. E.; and Burns, W. W. The effects of dental drill noise on the hearing of dentists. J Tenn Dent Assoc 52:305 Oct 1972.

[8] Information bulletin. Office of Naval Research, European Scientific Note no. 18–5, May 14, 1964.

[9] Guidelines to the Department of Labor's Occupational Noise Standards, bulletin 334, 1971.

[10] Dellheim, E. J. Dental air turbine noise. Dent Student 49:68 March 1971.

[11] Stockwell, C. L. Dental handpieces and rotary cutting instruments. Dent Clin North Am 15:219 Jan 1971.

[12] Goodman, A. Reference zero levels for pure tone audiometer. ASHA 7:262 July 1965.

[13] French, N., and Steinberg, J. Factors governing the intelligibility of speech sounds. J Acoust Soc Am 19:90 Jan 1947.

## STUDY QUESTIONS

1. Do you find the conclusions drawn in this article valid, considering the data presented? Do these writers examine an adequate number of samples?

2. Notice the stylistic differences between the section "Results" and the section "Discussion." Do you find this shift disturbing? How would you characterize the nature of this shift? Why do you suppose it exists?

3. The section entitled "Discussion" is considerably longer than the section "Materials and Method," which describes the experiments that were performed. What does this suggest to you about the relationship between the data and the interpretation of the data?

# Tentative Report on the Hudson River Bridge

## O. H. Ammann

We sometimes forget that even large projects have small beginnings. The George Washington Bridge, which spans the Hudson River between New York and New Jersey, was once a project that existed only on paper, a project that had to be "sold" to people who needed to be convinced both that they needed a bridge, and that the bridge being designed was in fact the bridge they needed. This report by O. H. Ammann convinced New York and New Jersey to commit themselves to the project, a project that the report itself claims would cost around $50,000,000.

THE PORT OF NEW YORK AUTHORITY

March 11, 1926

To the Governor of the State of New York:
To the Governor of the State of New Jersey:

Sirs: — We herewith transmit to you the Tentative Report of our Bridge Engineer dealing with a bridge across the Hudson River between Fort Washington and Fort Lee, which gives the engineers' tentative conclusions.

We send this report at this time in order that you may have the latest available engineering information on this matter. The Commission has not yet determined the design or location of the bridge.

We have the honor to remain,

Respectfully,

The Port of
New York Authority

{ JULIAN A. GREGORY, Chairman,
JOHN F. GALVIN, Vice-Chairman,
FRANK C. FERGUSON,
OTTO B. SHULHOF,
SCHUYLER N. RICE
HERBERT K. TWITCHELL,

Commissioners

O. H. Ammann, "Tentative Report of Bridge Engineer on Hudson River Bridge at New York Between Fort Washington and Fort Lee," Port of New York Authority, 1926.

The exhibits, tables, and appendices referred to in the text are not included here.

## THE PORT OF NEW YORK AUTHORITY

February 25, 1926

To the Commissioners of the Port of New York Authority:

DEAR SIRS: — The preliminary work necessary for the planning and construction of the Hudson River Bridge between Fort Washington and Fort Lee, with which The Port of New York Authority has been charged by the Legislatures of New York and New Jersey, has now advanced to a point where conclusions can be drawn regarding the physical and financial feasibility of this bridge, its necessity as a link in the local and interstate transportation systems, its location, size, type, method of construction, approximate cost and aesthetic merits.

Briefly the work so far accomplished embraces comprehensive traffic studies to determine the probable volume of traffic over the bridge and the revenues to be derived therefrom, topographical surveys, river borings and engineering design studies to determine the suitable site, size and type of crossing and its cost, and finally, architectural studies to determine the feasibility of rendering the bridge a befitting object in a charming landscape.

The project being of exceptional magnitude, and complex aspect, it was necessary that the preliminary studies be undertaken with great care and thoroughness. The appropriations by the two States for these preliminary studies, amounting to $200,000, became available only on July 1, 1925, and the time has not been sufficient to permit either the completion of the studies or the rendering of a comprehensive report on the project. However, it is believed that from the studies so far completed the following conservative conclusions may be drawn:

**Conclusions**

(1) The traffic studies reveal an urgent demand for a crossing for vehicular traffic in the vicinity of the proposed bridge to relieve the present intolerable traffic situation. The traffic volume is of more than sufficient magnitude to make it financially feasible to construct, operate and maintain, from tolls, such a crossing, not considering the broader benefits to the people of both States as well as to the local community.

(2) The general location of the bridge is well chosen with regard to topography in its vicinity and the feasibility of convenient connections to the important local and arterial highway routes on both sides of the river. A crossing at this point also appears to be the next logical step after construction of the vehicular tunnel at Canal Street, since the two crossings are far enough apart not to influence materially each other's traffic quota.

(3) From the engineering point of view the construction of the bridge is in every respect feasible and, while of unusual magnitude, will involve no extraordinary difficulties, nor hazardous or untried operations. The bridge will have a single river span of at least 3500 feet and a clear height above water of about 200 feet. The piers will be located within pier-head lines, as established by the War Department, and will therefore be no obstruction to navigation.

(4) The bridge is to be the suspension type, the most economical and

aesthetically superior type available. It will be of extremely simple construction, and its design is conceived so that it will be feasible to build the bridge at a minimum initial expenditure to serve present traffic needs, and to enlarge its capacity as the traffic volume increases.

(5) If funds for construction of the bridge shall become available in 1927, it is expected that not later than 1933 the bridge will be open for four-lane vehicular and bus passenger traffic and for pedestrians. It is estimated that this capacity will suffice to take care of the initial traffic and the expected increase until about 1943, and then it will probably become necessary to enlarge to an eight-lane vehicular capacity.

(6) While it is not possible, at the present time, to report definite cost figures, it is estimated, upon information so far available and upon such forecast of real estate values as may now reasonably be made, that the bridge can be opened for highway traffic at a cost of less than $50,000,000, inclusive of interest during construction.

(7) Depending upon traffic capacity finally to be decided upon, it is estimated that the bridge can later be enlarged at an additional cost of between $15,000,000 and $25,000,000, if, and when, the vehicular and passenger traffic will have grown in volume to pay for this additional cost.

(8) On the basis of conservative traffic analysis, and without counting upon the vehicular traffic which will be generated by the construction of the bridge, nor upon possible income from other than vehicular traffic, it is estimated that during the first year after completion the revenue will more than cover the annual interest charge, administration, maintenance, and amortization. The bridge will thus be self-sustaining in every respect from the first year without imposing unreasonable toll charges upon the traffic.

(9) On the basis of conservative assumptions for future growth of traffic, and counting upon revenue from vehicular traffic alone, it is estimated that within ten years after opening to traffic the bridge may be enlarged to eight-lane capacity, and that within twenty years thereafter the entire bond issue raised to cover construction cost can be amortized.

(10) The architectural studies so far made, while yet tentative, indicate clearly that the bridge may be so designed as to form an object of grace and beauty as well as utility, and to blend harmoniously with the grandeur of its natural setting.

(11) In view of this favorable aspect of the bridge, its urgent necessity to relieve traffic conditions and in order to derive the benefit of a complete investigation, it is recommended that the preliminary work be carried to completion, and that the States be asked to appropriate an additional sum of $100,000 to make that completion possible.

Following is a more complete and detailed account of the work so far accomplished:

**Traffic Studies**

Since the Legislative Acts provide that the Port Authority may levy charges for the use of the bridge and that the bridge shall be built and paid for in whole or in part by bonds of the Port Authority, or other securities, it has been necessary to ascertain whether or not the revenues from tolls for vehicles and pedestrians, and possibly franchise rights for rail passenger facilities, will be adequate to meet the cost of construction. This involves the study of a number of traffic factors, viz.:

First: The present volume of vehicular and pedestrian traffic over each of the seventeen ferries across the Hudson River.

Second: The volume of traffic the bridge will be expected to attract when it is opened to traffic. This requires an estimate of the effect on the bridge traffic of the opening of the vehicular tunnel in 1926.

Third: The volume of traffic that can reasonably be expected to be diverted to the bridge from each of the other crossings in that year.

Fourth: The volume of traffic over the bridge for each year, for twenty years subsequent to the opening of the bridge, proper allowance being made for the effect upon the bridge traffic of the possible construction of other crossings below 179th Street, Manhattan.

This necessitates the determination of the origin and destination of vehicles by types for the existing ferries and apportioning the divertible traffic to each of the proposed crossings in such a way as to take into account relative distances and ferry, tunnel, and bridge charges and the elimination of undue congestion on the approach streets to each of the proposed facilities.

Fifth: An estimate of the revenues for each year subsequent to the opening of the bridge, based upon an average toll per vehicle and per pedestrian.

In order to estimate the vehicular traffic, it was necessary to obtain the trend or rate of growth of the present-day traffic over seventeen ferries between the Battery and Tarrytown (for the most recent normal year). This required the records, by classes of vehicles, kept by each of the ferry companies from 1914 to date. Where revenues only are available for this traffic, average tolls for each class of vehicle must be applied to the revenues to estimate the number of vehicles. From these records the volume of traffic over each of the ferries can be forecast for each of the years subsequent to 1932.

Instead of forecasting the traffic for each of the ferries it is better to forecast the volume of traffic that will be diverted from the existing ferries to the bridge. To obtain this divertible bridge traffic it is necessary first to ascertain the distribution of the present-day traffic over each of the ferries for the most recent normal year. To do this the origin and destination of each vehicle is necessary for a sample period of time, so selected that the peak and the average traffic condition in the year will be reflected. These occur in the months of July and October. The variations of traffic between week-days and Sundays from hour to hour, or both, are necessary to estimate the peak traffic conditions to test out the roadway capacities on the bridge. Field clockings, therefore, were taken by placing inspectors on each of the ferry boats of every route to ride the boats throughout the day. The inspectors ascertained and recorded the following information respecting each vehicle crossing the river by ferry:

(a) Type of vehicle, that is, whether horse drawn or motor propelled. A division of motor vehicles was made as between commercial and pleasure, and again sub-divided to indicate the carrying capacity of the commercial vehicles and the seating capacity of the pleasure vehicles:

(b) Number of persons carried in each vehicle;

(c) State License;

(d) Origin and destination of each vehicle;

(e) Frequency of use of ferry route by each vehicle.

These clockings were made throughout the months of July, August, September and October, 1925. In carrying forward the clockings a field force of fifty-six men was employed on the seventeen ferry routes. The detailed information noted above was ascertained and recorded for a total of 242,000 vehicles.

Clockings were made of the vehicular traffic now passing over the streets and street intersections in the vicinity of the proposed location of the bridge, to determine the degree to which capacity of these streets is now used. Also a study was made to determine the volume of traffic carried at present by the East River bridges, particularly during the peak of travel; and the extent of saturation.

Examination of the records of the various ferry companies operating the seventeen ferry routes, for the purpose of ascertaining the volume of traffic and its classification handled by the ferries of each route for the past ten years, has required a force of three to four men constantly from July to the present date.

After having completed the field clockings, the next step was the tabulation and summarization of the data. The work of tabulating was carried on in part during the period of clocking and has proceeded since the clockings were completed in October, to bring it to a point to permit of detailed analysis.

These analyses are for the purpose of determining future distribution of vehicular traffic among the present crossings, the proposed 178th Street bridge, and any other crossings that might later be constructed and which might affect the future revenues of the 178th Street bridge.

One of the first determinations to be arrived at by analysis is the probable volume of traffic that may be expected to use the 178th Street bridge when it is opened, assuming that were the only highway across the Hudson River between Manhattan and New Jersey.

The second determination to be made is the volume of traffic which will be attracted to the vehicular tunnel, when it is opened, which otherwise might, in part at least, have used the 178th Street Bridge.

The third determination is the probable effect on the 178th Street bridge traffic by the opening of any additional highway crossing over the Hudson in the future.

Each of these steps involves a large number of intermediate steps. For example: highway access to the bridge; determination of a toll which will secure maximum traffic and maximum revenue; future crossings to be constructed by the City of New York across the East and Harlem Rivers; and traffic that will be generated by the stimulation of industrial and residential development, particularly on the Jersey side.

The results of all of these traffic studies are now being carefully recorded and will be included in a later report on the project.

Exhibit (A) illustrates the growth of the total trans-Hudson vehicular traffic as tentatively estimated from 1924 to 1960, inclusive, and the number of vehicles of this total traffic which would have been, or will be, diverted to the 178th Street bridge.

Below is recorded the first tentative estimate of total trans-Hudson vehicular traffic for all ferries from the Battery to and including Tarrytown, and the traffic that the bridge will divert from these ferries and the tunnel.

| Year | Hudson River Traffic | Bridge Traffic | Year | Hudson River Traffic | Bridge Traffic |
|------|---------------------|----------------|------|---------------------|----------------|
| 1924 | 11,706,000 | 3,208,000 | 1944 | 36,055,000 | 11,476,000 |
| 1925 | 12,912,000 | 3,596,000 | 1945 | 36,767,000 | 11,723,000 |
| 1926 | 14,185,000 | 4,017,000 | 1946 | 37,408,000 | 11,944,000 |
| . . . . . . . . . . . . . . . . . . . . . . . . . . . . . . . . . . . . . . . . . . . . . . . . |
| 1934 | 25,607,000 | 7,889,000 | 1953 | 40,841,000 | 13,144,000 |
| 1935 | 26,984,000 | 8,364,000 | 1954 | 41,172,000 | 13,263,000 |
| 1936 | 28,280,000 | 8,807,000 | 1955 | 41,478,000 | 13,369,000 |
|      |            |           | 1956 | 41,765,000 | 13,471,000 |

These figures must be revised as the analysis proceeds to take into account the effect of the opening of additional crossings. The above figures do not include traffic which will be generated from the adjacent territories, whose growth the bridge will stimulate. While this cannot be measured accurately, an analysis of the growth of population, intensity of realty development, and motor vehicle registration is in process to determine the effect of the East River bridges upon Brooklyn and Queens, in order to gauge roughly the effect that the Hudson River bridge will have upon Fort Lee and its contiguous communities. The amount of this traffic will be considerable and eventually will be added to the above estimates.

While the above traffic is the principal source of revenue, there are four other sources which will contribute to the income of the bridge. This revenue will come from passengers in vehicles, pedestrians, bus lines and rapid transit facilities. Studies are under way to ascertain the potential traffic which will give rise to this income and will be presented in a later report.

Tables (I-a), and (I-c), appended to this report, give the gross revenues estimated to date for a 50¢ rate, a 60¢ rate, a 70¢ rate, respectively, from vehicles only. It will be seen that for 1933, or the first year of operation, the income from vehicles alone is forecast as at least $3,700,000. Subtracting the charges for administration, maintenance and operation, the net operating income is close to 6½% on the $50,000,000, the probable maximum initial cost of the bridge. In addition, there will be revenue from passengers in vehicles, pedestrians, and bus lines, and from vehicular traffic which will be generated by the bridge. Consequently, it is safe to conclude at this time that the charges on the initial and ultimate cost of construction can be met out of the potential revenue from traffic, and that therefore the project is economically sound.

### Location Studies

The Legislative Acts of New York and New Jersey provide that the bridge shall be located at a point between 170th and 185th Streets in Manhattan, New York City, and a point approximately opposite thereto in the borough of Fort Lee, New Jersey.

After a general examination of the territory on both sides of the river, within these limits, three specific sites which appeared to offer possibilities were tentatively selected for more careful study. (See location map, Exhibit B.) The three sites chosen are those in close vicinity of 181st Street, 179th Street, and 175th Street, Manhattan, respectively. River borings, studies of approaches, grades, street connections, tentative designs and comparative cost estimates were made for these locations. These studies revealed the central location near 179th Street as being not only the most economical, but also the most desirable with respect to approach grades and street connections and natural setting, and it was therefore decided to confine the elaboration of more complete plans and estimates to this location.

In the selection of the locations, careful consideration was also given to the scenic effect of the bridge, more particularly with regard to the effect upon Fort Washington Park. While, by locating the bridge at 181st Street or 175th Street, encroachment upon this park by bridge piers might be avoided, the much longer river span required at these locations, and the consequent greater proportions of the bridge, would not be as favorable, aesthetically, as a bridge at 179th Street. Moreover, the location of a pier in the Park is not believed to curtail in any way the usefulness of the Park or to mar its beauty.

### Topographical Surveys, Mapping and Triangulation

Owing to lack of maps, sufficiently accurate and complete for preliminary planning and reliable estimates of cost, it has been necessary to undertake extensive and accurate topographical surveys extending over the territories on which the bridge approaches and street connections may be located. These surveys are now nearing completion and will form a valuable basis for the final planning and construction of the bridge. The results of these surveys have been embodied in a large map to the scale of $1'' = 100'$.

Owing to the lateness of the season it has been found impracticable to undertake an accurate triangulation across the river, but the necessary base lines have been established, and all other preparations for these measurements have been made, and it is expected that they can be accomplished in the Spring as soon as weather conditions permit. For the tentative studies, the triangulation made by the U.S. Coast and Geodetic Survey was considered to be sufficiently reliable.

### River Borings

In order to obtain reliable information on the character of the river bottom and to establish beyond question the surface of the solid bedrock upon which the bridge piers have to rest, it was necessary to undertake borings carried well into the solid rock.

In all, sixteen borings, at the three locations tentatively selected, have been sunk, the results carefully recorded and the rock cores preserved. These borings have established the fact that, outside of the pierhead lines established by the War Department, that is, within the width of river reserved for navigation, bedrock is too deep to permit of economical construction of bridge piers and that such piers must, and can, be placed between the pierhead lines and the shore, or on shore. Moreover, thus located, the piers will form no obstructions to navigation. . . .

Additional borings will have to be made when the location of the bridge is definitely established.

The character of rock revealed by these borings corresponds to that prognosticated by the U.S. Geological Survey. On the New Jersey side bedrock was found to consist partly of solid red sandstone and shale, known as the "Newark Formation," partly of the so-called "Stockbridge Dolomite" which forms the major portion of the rockbed under the Hudson River. The borings on the New York side revealed a solid bed of "Hudson Shist" (mica shist), which is the prevailing rock of Manhattan Island. All of these rock formations are sufficiently hard to constitute a solid and permanent foundation for the bridge piers and to safely sustain the great pressure from them.

The material overlying the rock is almost entirely river silt, unsuitable for foundation purposes.

### Engineering Design Studies

In order to determine the most economical and suitable type and general proportions of the structure, for various possible locations, it was essential to undertake extended comparative design studies and cost estimates, before any final planning could be undertaken. Complete tentative designs were made for a 3500 foot river span and a 3900 foot span, as required for the 179th and 181st Street locations, respectively. Comparative estimates of cost have also been prepared for various capacities for highway traffic and for combined highway and rail passenger traffic.

Various possible forms and materials for the individual parts of this structure were given most careful consideration, and all essential features of the structure have been studied in detail with a view to assure not only economy, but conformity to the most advanced standards of design and methods of fabrication and construction.

Tentative schemes of erection have been evolved, inasmuch as the method of erection of a large bridge not only has an important bearing upon its design and economy, but because in this case it involves operations of unprecedented proportions.

As a result of these studies a tentative design has been developed which, for the 179th Street location, may be briefly described as follows:

### Type and General Proportions of Bridge

Little study was necessary to determine the suspension bridge as the most suitable type, because its superior economy for such great spans and capacities is now generally recognized by engineers. Its superior aesthetic merits, when properly designed, further single it out as the best adapted type in this case.

A cantilever bridge, the nearest other possibility, would, with its dense and massive network of steel members, form a monstrous structure and truly mar forever the beauty of the natural scenery.

The general proportions of the bridge, as to length of spans and height above water, were sharply defined by the topographical and geological conditions of the site. As a result of the borings, heretofore described, the main pier on the New Jersey side was located well within the pierhead line at a point where rock can be reached at a depth of about 100 feet, which is the approximate limit for the pneumatic process, the safest and most reliable foundation method. On the New York side the logical and natural place for the pier is the rocky point of Fort Washington Park close to the pierhead line. This results in a central span of 3500 feet between centers of piers, or twice the span of the Philadelphia-Camden bridge, the longest suspension bridge so far built.

The rock cliffs of the Palisades form the natural abutment and anchorage on the New Jersey side and, for the sake of symmetry, which is an essential aesthetic requirement, the side span on the New York side is made the same, or approximately 700 feet.

The clear height of the bridge floor above water is approximately 200 feet, this height resulting from the elevations of the connecting streets on both sides of the river and the limiting grades of the approaches. Incidentally, this height is ample to permit passage of the largest vessels which are likely to go up the river beyond this point.

The general form and arrangement of the structure are of extreme simplicity. Essentially the floor deck is suspended throughout its length from simple cables or chains. The latter will pass over the two towers and are to be firmly anchored in rock or massive concrete blocks at their ends.

To enhance the gracefulness of the bridge, the cables are to have a comparatively small sag or flat catenary. Structurally, the cables are to be built either of steel wires or of high grade steel eyebars, both types of construction having reached a high degree of perfection in American bridge practice and a degree of safety superior to that of any other type of structural members.

Detailed studies have been made of two essentially different types of towers, a slender steel tower, as exemplified in the Manhattan bridge, and a combined steel and masonry tower of massive appearance. While the economic merits of the slender steel tower, and its justification in some localities, are recognized, it is felt that the conspicuous location of the proposed bridge in the midst of a bold and impressive landscape makes the selection of the aesthetically superior massive tower imperative.

### Traffic Capacity of Bridge

One of the most important and complex questions which had to be solved, and will involve further careful study in connection with the planning of this bridge, is the determination of its traffic capacity, as regards both kinds and volume of traffic.

The question is necessarily closed related to the study of the traffic situation and definite solution has to await the results of these studies. While the Legislative Acts do not specify the kind of traffic to be accommodated, existing conditions point clearly to the need of a crossing primarily for vehicular traffic. Furthermore, while the development of the territories contiguous to the bridge will, sooner or later, call for the accommodation of a considerable volume of passenger traffic, it is not likely that rapid transit or other rail passenger traffic facilities will be needed for many years to come.

It is also realized that the demand for passenger traffic in the immediate future, and possibly for many years to come, may be filled by passenger buses running over the bridge roadway. Any provision for the accommodation of rail traffic, which would involve a comparatively large outlay at present, would therefore not be warranted.

As a result of our studies it now appears quite feasible, however, to build the bridge initially for highway traffic only, but with provision, at a small extra expenditure, for the future accommodation of rail passenger, or additional bus passenger traffic.

In fact the design, as now developed (see Exhibit C), is exceptionally far-reaching in its provision for a gradual increase in traffic capacity with a minimum possible initial expenditure, and with the least possible time of construction before the bridge can be opened to traffic.

The plan provides for an initial capacity of two 24-foot roadways which will conveniently accommodate four lanes of vehicular traffic, two in each direction. Two footwalks for pedestrians are also provided for. It is estimated that these two roadways will be sufficient to fill the demand for highway traffic for about ten years after the opening of the bridge.

If and when justified by increased volume of vehicular traffic, another four-lane roadway can be added, and used for truck traffic, while the two initial roadways may be reserved for the faster passenger automobiles. It is estimated that the eight lanes will be ample to take care of all vehicular traffic which may be concentrated at this crossing. All of this highway traffic is to be accommodated on an upper deck of the structure.

If and when accommodation for rail passenger traffic, or for additional bus passenger traffic, across the bridge becomes necessary, two or four lanes, or tracks, of either form of such traffic can be added on a lower deck.

The question as to whether, and to what extent, rail passenger traffic should be provided for on the bridge is still under consideration, and the cooperation and advice of the transit authorities in the two States have been sought in order to arrive at a satisfactory solution.

### Approaches and Highway Connections

Tentative studies for the approaches and highway connections on both sides of the river have been made, but further studies in cooperation with the proper municipal and State highway authorities, are necessary. The studies so far completed indicate conclusively that direct connections of the bridge approaches with important highway arteries, such as Broadway and Riverside Drive in Manhattan, and Lemoine Avenue in New Jersey, are entirely feasible and involve no extensive changes in the street system, at least for many years after completion of the bridge.

It would be lacking in foresight, however, not to recognize the fact that when the bridge is to be completed to capacity the vehicular traffic will have grown to such an extent that new arteries will become necessary on both sides of the river, more particularly for that traffic which will flow to and from the bridge in an easterly and westerly direction. While such new arteries will not form part of the bridge project proper, studies are being made with respect to them and with a view to give the bridge a proper setting in the future net of highway arteries.

Regarding the structural arrangement of the approaches, more particularly that on the New York side, it should be mentioned that aesthetic considerations have been paramount in developing their design.

The New York approach is designed as a short viaduct of monumental appearance which will enhance rather than destroy the good character of the neighborhood (see Exhibit D). The New Jersey approach is designed as a cut through the top of the Palisades so marked at the face of the cliffs as not to destroy the appearance of the latter or to break their natural silhouette.

Tracks, if any are provided, will be hidden from view on the approaches.

### Architectural Studies

The commanding location of the bridge in a charming landscape made it imperative to give prominent consideration to the aesthetic side of the bridge design; in other words, to combine beauty with utility and strength. For this purpose the Port Authority has engaged an eminent architect, Mr. Cass Gilbert, to assist the engineering staff in the preparation of the plans. A statement by the architect on the architectural aspect of the project is appended.

### Estimates of Cost

In view of the incompleted state of the preliminary work and certain as yet unsettled questions, such as provision for passenger traffic, extent of architectural treatment of the bridge and approaches, more accurate appraisal of property and damages, etc., it is impossible to give at the present time reliable cost estimates. Making reasonable allowance for the uncertain features, it is estimated that the bridge can be constructed, ready for the initial highway capacity, at a cost of less than $50,000,000 inclusive of interest during construction, and that it can later be strengthened for the eight-lane highway capacity, and provision for from two to four electric railway tracks, at an additional cost of between $15,000,000 and $20,000,000.

### Financial Statement

The financial statement following gives, for the years 1933, 1943, 1953 and 1960, the gross revenue and net operating income from vehicles only, based upon average toll rates of 50¢, 60¢, and 70¢, respectively, an initial cost of $50,000,000 and an additional expenditure, ten years later, of $15,000,000, as required for the increased vehicular capacity.

| Year | Gross Revenue from Vehicles Only | Administration Operation Maintenance | Net Operating Income Available for Interest and Amortization | Per Cent of Net Operating Income to Estimated Cost |
|------|------|------|------|------|
| A. Average Toll Charge 50¢ | | | | |
| 1933 | $3,700,000 | $ 500,000 | $3,200,000 | 6.40% |
| 1943 | 5,608,000 | 750,000 | 4,858,000 | 9.72 |
| 1953 | 6,572,000 | 1,000,000 | 5,572,000 | 8.57 |
| 1960 | 6,910,000 | 1,000,000 | 5,910,000 | 9.09 |
| B. Average Toll Charge 60¢ | | | | |
| 1933 | 4,441,000 | 500,000 | 3,941,000 | 8.76 |
| 1943 | 6,730,000 | 750,000 | 5,980,000 | 13.29 |
| 1953 | 7,886,000 | 1,000,000 | 6,886,000 | 10.59 |
| 1960 | 8,293,000 | 1,000,000 | 7,293,000 | 11.22 |
| C. Average Toll Charge 70¢ | | | | |
| 1933 | 5,181,000 | 500,000 | 4,681,000 | 10.40 |
| 1943 | 7,851,000 | 750,000 | 7,101,000 | 15.78 |
| 1953 | 9,201,000 | 1,000,000 | 8,201,000 | 12.62 |
| 1960 | 9,675,000 | 1,000,000 | 8,675,000 | 13.35 |

Note: — Additional revenue from generated vehicular traffic and from bus and rail-passenger traffic is expected to increase materially the potential net operating income.

**Acknowledgement**

The Engineering Staff has been aided in its studies so far made by valuable advice and information from various individuals and organizations to whom due credit will be given at the proper time.

The undersigned also take this occasion to express their acknowledgement for the valuable services so far rendered by other members of the engineering staff, more particularly, R. A. Lesher, Traffic Engineer, in charge of traffic studies; W. J. Boucher, Engineer of Construction, and R. Hoppen, Jr., Resident Engineer, in charge of surveys and borings; W. A. Cuenot and A. Andersen, Assistant Engineers, in charge of design studies.

Respectfully submitted

(Signed) O. H. AMMANN,
Bridge Engineer

# STUDY QUESTIONS

1. This report is directed to the governors of the states to be connected by the bridge. What kinds of evidence might interest such an audience most? What evidence might interest them least? Do you find that Ammann has slanted his report to this special audience in any way?

2. Ammann is an engineer. Does his profession seem to influence the language or style of this report?
3. Notice the arrangement of Ammann's report. Traffic studies, for example, come first; costs come last. What were Ammann's reasons for arranging these sections in this particular order? The individual sections are not of equal length, and some are obviously more detailed than others. Why?
4. Does selling a project as large as this one always depend on the use of factual data rather than emotional appeals? (Before you answer consider the United States' Apollo projects, or the collaboration of France and Britain on the Concorde.) Does Ammann use any emotional appeal in his presentation?

# The Stylists: It's the Curve that Counts

## Ralph Nader

Today almost everyone is familiar with the work of Ralph Nader. At the time that this selection was first published, however, Nader was still relatively unknown, a writer and critic whose power was yet to be discovered. His book *Unsafe at Any Speed* (which is sometimes credited with forcing General Motors to stop production of the Chevrolet Corvair) focused public attention on the way automobiles are designed, manufactured, and marketed.

The importance of the stylist's role in automobile design is frequently obscured by critics whose principal tools are adjectives. The words are familiar: stylists build "insolent chariots," they deal with tremendous trifles to place on "Detroit Iron." Or, in the moralist's language, the work of the stylists is "decadent, wasteful, and superficial."

The stylists' work cannot be dismissed so glibly. For however transitory or trivial their visible creations may be on the scale of human values, their function has been designated by automobile company top management as *the* prerequisite for maintaining the annual high volume of automobile sales — no small assignment in an industry that has a volume of at least twenty billion dollars every year.

It is the stylists who are responsible for most of the annual model change which promises the consumer "new" automobiles. It is not surprising, therefore, to find that this "newness" is almost entirely stylistic in content and that engineering innovation is restricted to a decidedly secondary role in product development.

In the matter of vehicle safety, this restriction has two main effects. First, of the dollar amount that the manufacturer is investing in a vehicle, whatever is spent

Ralph Nader, *Unsafe at Any Speed* (New York: Grossman Publishers, 1972).

for styling cannot be spent for engineering. Thus, the costs of styling divert money that might be devoted to safety. Second, stylistic suggestions often conflict with engineering ideas, and since the industry holds the view that "seeing is selling," style gets the priority.

Styling's precedence over engineering safety is well illustrated by this statement in a General Motors engineering journal: "The choice of latching means and actuating means, or handles, is dictated by styling requirements. . . . Changes in body style will continue to force redesign of door locks and handles." Another feature of style's priority over safety shows up in the paint and chrome finishes of the vehicle, which, while they provide a shiny new automobile for the dealer's floor, also create dangerous glare. Stylists can even be credited with overall concepts that result in a whole new variety of hazard. The hard-top convertible and the pillarless models, for example, were clearly the products of General Motors styling staff.

Engineering features that are crucial to the transportation function of the vehicle do exert some restraining influence on styling decisions. A car must have four tires, and though the stylists may succeed shortly in coloring them, it is unlikely that aromatic creampuffs will replace the rubber. But conflicts between style and traditional engineering features are not often resolved in the latter's favor. For example, rational design of the instrument panel does not call for yearly change or recurring variety. Yet the stylists have had their way and at the same time have met management's demands for the interchangeability of components between different car makes. In one instance the 1964 Oldsmobile used exactly the same heater control as the 1964 Buick. In one brand it was placed in a horizontal position; in the other it was used vertically. With this technique, four separate and "different" instrument panels were created for each division.

This differentiating more and more about less and less has reached staggering proportions. In 1957 the Fisher body division produced for the five General Motors car divisions more than 75 different body styles with 450 interior soft trim combinations and a huge number of exterior paint combinations. By 1963 this output proliferated to 140 body styles and 843 trim combinations.

Different designs for what General Motors styling chief Harley Earl called "dynamic obsolescence" must be created for many elements of the car: front ends, rear ends, hoods, ornaments, rear decks and rear quarter panels, tail lamps, bumper shades, rocker panels, and the latest items being offered in an outburst of infinite variation — wheel covers and lugs.

These styling features form the substance of sales promotion and advertising. The car makers' appeals are emotional; they seek to inspire excitement, aesthetic pleasure, and the association of the glistening model in its provocative setting with the prospect's most far-reaching personal visions and wish-fulfillment. This approach may seem flighty, but the industry has learned that the technique sells cars to people who have no other reason to buy them with such frequency.

In recent years, campaigns saturated with the "style sell" have moved on to bolder themes. A 1964 advertisement for the Chevrolet Chevelle said, "We didn't just make the Chevelle beautiful and hope for the best. . . . If you think all we had in mind was a good-looking car smaller than the Chevrolet and bigger than Chevy II, read on." Curved side windows, the ad continued, are not just for appearance, "they slant way in for easy entry and don't need bulky space-wasting doors to roll down into." In addition, Chevelle's "long wide hood looks nice, too," because of all that goes under it — "a wide choice of Six and V-8 engines."

A Buick advertisement listed a number of regular vehicle features and commented, "You don't really need these, but how can you resist them?". . .

General Motors has been the most aggressive advocate of styling. The first distinct styling section was organized in 1927 under Harley Earl. It was called "The Art and Colour Section." At first, the stylists' position was not secure when it came to disagreements with engineers. Earl's first contributions, slanted windshields and thin corner-pillars, had to be justified as "improving visibility." But by the late thirties Earl's group became the "General Motors styling section," and he was elevated to a vice presidency, indicating that the stylist's function was equal in importance to the work of the engineering, legal, company public relations, and manufacturing departments. The styling departments went through similar developments in other automobile companies. The engineer's authority over the design of the automobile was finished. As Charles Jordan of General Motors said, "Previously, functional improvement or cost reduction was a good reason for component redesign, but [in the thirties] the engineer had to learn to appreciate new reasons for redesigns." In a paper delivered before the Society of Automotive Engineers in 1962, Jordan demonstrated how the importance of the stylist has continued to grow when he urged that the word "styling" be replaced by the word "designer." Jordan said that the "designer" (that is, the stylist) is "the architect of the car, the coordinator of all the elements that make up the complete car, and the artist who gives it form. He stands at the beginning, his approach to and responsibility for the design of the vehicle is parallel to that of an architect of a building." An observer might wonder what was left for the engineer to do but play the part of a technical minion. Jordan ended his address by looking into the future. He foresaw changes in the automobile industry that he described as "drastic and far-reaching." He listed eleven questions in advanced research inquiry for which the styling research and the advanced vehicle design sections were working to find answers. Not one concerned collision protection. . . .

The callousness of the stylists about the effects of their creations on pedestrians is seen clearly in the case of William Mitchell, chief stylist at General Motors and the principal creator of the Cadillac tail fin. This sharp, rising fin was first introduced in the late forties, soaring in height and prominence each year until it reached a grotesque peak in 1959 and gradually declining thereafter until it was finally eliminated in the 1966 models. To understand how a man could devise and promote such a potentially lethal protuberance, it is necessary to understand the enthusiasm of Mr. Mitchell, who frequently confides to interviewers that he has "gasoline in his blood." His vibrancy in conversation revolves around the concepts of "movement," "excitement," and "flair." Samples of his recent statements are illustrative: "When you sat behind the wheel, you looked down that long hood, and then there were two headlight shapes, and then two fender curves — why, you felt excited just sitting there. A car *should* be exciting." Or, "Cars will be more clearly masculine or feminine," and "For now we deal with aesthetics . . . that indefinable, intangible quality that makes *all* the difference." Mr. Mitchell's reported view of safety is that it is the driver's responsibility to avoid accidents, and that if cars were made crashworthy, the "nuts behind the wheel" would take even greater chances.

The world of Mr. Mitchell centers around the General Motors technical center, where in surroundings of lavish extravagance he presides over a staff of more than 1,400 styling specialists. It is a world of motion, color, contour, trim, fabric. To illustrate the degree of specialization involved, one color selector holds 2,888 metal samples of colors; glass-enclosed studios, surrounding verdant roof gardens, are specially designed so that colors may be matched under varying lighting conditions. In such an environment, it is easy for Mr. Mitchell to believe that "Eighty-five per cent of all the information we receive is visual." His two favorite sayings are, "Seeing is selling," and "The shape of things shape man."

The matter of Cadillac tail fins, however, transcends the visual world of Mr. Mitchell. Fins have been felt as well as seen, and felt fatally when not seen. In ways that should have been anticipated by Mr. Mitchell, these fins have "shaped" man.

In the year of its greatest height, the Cadillac fin bore an uncanny resemblance to the tail of the stegosaurus, a dinosaur that had two sharp rearward-projecting horns on each side of the tail. In 1964 a California motorcycle driver learned the dangers of the Cadillac tail fin. The cyclist was following a heavy line of traffic on the freeway going toward Newport Harbor in Santa Ana. As the four-lane road narrowed to two lanes, the confusion of highway construction and the swerving of vehicles in the merging traffic led to the Cadillac's sudden stop. The motorcyclist was boxed in and was unable to turn aside. He hit the rear bumper of the car at a speed of about twenty miles per hour, and was hurled into the tail fin, which pierced his body below the heart and cut him all the way down to the thigh bone in a large circular gash. Both fin and man survived this encounter.

The same was not true in the case of nine-year-old Peggy Swan. On September 29, 1963, she was riding her bicycle near her home in Kensington, Maryland. Coming down Kensington Boulevard she bumped into a parked car in a typical childhood accident. But the car was a 1962 Cadillac, and she hit the tail fin, which ripped into her body below the throat. She died at Holy Cross Hospital a few hours later of thoracic hemorrhage.

Almost a year and a half earlier, Henry Wakeland, the independent automotive engineer, had sent by registered mail a formal advisory to General Motors and its chief safety engineer, Howard Gandelot. The letter was sent in the spirit of the Canons of Ethics for Engineers, and began with these words: "This letter is to insure that you as an engineer and the General Motors Corporation are advised of the hazard to pedestrians which exists in the sharp-pointed tail fins of recent production 1962 Cadillac automobiles and other recent models of Cadillacs. The ability of the sharp and pointed tail fins to cause injury when they contact a pedestrian is visually apparent." Wakeland gave details of two recent fatal cases that had come to his attention. In one instance, an old woman in New York City had been struck by a Cadillac which was rolling slowly backward after its power brakes failed. The blow of the tail fin had killed her. In the other case, a thirteen-year-old Chicago boy, trying to catch a fly ball on a summer day in 1961, had run into a 1961 Cadillac fin, which pierced his heart.

Wakeland said, "An obviously apparent hazard should not be allowed to be included in an automobile because there are only a few circumstances under which the hazard would cause accident or injury. When any large number of automobiles which carry the hazard are in use, the circumstances which translate the hazard into accident or injury will eventually arise. Since it is technically possible to add [fins to automobiles] it is also technically possible to remove them, either before or after manufacture.

Howard Gandelot replied to Wakeland, saying that only a small number of pedestrian injuries due to fins or other ornamentation had come to the attention of General Motors, adding that there "always is a likelihood of the few unusual types of accidents."

The lack of complaints is a standard defense of the automobile companies when they are asked to explain hazardous design features. Certainly no company has urged the public to make complaints about such injuries as described by Wakeland. Nor has any company tried to find out about these injuries either consistently or through a pilot study. Moreover, the truth of the statement that "very few complaints" are received by the automobile companies is a self-serving one that is not verifiable by any objective source or agency outside the companies. Also, it must be remembered that since there is no statistical reporting system on this kind of accident — whether the system is sponsored by the government or the insurance industry — there is no publicly available objective source of data concerning such accidents.

As an insider, Gandelot knew that the trend of Cadillac tail fin design was to lower the height of the fin. He included in his reply to Wakeland this "confidential information" about the forthcoming 1963 Cadillac: "The fins were lowered to bring them closer to the bumper and positioned a little farther forward so that the bumper face now affords more protection.

Gandelot's comment touches on an important practice. The introduction, promotion, and finally the "phasing out" of external hazards is purely a result of stylistic fashions. For example, a few years ago sharp and pointed horizontal hood ornaments were the fad. Recent models avoid these particular ornament designs, not for pedestrian safety but to conform to the new "clean look" that is the trademark of current styling. The deadly Cadillac tail fin has disappeared for the same reason. New styles bring new hazards or the return of old ones.

Systematic engineering design of the vehicle could minimize or prevent many pedestrian injuries. The majority of pedestrian-vehicle collisions produce injuries, not fatalities. Most of these collisions occur at impact speed of under twenty-five miles per hour, and New York City data show that in fatality cases about twenty-five per cent of the collisions occurred when the vehicles involved were moving at speeds below fourteen miles per hour. It seems quite obvious that the external design and not just the speed of the automobile contributes greatly to the severity of the injuries inflicted on the pedestrian. Yet the external design is so totally under the unfettered control of the stylist that no engineer employed by the automobile industry has ever delivered a technical paper concerning pedestrian collision. Nor have the automobile companies made any public mention of any crash testing or engineering safety research on the problem.

But two papers do exist in the technical literature, one by Henry Wakeland and the other by a group of engineers at the University of California in Los Angeles. Wakeland destroyed the lingering myth that when a pedestrian is struck by an automobile it does not make any difference which particular design feature hits him. He showed that heavy vehicles often strike people without causing fatality, and that even in fatal cases, the difference between life and death is often the difference between safe and unsafe design features. Wakeland's study was based on accident and autopsy reports of about 230 consecutive pedestrian fatalities occurring in Manhattan during 1958 and early 1959. In this sample, case after case showed the victim's body penetrated by ornaments, sharp bumper and fender edges, headlight hoods, medallions, and fins. He found that certain bumper configurations tended to force the adult pedestrian's body down, which of course greatly increased the risk of the car's running over him. Recent models, with bumpers shaped like sled runners and sloping grill work above the bumpers, which give the appearance of "leaning into the wind," increase even further the car's potential for exerting down-and-under pressures on the pedestrian.

The UCLA study, headed by Derwyn Severy, consisted of experimenting with dummies to produce force and deflection data on vehicle-pedestrian impacts. The conclusion was that "the front end geometry and resistance to deformation of a vehicle striking a pedestrian will have a major influence on the forced movement of the pedestrian following the impact." These design characteristics are considered crucial to the level of injury received, since subsequent contact with the pavement may be even more harmful than the initial impact. As additional designs for protections the Severy group recommends the use of sheet metal that collapses, greater bumper widths, and override guards to sweep away struck pedestrians from the front wheels.

If the automobile companies are seeking more complaints about the effects of styling in producing pedestrian hazards, they might well refer to a widely used textbook on preventive medicine written by Doctors Hilleboe and Larimore. Taking note of the many tragic examples of unnecessarily dangerous design, the results of which "are seen daily in surgical wards and autopsy tables," the authors concluded that "if one were to attempt to produce a pedestrian-injuring mechanism, one of the most theoretically efficient designs which might be developed would closely approach that of the front end of some present-day automobiles."

The ultimate evidence that the work of the stylist is anything but trivial is to be found in the effect styling has had on the economic aspects of the automobile industry.

General Motors, which controls over fifty per cent of the automobile market, whenever it introduces and promotes a particular styling feature can compel the other companies to follow suit. The history of the wrap-around windshield, the tail fin, and the hard-top convertible confirms this point. For although the wrap-around windshield created visual distortion that shocked the optometry profession, and the tail fin and hard-top designs engendered the dangers discussed earlier, every one of the other automobile companies followed the lead of General Motors in order not to be out of date.

Economists call this phenomenon "protective imitation," but under any name, following suit involved tremendous tooling costs, the curtailment of engineering diversity and innovation, and most important, the wholesale adoption of features that were intended to please the eye of the driver rather than to protect his life.

George Romney, then the president of American Motors, described the situation aptly when he told the Kefauver Senate antitrust subcommittee in 1958, "It is just like a woman's hat. The automobile business has some of the elements of the millinery industry in it, in that you can make style become the hallmark of modernity . . . A wrap-around windshield, through greater sums of money and greater domination of the market, can be identified as being more important than something that improves the whole automobile. . . . In an industry where style is a primary sales tool, public acceptance of a styling approach can be achieved by the sheer impact of product volume."

Still the industry has persisted in declaring that it merely "gives the customer what he wants." This hardly squares with Mr. Romney's statement or with the facts. The history of every successful style feature is that it was conceived in one of the automobile company style sections — often without reference to company engineers, let alone considerations of safety — and then turned over to marketing specialists for repetitive, emotional exploitation until it was an entrenched, accepted "fashion."

Entrenched, that is, until the need to make the customer dissatisfied with that fashion sent the styling staffs back to their drawing boards. The principle that governs then is in direct contradiction to the give-them-what-they-want defense. In the words of Gene Bordinat of Ford, the stylist at work must "take the lead in establishing standards of taste." That, in fact, is what they have done.

The follow-the-leader spiral of styling innovations has had other profound effects. One of the most important results is that by concentrating model "changes" in the area of styling, the manufacturers have focused consumer attention on those features of the automobile that are the most likely subject of "persuasive" rather than "informational" appeals. As in the fashion industry, dealing with emotions rather than dealing with the intellect has had the result that the car makers have rarely been threatened with consumer sovereignty over the automobile. On the contrary, car manufacturers have exerted self-determined control over the products they offer. This control is reflected in another statement from Mr. Mitchell, who said, "One thing today is that we have more cars than we have names. Maybe the public doesn't want all these kinds, but competition makes it necessary."

The narrowing of the difference between automobiles to minor styling distinctions is not the only unhealthy result of the stylists' dominance. Even more discouraging has been the concomitant drying-up of engineering ingenuity. As the stylists have steadily risen to pre-eminence, the technological imagination of automotive engineers has slowed to a point where automobile company executives themselves have deplored the lack of innovation. Ford vice president Donald Frey recognized the problem clearly when he said in an address delivered in January 1964, "I believe that the amount of product innovation successfully introduced into the automobile is smaller today than in previous times and is still falling. The automatic transmission [adopted in 1939 on a mass-production basis] was the last major innovation of the industry."

The head of Mr. Frey's company, Henry Ford II, seemed troubled by the same question in his address to the same group. He said, "When you think of the enormous progress of science over the last two generations, it's astonishing to realize that there is very little about the basic principles of today's automobile that would seem strange and unfamiliar to the pioneers of our industry. . . . What we need even more than the refinement of old ideas is the ability to develop new ideas and put them to work."

Neither of these automobile executives, of course, makes the obvious connection that if an industry devotes its best efforts and its largest investment to styling concepts, it must follow that new ideas in engineering — and safety — will be tragically slow in coming.


## STUDY QUESTIONS

1. In the years since this essay was written Nader has obviously convinced many people of the validity of his opinions. Are you persuaded by this essay? What elements of it do you think are the most persuasive?
2. Nader's book has drawn a good deal of criticism from the automobile industry (naturally), and from some automotive writers who contend that his arguments are based on insufficient evidence and misunderstandings. The result, these critics claim, is a one-sided distortion of the truth. Is Nader fair in this essay? At any time do you feel as though Nader is pushing his argument too strongly, or making too much of the evidence he presents?
3. *Writing Assignment:* Choose a product that you own that seems to have a design flaw. (Hint: many automobiles, toys, and household items with flaws are recalled every year.) Write a report that explains why this faulty product needs to be redesigned.

# The Quality Profession: Today and Tomorrow

## Edward A. Reynolds

Quality control can be a career in itself, although it is an occupation many students may not have even heard of. The following essay, taken from a journal for quality assurance specialists, attempts to forecast what the job possibilities may be like in this field in the coming years.

As Walter Shewhart stated half a century ago, we must prevent defects, not simply inspect them out after production, if we are to have economic control of quality. Thus the control of quality has become a more sophisticated engineering-statistical science, planning for quality from design to field application, and with much emphasis on procurement and in-process controls.

That lesson, however, has not been completely learned. Some of our "quality departments" are the old inspection or test, given new names, but still with prime emphasis on attempting after-the-fact segregation of bad from passable.

Fortunately, the number and industrial significance of these companies decreases each year, through education and attrition of those with less alert management. Today, the concepts and methodology of preventative quality assurance is well established in the U.S. and other countries, and the QA methods do cover the entire spiral from concept to application.

What lies ahead? One lesson we must take to heart is that change is not a probability; it is one of the few certainties. The careers of all of us now in quality control were born only a generation ago, and today few careers last a lifetime without major change. For example, the emphasis moved to inspection after centuries of hand craftsmanship. It moved to statistical quality control in our lifetime, and is already moving from that to automation and complex man-machine systems utilizing the sensing ability of instruments, the calculation and display of computers, all coupled to the pattern recognition and judgement of man. Where the emphasis will move, when, or how, we do not know, but we know there will be further change.

Let's take a close look at some of the changes that may occur in the quality control field within the near future.

The statistical "tools" for quality grow each year in sophistication and complexity, but, perhaps even more important, they are being improved in simplicity for wider use and better understanding. Ellis Ott is one of the leading statisticians who has advocated this and, in his fine text *Process Quality Control Troubleshooting and Analysis of Data* (published by McGraw-Hill, 1975), emphasizes the use of simple graphical methods. In my own consulting and teaching for many years I have been using visual examination of frequency distributions (I call them

Edward A. Reynolds, "The Quality Profession: Today and Tomorrow," *Quality Progress*, September 1978, pp. 18–20.

"eyeball" statistics) in place of t or F tests of significance; combined graphs of individual test results and lot averages (visually determined) in place of X bar and Range control charts; and graphical, arithmetic, summations of designed experiments in place of variance analysis.

## Elephant Gun or Fly Swatter

The result has been better understanding and wider use. As one head of a large engineering organization said to me: "I studied statistics in college, but I never knew I could use them!" Modern statistics have a wide variety of techniques — fly swatters for little bugs of problems, elephant guns for great beasts. The quality professional must have knowledge of both, but resist the tendency to use the elephant gun on flies simply to show his prowess!

An exciting indicator toward the more common use of quality statistics has been the leadership of Japanese industry in teaching data gathering and evaluation at the worker level and encouraging their use through the quality circles programs. Some American corporations are experimenting with this method and the American Statistical Association has been attempting to encourage statistical education at the elementary school levels.

For the understanding and intelligent usage of statistical quality methods to be widespread among design engineers, manufacturing supervisors, and general managers, the quality profession must develop language that is understandable, and techniques that are quick and simple . . . certainly no more complex than needed for any problem at hand. I believe this trend is already well under way — less mystery, less complexity, more day-to-day use of quality statistical techniques by those not directly in the quality profession.

Many current leading practitioners in quality assurance have education and experience foundations primarily in the statistical methods. This is certainly understandable, since statistical quality control was the cornerstone on which our work was started. However there are many highly important techniques for controlling and improving quality, and the professional of the future, unless he is to be a narrow specialist (which, of course, we shall continue to require), must be able to help his organization improve quality through such methods as quality cost systems to locate major needs, personnel training and motivation for quality, vendor evaluation and certification programs, better customer or field feedback on quality conditions, and long-range "breakthrough" programs for achieving new levels of quality through combined management-diagnostic investigations.

In much of modern industry, the demands for near-perfect reliability and the speed and complexity of both process and piece parts operations have made sampling and control by manual charting ineffective. Increasingly, we are controlling processes and products by again 100 percent checking — not by hand inspections, but automatically — often with process control feedback through on-line instrumentation and computers, or with automated segregation of discrepant materials. Electronic, air flow and gamma gages; nondestructive tests using eddy currents, sonics or ultrasonics, microwaves, infra-red heat sensing, fluorescent particle, radiography, etc. are all in increasing use.

## Continuing Automation

It is my opinion that this tendency toward automation in both production and quality controlling will certainly continue. Unfortunately it has not been given as much attention as it deserves. Too many American corporations have been content to achieve cost reductions by moving about the country, and then out of it, to reduce labor costs, rather than developing labor saving methods for routine tasks. (If Eli Whitney had been of this mentality, we should now be importing hand fitted muskets!) Regardless of this, automation and automatic quality controls will continue to grow in the U.S.

The QA specialist, who studied statistics a decade or more ago, should now learn instrumentation, computers and NDT. He might recall the example of the old-time chief inspector who resisted SQC training and use some years ago, only to be displaced by the SQC engineer as the key person for control of quality.

To date, in the U.S., the great majority of modern QA applications has been in large manufacturing organizations where both the demonstrated needs and the facilities, in technical manpower and funds, encouraged such work. However there are important and unfilled potential applications in small companies, often individually owned and directed. Here the same concepts apply, but the organization for quality and the administrative techniques must be quite different. The small organization usually has very limited technical-supervisory personnel — often only the owner-manager — with many responsibilities in addition to quality control. We must either supply him with easily assimilated instruction (here the simplification of SQC will help) or provide convenient and economical consulting services to help in installation of effective QA techniques — simple ones aimed at his particular needs and abilities.

It would seem this need could best be filled by technical societies, government agencies; small business associations, or sparetime contribution of retirees or highly competent, and altruistic QA engineers and managers. In any event, it is an important need and especially necessary in those areas where there are a large number of small manufacturing establishments. The solution offers a challenge to QA leaders.

The trend for effective QA in such service organizations as banks, insurance companies, airlines, car rental agencies is well established and we are beginning to see use of QA programs in hospitals and other areas of health care. This trend seems certain to continue, including QA of government activities — fire and police departments, etc.

Until now, the quality engineer has usually been a person with specialized education (statistics, NDT, instrumentation, computers, cost systems, motivation methods, etc.) built on a foundation of some particular discipline — most commonly mechanical engineering, electronic engineering, metallurgy, chemistry, etc. — with some having a more general mathematical-statistical or industrial engineering foundation. For the most part, the past practice was to require basic education germane to the process or product to be quality controlled. To date, this is less so, and the tendency seems to be that soon the quality professional will be akin to the accounting or personnel specialist, able to use his skills almost regardless of process or product (almost — not entirely).

As we consider this, it seems quite logical. The statistical skills are not dependent upon product, but rather on complexity of the quality problems. The administrative techniques depend largely upon size of the operation, and may be similar for a large bank or hospital as for a machine shop or textile factory. Only the engineering techniques differ, and even here we find mutuality in many instances. For example, infra-red NDT may be used in tire testing, in the automated 100 percent checking of complex electronic assemblies, in evaluation of home construction, utility maintenance control, or in hospital diagnostics. Human resources engineering, value engineering, industrial engineering, and inspection engineering are common to control of quality in many different applications. The ASQC Certification as quality engineer makes no distinction as to type or product, nor does the first of the U.S. states to grant a professional license to quality engineers (California). Today it is quite common to see ads for quality engineers which make no mention of type operations or products (however, these are still in the minority).

## Product Generalist

The general trend is for the quality specialist to be a product generalist, able to ply his professional skills in many fields of endeavor such as research, commerce, service organizations, even government and education. Those who plan careers in quality may find it advantageous to broaden both their outlooks and skills.

Growth in the quality controlling sciences seems to assure many future opportunities. Recently, one issue of the *New York Times* contained 40 ads for quality or reliability engineers or managers, and salaries for the quality specialist are higher than for several other technical areas with the same foundation degree and years of experience. The demand, and shortage, seems certain to continue in view of consumer movement and government quality, safety and reliability requirements, as well as competitive demands for product and service performance.

American universities have not kept up with the demand, and three years ago ASQC published a listing showing only 23 colleges and universities in the U.S. offering degrees (at the BS or MS levels) in quality sciences — although many more offered courses or majors in other programs. Today, this number is higher, but still inadequate to fill the future needs. However, there are many short courses or seminars to help in training. In addition, the level of training is undoubtedly increasing and several colleges now offer PhD programs majoring in quality assurance, and the level of seminar training is generally higher.

Beyond the quality control function, careers in quality offer good prospects for advancement to general, high management. There are few areas of industry which as well permit access to all activities and problems as does the quality function, and this alone gives the quality professional well rounded managerial exposure. Some 20 years ago, the head of one of America's most respected psychological testing services said that the quality manager had characteristics well suited to top management, being more technical and innovative than the production or sales manager and more practical and cost minded than the engineering manager. Today, I see many examples of good QA managers moving to high executive positions and suggest this trend will intensify.

With these opportunities, come responsibilities. The quality professional must, of course, have the high standards of integrity that should apply to all in responsible positions. And few areas of modern activity are subject at times to more pressure than the quality head to bend the rules, ignore the unfavorable fact, "make yet another retest!" Trade-offs of quality-cost-delivery must be made, but always openly, objectively, honestly, or not at all. Certainly all decisions and actions must be unbiased by fear or personal gain.

But more than that, the quality professional has great responsibilities to work for improvement — products or services that are not only better, but more available through lower costs or higher yields. We cannot be content to "control." Tomorrow must be better than today, and we have a responsibility to make it so.

## STUDY QUESTIONS

1. Reynolds makes some predictions about supply and demand in the field of quality control, yet he includes no specific figures in his predictions. Is this a serious drawback? Do terms such as "good prospects" and "many future opportunities" seem to be proof enough in themselves?

2. What kind of research has gone into this report? Does this amount of research seem adequate? Is reading the ads in the *New York Times* a reliable method of determining salary ranges?

3. What conclusions are you asked to accept in this article? Have you been persuaded to agree with Reynolds on these points? How do you feel the readers of this journal might react?

4. *Writing Assignment:* Study the employment opportunities for your prospective career in a large classified section (your local newspaper may not be appropriate if it has a small circulation). Write an article similar to this one, entitling it "The _____ Profession: Today and Tomorrow."

# Section 3

# Implementing

We have seen writing that has the primary purpose of presenting information, and we have seen writing that has the primary objective of using information to influence decisions. Now we will consider writing as an instrument of task performance. In this form, writing is an instrument without which a job cannot be completed readily and correctly — or perhaps cannot be completed at all. Such writing is likely to be in one hand while a wrench is in the other. Without the writing there is no using the wrench. Writing is now more than a supplement to the tool; it is itself a tool.

Such writing includes *manuals* and other less formal versions of *"how-to" instructions.* Although these are perhaps the most obvious categories of instrumental writing, they aren't the only ones possible. In addition to manuals and "how-to" instructions we will include *specifications* and some forms of *laboratory reports.* This might seem a diverse set of documents to group together, but we must remember that one central objective underlies them all. These documents are not for the person who is seeking only to enlarge his knowledge of a topic. They are not for the person who is looking to be convinced that a certain approach is proper. These documents are for the person who has a given task to complete and who must have the proper guidelines, instructions, and data for completing this task.

Let us look briefly at each type of document that might come under this heading:

*Manual* — a written tool that accompanies a product or guides an activity. In the first sense, manual is a term commonly used to denote the written document that gives necessary information on the care and use of a manufactured product. A comprehensive manual might include data on product design, function, and performance capabilities, as well as directions on unpacking, installation, maintenance, trouble-shooting, repair, and overhaul. Manuals of this scope are generally found with highly complex products intended for industrial or commercial use.

The manual that accompanies a portable hair dryer, a consumer product, would not be nearly as comprehensive. Owners' manuals might include little more than a description of operating controls, operating instructions, safety tips (which might be in the form of a highly simplified list of DO's and DON'Ts), and directions for returning the product to the factory in case of malfunction.

Manual is also a term that is commonly used to describe written instructions that give a comprehensive approach to a skill or activity. Examples could include *A Manual for Wood-Carvers* and *The Jogger's Manual*. Strictly speaking, such works are better described as *instruction manuals*.

*How-To Instructions* — a step by step guide for performing a given task. Instructions are included in manuals, but a set of instructions alone does not make a manual. A helpful distinction is to understand the manual as a format that provides for several categories of information and instruction for using and maintaining a product. On the other hand, a set of "how-to" instructions is usually less ambitious in its objective. Typically, it isolates a task to be accomplished, then proceeds to mark off the appropriate steps necessary for the accomplishment of that task.

Although we identify instructions with the rigidity of commands — "insert tab A into slot B" — not all activities can be approached so directly. Instructions must always involve "how" but sometimes need the "what" and "why" to provide a necessary perspective. Sometimes the "what" and "why" are isolated in an introduction, then followed by a rigid set of "how" steps. Other times the "what" and the "why" are not readily separated from the "how," thus dictating a looser, less rigid approach to instructions.

*Specification* — the word *specification* is in the terminology of several types of related documents. For example, *The Handbook of Technical Writing Practices* lists four forms:

1.) In the singular form, *specification* denotes a clear and accurate description that a company draws up to serve as a guide in the purchase of materials and services.
2.) A *specification control drawing* includes configuration, design, and test requirements for a manufactured item.
3.) A *specification sheet* has details which define product functions, appearance dimensions, performance capacities, and the like.
4.) *Specifications* are a form of sales literature that give a detailed listing of the characteristics of a product that are guaranteed by the seller.

With all forms of the specification we move into a truly "technical" aspect of report writing. They are highly specialized documents and are usually handled by writers who have an extensive background in the fields for which they write. As working documents with a high degree of practical applicability, their implemental quality should be obvious.

*Laboratory Notes and Reports* — the researcher must keep clear and accurate notes on his laboratory experiments. The raw data in the notes will be used to support the conclusions drawn in the laboratory report. The notes and the report are instrumental documents to the extent that other researchers can draw their own conclusions from the data, or use the data and conclusions as the basis for future experiments.

The documents that we have discussed are not meant to entertain, so we cannot expect them to be "fun" to read. We expect them to help us complete the task at hand, and they are successful to the extent to which they do just that. Nor are they necessarily "fun" to write, although some instrumental documents do call upon a writer's creative abilities. For example, translating your expert knowledge of a complex technical field into instructions that the layman can readily follow requires a high order of creativity.

These documents stress two main virtues: clarity and accuracy. For some documents these virtues are mandatory. In the airline evacuation chart, there can be no room for misinterpretation. Anything less than immediate and total understanding of the instructions could be fatal. Clarity and accuracy in specifications are important because ambiguities of language and mistakes can, and do, lead to costly lawsuits. Inaccuracies in a laboratory report will lead to faulty conclusions.

A key principle in writing such documents is to include enough without including too much. Of course, "enough" and "too much" are, depending on the task, highly variable terms. "Too much" information in the evacuation chart might be dangerously misleading. However, too little in the instructions for changing a flat tire might leave the motorist stranded far from help.

Or perhaps we are considering instructions aimed at the young executive who wants to learn to get ahead in the corporate world. Seemingly, the writer would have room for more flexibility in the categories of information that he includes, because he must keep his reader interested. Becoming successful is important to many people, but rarely does success have the immediacy of, say, a busted water pipe.

So there must be an appropriate balance of information, and there is, unfortunately, no rule book that can tell you how to establish that balance. The answer, as much as there is one, lies in careful analysis on the part of the writer. He must think through the task and determine what the reader must know, and what might be useful, and what might be discarded. As in all writing tasks, the writer must plan ahead. As always, the written document is but the end product of a carefully planned and thought-out project.

Most of the examples included in this section are valuable for their content; but the guiding principle in assembling these pieces was to introduce students to a wide variety of formats for implemental documents. In some cases, the documents are very rigid in their format, in others there is room for looseness. You might try analyzing each one as to its success in finding the appropriate balance for achieving its implemental objective.

# Replace Thermostat

## From the *1978 Owner Maintenance and Light Repair Manual*

Many new cars come with an owner's manual that does little more than tell what knob does what and when to take the car in for servicing. Ford, among others, provides another manual for those owners who want to undertake simple repair jobs. Usually you must request these manuals from the company — and sometimes they are free.

> Tools Needed:
>     Socket
>     Extension Handle
>     Pliers
>     Drain Pan
>     Putty Knife
> Materials Needed:
>     New Thermostat
>     Gasket for Housing
>     Gasket Sealer
>     Ford Cooling System Fluid or Equivalent

In most cases the thermostat (Figure 1) is located at the engine end of the radiator top hose. On the 2.8 L, V-6 engine, the thermostat is located at the bottom of the water pump housing (Figure 2).

The thermostat is designed so that it will go to the full-open position after the engine reaches normal operating temperatures. If you suspect a thermostat malfunction, put the heater controls to maximum heat. With the heater blower on, the air discharging from the heater should be hot after you have driven 2 or 3 miles from a cold start. If the air is not hot, you have reason to suspect the thermostat.

1. Remove the radiator cap. (See radiator cap warning in this section.)
2. Place a clean container under the radiator draincock. With a pair of pliers, open the draincock by turning it counterclockwise and drain the cooling system.

---

**NOTE: The draincock can be located on either side of the radiator at the lower end, depending on the vehicle.**

---

From the *1978 Owner Maintenance and Light Repair Manual* (FPS 365–13078, January, 1978), pp. 4–22 through 4–23.

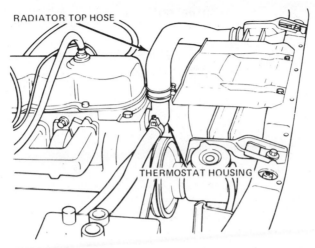

Figure 1. Typical Thermostat Housing — Except 2.8 L, V-6 Engine.

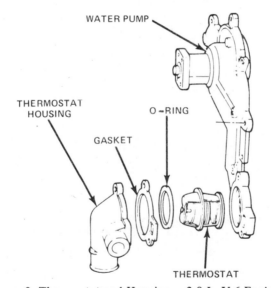

Figure 2. Thermostat and Housing — 2.8 L, V-6 Engine.

3. Remove the bolts that attach the thermostat housing to the engine.
4. Pull the thermostat housing away from the engine.
5. Remove the thermostat from the housing (Figures 2 and 3). **Note the position of the thermostat upon removal.**

**Thermostat Test:**

a. Immerse the thermostat in boiling water. Replace it if the thermostat valve doesn't open at least ¼ inch.

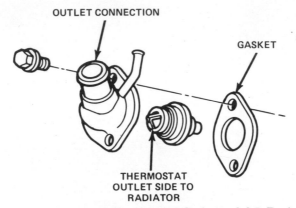

Figure 3. Thermostat Housing and Gasket — 2.3 L Engine.

b. Let the thermostat cool.
c. Check the thermostat for leakage by holding it up to a lighted background. If excessive light is visible around the thermostat valve when the thermostat is at room temperature, it should be replaced. Some light leakage is normal.

---

NOTE: Do not attempt to repair a thermostat; replace it.

---

6. Remove the thermostat housing mounting gasket(s) with a putty knife, and clean the adjacent areas with a clean rag.
7. Coat the new gasket with water-resistant sealer and position gasket(s) properly before installing the thermostat. On 5.8 L, 6.6 L, and 7.5 L (351 M, 400, and 460 CID) engines, install the thermostat before installing the gasket.

---

NOTE: Do not install the thermostat backwards. Inspect the thermostat for a stamping to indicate which end goes toward the radiator.

---

8. Carefully position the thermostat housing against the engine.

---

NOTE: To prevent incorrect installation of the thermostat, the outlet on some engines contains a locking recess into which the thermostat is turned and locked. Install the thermostat with a bridge section (Figure 4) in the outlet casting. Turn the thermostat clockwise to lock it in position on the flats cast into the outlet elbow.

---

9. Install the attaching bolts. Tighten the attaching bolts to 12 to 15 ft-lb (17 to

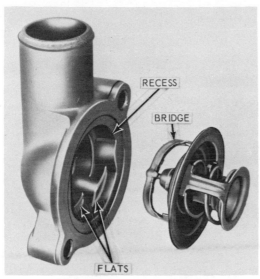

Figure 4. Thermostat Installation Typical in Line 6 and V-8 Engines.

20 N·m) torque. On 2.3 L engines, tighten the bolts to 14 to 21 ft-lb (23 to 28 N·m) torque.
10. Close the radiator draincock.
11. Fill the cooling system. Follow procedures under Refill Cooling System in this section.
12. Reinstall the radiator cap.
13. Start and run the engine until it reaches normal operating temperature.
14. Stop the engine.

---

**WARNING: Be careful not to touch the hot engine when checking for leaks. Also, be careful when removing the radiator cap.**

---

15. Inspect the thermostat for leaks.
16. Check the radiator for proper coolant level after the engine has cooled. If additional coolant is needed, add a mixture of half water and half permanent antifreeze.

## STUDY QUESTIONS

1. In *Reporting Technical Information,* Houp and Pearsall list six possible categories of information that might be included in "how to" instructions: 1.) introduction, 2.) basic principles of operations, 3.) equipment and materials needed for the operation, 4.) a description of the mechanism, 5.) step-by-step perfor-

mance instructions, and 6.) a conclusion. Which of these does the author use in giving instructions for replacing a thermostat? Why were some categories omitted?

2. In three of the four illustrations, the author uses line drawings rather than photographs. What are the advantages of line drawings?

3. What is the purpose of numbering the steps?

4. Step number 12 instructs you to "Reinstall the radiator cap." Is this level of instruction necessary? Could something so simple be included as part of another step? Discuss.

# Fish

## Julia Child, Louisette Bertolle, and Simone Beck

One of the most common forms of written instructions is the recipe. In a world in which women are (currently) more likely to do the cooking, many men are unaware of the importance of a good recipe. The selection below is taken from Child, Bertolle, and Beck's *Mastering the Art of French Cooking.* The value of their cookbook lies not in the fact that they were able to translate French into English — any French major could do that — but in the fact that they could translate the practice of the French kitchen into an American practice.

THE FRENCH are magnificent with fish. Not only is fresh fish abundant all year round, but the art of its cooking and saucing is accomplished with great taste and skill.

This chapter includes two fine recipes for scallops, one for tuna or swordfish, three for lobster, and a group for mussels. But the main emphasis is on the important and typically French method of poaching *filets* of fish in white wine and serving them in a wine sauce, starting with the simplest type of sauce and ending with several of the most famous of *la grande cuisine.* These last, as you will observe, are fish *veloutés* (flour and butter *roux* simmered with the fish cooking liquid), which are then enriched with cream and egg yolks. They are all the same basic sauce described in detail on page 60 in the Sauce chapter. Under numerous disguises and with various flavorings, this sauce appears throughout almost every phase of French cookery.

### A NOTE ON BUYING FISH

Fish must be fresh smelling and fresh tasting. If it is whole, its eyes are bright and full, not filmed, opaque, and flat. Its gills are bright red, its flesh firm to the touch, its skin fresh and glistening.

Frozen fish should be bought from a dealer who has the proper facilities to ship and store it at a constant temperature of zero degrees. It should be solidly frozen. A block of frozen juices at the bottom of the package is proof that it has been thawed and re-frozen. Before cooking, defrost it in the refrigerator, or under cold running water.

From Julia Child, Louisette Bertolle, and Simone Beck, *Mastering the Art of French Cooking* (New York: Knopf, 1971), pp. 207–209.

## SERVING SUGGESTIONS

A beautifully sauced fish can well be considered as a separate course and needs nothing but French bread and a good wine to go with it. If it is a main course, include *risotto* or steamed rice for shellfish, boiled potatoes for other fish. A salad or vegetable should come afterward, so as not to disturb the harmony of the fish, the sauce, and the wine.

---

## FISH FILETS IN WHITE WINE SAUCE

### THE FISH FILETS

Most of the famous French dishes involving *filet* of sole center around fish poached in white wine and coated with a lovely, creamy sauce made from the poaching liquid. Although many types of American flat fish and fish *filets* are called sole, they are usually flounder because the true sole is not a native American fish. European sole is flown over to America, and can be bought, but it is rarely seen in the usual American market. The sole's ease of skinning and filleting, and its close-grained yet delicate flesh make it ideal for poaching. The best American substitutes for European sole are the Atlantic winter flounder, *Pseudopleuronectes americanus,* and the Pacific Petrale sole or brill, *Eopsetta jordani.* Each of these is a common sole *filet* in America, depending on where you live; it is doubtful, however, if their technical names would be known at your fish market. Whiting or silver hake; pollack or Boston blue-fish; summer flounder or fluke; dab; gray, lemon, or English sole; and fresh-water trout are other types of fish you may fillet and poach. If you notice that any of these flake during their poaching, they should be sauced and served in the poaching dish. All of the fish mentioned in this paragraph, including the true European sole, may be used interchangeably in any of the following recipes.

---

## * FILETS DE POISSON POCHÉS AU VIN BLANC

[Fish *Filets* Poached in White Wine]

*For 6 people*

---

Preheat oven to 350 degrees.

---

A buttered, 10- to 12-inch, fireproof baking and serving dish, 1½ to 2 inches deep
2 Tb finely minced shallots or green onions
2½ lbs. skinless and boneless sole or flounder filets cut into serving pieces
Salt and pepper
1½ Tb butter cut into bits
1¼ to 1½ cups cold, white-wine fish stock made from heads, bones, and trimmings
OR ¾ cup dry white wine or ⅔ cup dry white ver-

Sprinkle half the shallots or onions in the bottom of the dish. Season the *filets* lightly with salt and pepper and arrange them in one slightly overlapping layer in the dish. If *filets* are thin, they may be folded in half so they make triangles. Sprinkle the *filets* with the remaining shallots or onions, and dot with butter. Pour in the cold liquid and enough water so fish is barely covered.

mouth plus ¼ cup bottled
clam juice, and water
OR 1½ cups wine and water
mixed

---

Buttered brown paper or
waxed paper (do not use
aluminum foil—it will
discolor the wine)

Bring almost to the simmer on top of the stove. Lay the buttered paper over the fish. Then place dish in bottom third of preheated oven. Maintain liquid almost at the simmer for 8 to 12 minutes depending on the thickness of the *filets.* The fish is done when a fork pierces the flesh easily. Do not overcook; the fish should not be dry and flaky.

---

An enameled saucepan

Place a cover over the dish and drain out all the cooking liquid into an enameled saucepan.
(*) The fish is now poached and ready for saucing. It may be covered and kept warm for a few minutes over hot, but not simmering, water. Or set it aside, covered with its piece of paper, and reheat later for a few minutes over simmering water. Be very sure the fish does not overcook as it reheats. Before saucing the fish, drain off any liquid which may have accumulated in the dish.

## STUDY QUESTIONS

1. Identify the intended audience for this recipe. Is it for the beginning cook, the intermediate cook, or the advanced cook? Explain.
2. The authors use a two-column format for presenting the steps of the operation. What are the virtues of this method?
3. The first six paragraphs are concerned not so much with the actual cooking as with general information. What is the value of the introduction? Could this introductory material be omitted?

# The Mile-High Illinois

## Frank Lloyd Wright

In America, modern architecture is synonymous with the name Frank Lloyd Wright, whose designs — residential and public — have influenced generations of architects. Although Wright was generally anti-skyscraper, he drew up the specifications for a building that would be about a mile high. It was never built.

CANTILEVER SKY-CITY · 528 STORIES · TRIPOD IN PLAN · ONE MILE HIGH FROM GRADE TO TOP FLOOR, DIVIDED INTO FOUR SECTIONS · EXPOSED MEMBERS ALUMINUM OR STAINLESS STEEL · ELEVATORS ESPECIALLY DESIGNED TANDEM-CABS RATCHET-GUIDED TYPE, ATOMIC POWER · ESCALATOR SERVICE BASEMENTS AND FIRST FIVE FLOORS · FOUR QUADRUPLE-LANE APPROACHES TO EACH OF THE FOUR ENTRANCES · ONE ENTRANCE AT EACH CORNER · PARKING FOR ABOUT 15,000 CARS AND LANDING DECKS FOR 150 HELICOPTERS

### Specification

1. Of all forms of upright structure, most stable is the tripod: pressures upon any side immediately resisted by the other two. For general stability at great height this form of the *Illinois* is planned to employ the new principles of cantilever — steel in suspension — as in the Imperial Hotel, Tokyo, the Johnson Heliolaboratory at Racine, Wisconsin, and Price Tower at Bartlesville, Oklahoma. The exterior of the *Illinois* is entirely metal. The exterior wall screens are suspended from the edges of the rigid upright steel cores — cores buried in light-weight concrete: this building thus designed from the inside outward instead of the dated steel-framed construction from outside inward. The entire structure is thus more airplane in character (twentieth century construction) than the usual heavy nineteenth century building. For instance, the support of the outer walls and sixteen feet of the outer area of the floors is pendent, and the science of continuity is employed everywhere else. From inside outward always. Floor-slabs are extended across the vertical central core. All floor-loads are balanced against each other over central supports, even the outer perimeter of floor slabs and exterior walls which are suspended from the cores. The science of continuity thus employs the type of construction similar to that of the airplane and ocean liner. The Imperial Hotel and the Price Tower, the National Life Insurance Building, the Johnson Heliolaboratory, Fallingwater, etc. were all of this type of structure: a "natural" for either great spans or great heights. Throughout the *Illinois* typical weights of this

From Frank Lloyd Wright, *A Testament* (New York: Bramhall House, 1957), pp. 239–240.

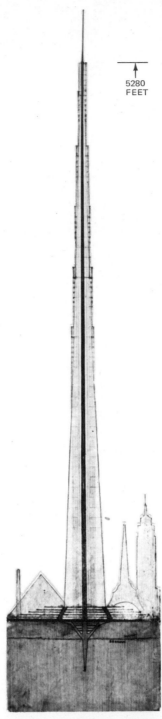

5280
FEET

1956. Project, The Mile-High Illinois, Chicago, Illinois Section Showing Taproot
Foundation in Bedrock.

structure are little more than half those of nineteenth century building practice: the customary bridge-engineers' welded steel framing.

2. This interior system of building-construction is new: twentieth century. Tension as here involved in the *Illinois* (upright) was first used by myself in the Imperial Hotel (horizontally), 1915, and proved earthquake-proof, 1922. The same general system has now been repeated vertically — and successfully — in the nineteen story, forty-foot square, Price Tower — steel-in-tension there used as years before in the Heliolaboratory.

   The same principle centralizes loads over a giant core of properly designed steel fabrication in the *Illinois,* cast in appropriate masses of light-weight concrete. As all floor loads balance each other over staunch rigid cores, with the outer portion of the slabs and the outer wall-screens suspended from these cores, the framework of the *Illinois* is like a tree — the horizontal floor slabs integral with the vertical cores, making the total structure light and rigid, not rickety.

3. The 528 light floor slabs are hollow, tapered from the cores to carry air-conditioning, lighting and appurtenance systems. These cantilevered floor slabs are formed by special high-tension steel, diamond-mesh reinforced, and cast into light concrete slabs. Excepting vertical elevator enclosures, which issue from the sloping sides of the tripod, all exterior surface features of this structure and certain outer areas of the floors also are suspended by steel strands from the sloping corners of the core as already described. Outer glass surfaces are set four feet back under the metal parapets to avoid glare of glass and afford a human sense of protection at such enormous heights as characterize the *Illinois Sky-City.*

4. Elevator transit is by atomic power; especially designed elevators, each five stories high, serving in series the five divisions of 100-floor heights. A group of 76 tandem-cab elevators five units high begin to load where the escalators leave off at the fifth floor. These elevators entirely independent of ordinary suspension systems. As motorized they rise and emerge on ratchet-guides, soaring into the air, independent of, yet an integral feature of the tripod, and appear outside the tripod as graceful vertical features of the *Illinois.* The entire elevator system thus rises perfectly upright to five different story-heights. Special through-service is provided to the upper stories and to the very top floor at various speeds all the way to one mile above the ground floor level. All elevators are motivated by atomic power, engines on the cabs engaging ratchet-tracks, cabs moving vertically at various speeds much as an automobile runs on the level. Approximate speed: say, a mile per minute; appropriate automatic stop-and-go controls without attendance are to be provided. Additional private lifts may connect various departments independently of main elevators. Cars are set aside for non-stop emergency service. As escalators from the lower parking levels serve the first five stories, the main floor of the *Illinois* is thus practically the fifth floor. This combination escalator-elevator service

should empty the entire building within the hour by day and the various occupations by night in half the time.

5. The *Illinois* employs the now proved system of "tap-root" foundation sloping to hard-pan or bedrock, again similar to the foundations of the Heliolaboratory and the Price Tower; all similar in principle to the foundation system that saved the structure of the Imperial Hotel in the 1922 temblor. To make rigidity possible at the extreme heights of the *Illinois,* this type of foundation continues the main core into the ground to reach rock or hard-pan formation beneath. The foundation has available spaces within it for utilities, owing to its tapered form. Final drilling into bedrock for insert of the spinal cores is not difficult to construct.

6. Finally — throughout this light-weight tensilized structure, because of the integral character of all members, loads are in equilibrium at all points, doing away with oscillation. There would be no sway at the peak of the *Illinois.*

   A rapier, with handle the breadth of the hand, set firmly into the ground, blade upright, as a simile, indicates the general idea of the *Illinois* five times the height of the highest structure in the world.

7. Exterior features: the elevators, outer parapets, all exposed horizontal or vertical members are of gold-colored metal and, with the windows set back under steel parapets to avoid glare, give the building emphasis as an all metal structure.

8. Covered parking for about 15,000 cars may be reached by ramps connected to one central level below grade and four levels above. These lower levels and the sub-floor parking beneath the building itself have direct access to and from escalators; there above, beside the main building, are two decks, each for 75 helicopters.

IN GENERAL: The *Illinois* is divided into four parts, and is reached at four points by four four-lane approaches. Fountain features and green-planted parterres are thus related to the tripod entrances, each independent of the other.

The riveted or welded steel framing of the nineteenth century  has been abandoned. All this well done, this great twentieth-century edifice will be more permanent than the Pyramids.

| Areas of the Illinois | Gross Area | Net Rentable Area |
|---|---|---|
| From Grade level to entrance level | 490,000 sq. ft. | 374,000 sq. ft. |
| Entrance level to 1st elevator floor | 480,000 sq. ft. | 364,000 sq. ft. |
| 1st elevator floor to 110th floor | 8,050,000 sq. ft. | 5,250,000 sq. ft. |

| | | |
|---|---|---|
| 111th floor to<br>328th floor | 7,800,000 sq. ft. | 5,840,000 sq. ft. |
| 329th floor to<br>428th floor | 1,150,000 sq. ft. | 850,000 sq. ft. |
| 429th floor to<br>528th floor | 492,000 sq. ft. | 369,000 sq. ft. |
| TOTALS | 18,462,000 sq. ft. | 13,047,000 sq. ft. |

IN SPACIOUS COMFORT APPROXIMATE TOTAL POPULATION 130,000 INHABITANTS

**STUDY QUESTIONS**
1. Wright calls his description a *specification*. Review pages 319–321 for the different types of *specification*. Which type has Wright used?
2. A mile-high building will not be constructed in the foreseeable future — maybe never — a fact that the very shrewd Wright must have been aware of. Why do you suppose he wrote up this fanciful project? Why did he put it in the matter-of-fact format of a specification?
3. Why is at least half of the specification concerned with the construction principles of the building? Is this appropriate for a specification?
4. The illustration that accompanies the specification is not precise. Despite its lack of precision, what useful purpose does it serve?
5. *Writing Assignment:* Use Wright's format to write a specification for a product or tool in your job field. Or, use the format for drawing up the specifications of a proposed project.

# The Laboratory Notebook

## Walter E. Harris and Byron Kratohvil

The following selection includes "how-to" information on keeping a laboratory notebook as well as a sample of laboratory notes. Although the sample notes present only raw data, the precise entries will aid the experimenter and his colleagues in making sound conclusions. Note the usefulness of the tabular format for recording data.

A laboratory notebook is designed to provide a permanent record of firsthand

From Walter E. Harris and Byron Kratohvil, *Chemical Separation and Measurements* (Philadelphia: W. B. Saunders, 1974), pp. 3–5.

observations. The criterion for sound record keeping is that someone else can readily locate pertinent data and results for an experiment. Although reasonable legibility and neatness are desirable, the usefulness of a record is determined largely by whether it is original, systematic, and complete — not by whether it is a work of art.

The notebook should be hard-covered and bound. A 7- by 9-in. size is convenient. Number the pages consecutively, leaving two or three pages at the front for an index.

To keep a proper experimental record:

1. Enter data directly in ink as soon as taken. Never recopy numbers or use loose sheets of paper. Cancel errors or rejected data by drawing a single line through them. Do not erase or remove pages. The notebook should be a permanent record of the original laboratory work.
2. Enter data only on the right side of the page. Use the left side for recording observations that may be useful in evaluating results, for calculations, and so on.
3. Clearly label all entries. To facilitate direct entry of experimental work, set up a data page for each experiment before starting. Examples are shown below.
4. Make the record of experiments complete but concise. Date each entry. Give each experiment a title and list it in the index.

## Examples of Summary Data Pages

| Standardization of HCl against $Na_2CO_3$ | Date *October 10, 1973* | | | |
|---|---|---|---|---|
| | 1 | 2 | 3 | 4 |
| Wt weighing bottle + sample, g | *19.2750* | *18.8346* | *18.4640* | *18.0229* |
| Wt weighing bottle, g | *18.8346* | *18.4640* | *18.0229* | *17.6317* |
| Wt $Na_2CO_3$ sample, g | *0.4404* | *0.3706* | *0.4411* | *0.3912* |
| Final buret reading | *40.22* | | *43.41* | *40.97* |
| Initial buret reading | *0.14* | *1.03* | *3.23* | *3.48* |
| Difference | *40.08* | *End point missed* | *40.18* | *37.49* |
| Net buret correction | *+0.02* | | *+0.01* | *+0.02* |
| Volume HCl (corrected) | *40.10* | | *40.19* | *37.51* |
| Molarity of HCl | *0.2072* | | *0.2073* | *0.2069* |
| Average molarity of HCl | | *0.2071* | | |
| Relative average deviation | | *0.8 ppt* | | |
| Median | | *0.2072* | | |
| Relative probable deviation | | *0.5 ppt* | | |

| Determination of $Na_2CO_3$ in a Sample (Sample Code Number B-231) | Date *October 10, 1973* | | | |
|---|---|---|---|---|
| | 1 | 2 | 3 | 4 |
| Wt weighing bottle + sample, g | 16.4361 | 16.0632 | 15.6321 | 15.2207 |
| Wt weighing bottle, g | 16.0632 | 15.6321 | 15.2207 | 14.8100 |
| Wt sample, g | 0.3729 | 0.4311 | 0.4114 | 0.4107 |
| Final buret reading | 33.96 | 43.90 | 38.31 | 36.44 |
| Initial buret reading | 1.23 | 5.84 | 1.86 | 0.16 |
| Difference | 32.73 | 38.06 | 36.45 | 36.28 |
| Net Buret correction | +0.04 | +0.01 | +0.03 | +0.03 |
| Volume HCl (corrected) | 32.77 | 38.07 | 36.48 | 36.31 |
| % $Na_2CO_3$ from average HCl $M$) | 96.45 | 96.93 | 97.34 | 97.02 |
| %$Na_2CO_3$ (from median HCl $M$) | 96.50 | 96.98 | 97.38 | 97.07 |
| Average % $Na_2CO_3$ | 96.94% | | | |
| Relative average deviation | 2.3 ppt | | | |
| Median % $Na_2CO_3$ | 97.02% | | | |
| Relative probable deviation | 2.0 ppt | | | |

## STUDY QUESTIONS

1. The authors state that the "criterion for sound record keeping is that someone else can readily locate pertinent data and results for an experiment." Does this criterion suggest an implemental purpose for laboratory notes? Explain.
2. Why must laboratory notes be systematic? Are all implemental documents necessarily systematic? Why or why not?

# How to Write a Letter
# That Will Get You a Job

## Nona Aguilar

Companies receive a constant stream of letters asking for employment — so many that perhaps only the very exceptional letter catches anybody's attention. Nona Aguilar gives some good tips on writing a letter that will get attention, and maybe a job.

Whether you're just getting back into the job market after years out of it or you're looking for a better job to advance your career, you can double your chances of success by using a "tailored" letter.

What's a tailored letter? It's simply a brief letter highlighting background elements which most relate to the needs of a prospective employer. In other words, you "tailor" your experience to meet the needs of the person or company you want to work for. By following our simple guidelines, you can write a persuasive, concise letter that gets results.

Here's an example of the power of a tailored letter: My friend's mother, Mrs. Kinley, had been widowed for almost three years. She was 54 years old, her children were grown and she hadn't worked during the 29 years of her marriage. Now, with time on her hands, she wanted a job, but employment agencies discouraged her because, she was told, she didn't have skills or work experience.

Since she knew that I had always managed to rustle up a job no matter what or where, she talked to me about her problem. She realized that she didn't really want a full-time job; she had looked for one because it was the only type of work available through the agencies. And her only work experience was the hospital and Red Cross volunteer work she'd done throughout her marriage. We used that experience in composing her tailored letter, which she sent to 30 doctors (found in the "Physicians" section of the Yellow Pages).

## What Were the Results?

Within three days of mailing the letter, Mrs. Kinley had received four telephone calls. One wanted someone to work full time for five doctors in practice together; she declined that interview request but went to the other three.

While trying to decide which of two opportunities she might take — one of the positions wasn't offered to her after the interview — the mail brought a written reply asking her to call for an interview. On a hunch, she decided to make the call. That last interview turned out to be THE job: four days a week, from 9 to 1:30.

A postscript to the story: She received two more calls after she started work-

Nona Aguilar, "How to Write a Letter That Will Get You a Job," *Family Circle,* 1977, pp. 14 and 20.

ing. She also got a few P.B.-O.s (Polite Brush-Offs) in the mail plus a "not right now but maybe in six months" letter. *That* is what I mean about the power of a tailored letter!

## Measurable Accomplishments

Take a look at the letter Mrs. Kinley wrote. She used the Four Elements that form the basic structure of a good tailored letter: (1) an opening grabber, (2) an appeal to the self-interest of the reader, (3) a number of examples of her experience and (4) a good closing. The Four Elements are detailed . . . here, but as you can see, specific accomplishments are at the letter's heart. If your accomplishments are measurable in any way, you will look that much more impressive.

Here's what I mean by measurable. I landed a job teaching English in a language school in Italy. I was not a professional teacher; I had never taught in a school — obviously I had never been certified! However, one summer while I was still in high school, I started a little brush-up school in our family dining room to help three of my kid brothers and sisters. Four neighborhood kids joined my "class" and, through tutoring, I literally boosted English grades by about 13%.

The opening line of my tailored letter to the directress of the language school — the "grabber" — read: "I raised students' grades in English an average of 13% during a summer-school program which I began in my neighborhood." The letter made an impression, sailing past almost 100 weighty epistles and résumés sent by teaching professionals listing schools, courses, degrees and experience in abundance. When I came in for my interview, the directress was already anxious to meet me. My letter had shown an awareness of her major problem: Finding teachers who could actually teach — and could then prove it.

Mrs. Kinley's letter

March 1, 197-

Marvin Willis, M.D.
488 Madison Avenue
New York, NY 10022

Dear Dr. Willis:

In the past several years, I have worked over 6,000 hours in hospitals    ①
handling bookeeping and billing.

I am writing to you because your office may be in need of a woman
② with my background and experience to work on a part-time basis.
If so, you may be interested in some of the things I have done.

For example, I was responsible for handling Wednesday receipts for a
volunteer-operated hospital gift shop.

I sorted 500 pieces of patient mail per week.    ③

I handled all bookkeeping for the gift shop, insuring payment of suppliers and disbursement of profits to the hospital by the 30th of each month.

④ If such experience would be valuable to your office to help with book-keeping or billing, I would be happy to talk to you in more detail. My telephone number is EL 6–0000.

Sincerely yours,

## The Four Elements

1. **An Opening Grabber**

   Mrs. Kinley's letter begins with a short sentence listing a memorable figure: 6,000 hours of work experience in a hospital. This grabs the reader's interest.

2. **Self-Interest Appeal**

   She appeals to his self-interest right away in her second sentence by letting the doctor know that those 6,000 hours are part of valuable experience which might be useful to him.

3. **Examples**

   Mrs. Kinley gives three specific examples of accomplishments to further appeal to the self-interest of a would-be employer.

4. **The Closing**

   Mrs. Kinley does not plead for an interview. She doesn't even ask for one. Rather, she lets the recipient know she's a worthwhile professional person and that "if such experience would be valuable to your office . . ." she'd be happy to discuss it in an interview.

### Specific Accomplishments

So I can't stress it enough: *The heart of a successful tailored letter is specific accomplishments.* When your accomplishments are measurable, you look even more impressive — but don't equate *paid* with *measurable.* Mrs. Kinley does not apologize for her lack of paid business experience. That isn't even mentioned, nor is the fact that her work had been on a volunteer basis. Instead she casts all her experience in terms of *accomplishment.* Each separate accomplishment that relates to working in a medical environment is placed in its own brief, one-sentence paragraph. Indeed, the whole letter is brief, only eight sentences in all, so busy recipients — in this case doctors — are more inclined to read the letter straight through to the end.

It's important that your letter be short and crisp. Work and rework the letter so that your grabber is brief and punchy. Appeal right away to the self-interest of the recipient. In the Examples section of your letter, cover each accomplishment in one short sentence in its own paragraph. The succession of short, accomplishment-laden paragraphs makes a greater impact on the reader than long, cumbersome prose. Make your closing sharp and clean. And *don't beg* for an interview.

### Finding Job Prospects

Of course, you have to find prospects for your letter!

It was easy with Mrs. Kinley: We just opened the phone book and picked physicians whose offices were convenient to her home.

If you already have a job but want a better one, you're probably aware of where and for whom you want to work. All you have to do is send letters to the companies on your "list".

If you use the help-wanted ads in the paper, send a tailored letter *instead* of a résumé, even if a résumé is asked for. All résumés tend to look alike, so your letter will stand out, considerably increasing your chances of getting an interview — *the* crucial first step toward getting a new job.

If you're interested in a particular business or industry, check with your librarian to see if a directory exists for it. There you'll find listings complete with spellings, business titles and addresses. You can also pick up your telephone and call companies or businesses. Ask for the name and correct spelling of the owner or president — if it's a small company — or the district manager, if it's a large company and you're calling the regional office in your city. If a secretary insists on knowing why you're calling before she gives the information, simply say that you're writing the man or woman a letter and need the information.

How many letters will you have to send out? That's hard to say.

When you send letters to companies that aren't specifically advertising or looking for someone, you can expect to send a lot. I did that some years ago; I sent over 60 letters to advertising agencies. Some of the letters drew interviews; only one interview finally resulted in a job. But I only needed *one* job, and I got the job I wanted!

As a general rule, your letter is a good one when requests for an interview run about 8% to 10%. If Mrs. Kinley had received just two or three interview requests, she would have been doing fine. If she had received only one reply, or none, we'd have reworked the letter. As it turned out, she got six interview requests out of 30 letters — that's an exceptionally high 20%.

If you send a tailored letter when you know a company is hiring — for instance, in reply to a help-wanted ad — you will increase your chances of being called for an interview at least 30% to 50%, sometimes more. Once I was the only person called for an interview for an advertised editorial job, even though I had never worked on either a newspaper or a magazine in my life — there's the power of a tailored letter!

**Look Professional!**

Once you've composed your short, punchy, accomplishment-laden letter and decided who's going to get it, make sure you're careful about three things:

First, *type* — don't handwrite — the letter, following standard business form; a secretarial handbook in your library will show you examples. Or follow the form Mrs. Kinley used.

Second, use plain white or ivory-colored stationery. Very pale, almost neutral, colors are okay too, but nothing flashy or brightly colored. I've found that it is helpful to write on monarch-sized stationery, which is smaller than the standard 8½" X 11" paper; the letter looks much more personal and invites a reading.

Finally, do *not* do anything gimmicky or "cute." I remember the laughter that erupted in an office when a job-seeking executive sent a letter with a small, sugar-filled bag carefully stapled to the top of the page. His opening line was "I'd like to sweeten your day just a little." He came across looking foolish . . . and the boss didn't sweeten his day by calling him for an interview.

These are the basics that add up to a professional business letter. I've worked in a lot of offices and seen some pretty silly letters tumble out of the mail bag. Don't let yours be one of them; especially not if you're a woman of specific accomplishments who's ready for a job!

## STUDY QUESTIONS
1. Of course, not all business letters are employment applications. Could you apply Aguilar's advice to other forms of business correspondence? Discuss.
2. Does she follow her own advice by writing instructions that grab — and hold — the reader's attention?
3. Where and how does she use graphics and layout for purposes of clarity?
4. *Writing Assignment:* Use Aguilar's directions for writing a letter applying for a summer job. If you are near graduation, write a letter applying for a permanent position in your job field.

# Program Planning and Proposal Writing

## Norton J. Kiritz

Much of the world's work starts as written proposals that are then evaluated, refined, rewritten, and finally — perhaps — acted upon. The following selection presents a guide for program planning and a flexible format for organizing the proposal.

From Program Planning and Proposal Writing, *Grantsmanship Center News,* No. 6, © 1974, 1978, pp. 11–14.

Proposals written for foundations and those written for federal grants will differ markedly in final form. Foundations usually require a brief letter; federal agencies usually require you to complete an extensive array of forms and possibly attach your own narrative.

We suggest the following format as a basic planning format for all proposals. Thinking through the various sections as we suggest will enable you to draw from the content virtually all that either a private or public funding source will ask from you. Thinking through the various components will also enable you to develop a logical way to approach your plans and programs. And hopefully this planning will make your programs more effective. The proposal format looks like this:

     I Introduction
    II Problem Statement or Assessment of Need
   III Program Objectives
   IV Methods
    V Evaluation
   VI Budget
  VII Future Funding

## Proposal Summary

The summary is a very important part of a proposal — not just something you jot down as an afterthought. There may be a box for a summary on the first page of a federal grant application form. In writing to a foundation, the summary may be presented as a cover letter, or the first paragraph of a letter-type proposal. The summary is probably the first thing that a funding source will read. It should be clear, concise and specific. It should describe who you are, the scope of your project, and the projected cost.

Some funding sources may screen proposals as a first step in grant-making. That is, they briefly examine each proposal to see if it is consistent with their priorities, if it is from an agency eligible to apply for their funds, etc. As a further step, the "screeners" may draw up a summary of their own and these proposal summaries may be all that are reviewed in the next step of the process. It is much better to spend the time to draw up a summary of your own that the funding source can use, than to hope that the reviewer sees the importance of your program in his brief initial look at your proposal. So do a good job!

## I Introduction

This is the section of a proposal where you tell who you are. Many proposals tell little or nothing about the applicant organization and speak only about the project or program to be conducted. More often than not proposals are funded on the basis or the reputation or "connections" of the applicant organization or its key personnel rather than on the basis of the program's content alone. The Introduction is the section in which you build your credibility as an organization which should be supported.

## Credibility

What gives an organization credibility in the eyes of a funding source? Well, first of all, it depends on the funding source. A traditional, rather conservative funding source will be more responsive to persons of prominence on your Board of Directors, how long you have been in existence, how many other funding sources have been supporting you and other similar characteristics of your organization. An "avant garde" funding source might be more interested in a Board of "community persons" rather than that of prominent citizens and in organizations that are new, rather than established, etc.

Potential funding sources should be selected because of their possible interest in your type of organization or your type of program. You can use the introduction to reinforce the connection you see between your interests and those of the funding source.

What are some of the things you can say about your organization in an introductory section?

- How you got started.
- How long you have been around.
- Anything unique about the way you got started, or the fact that you were the first thus-and-so organization in the country, etc.
- Some of your most significant accomplishments as an organization, or, if you are a new organization, some of the significant accomplishments of your Board or staff in their previous roles.
- Your organization's goals — why you were started.
- What support you have received from other organizations and prominent individuals (accompanied by some letters of endorsement which can be in an appendix).

We strongly suggest that you start a "credibility file" which you can use as a basis for the introductory section of future proposals you write. In this file you can keep copies of newspaper articles about your organization, letters of support you receive from other agencies and from your clients. Include statements made by key figures in your field or in the political arena that endorse your kind of program even if they do not mention your agency.

For example, by including a presidential commission's statement that the type of program which you are proposing has the most potential of solving the problems with which you deal, you can borrow credibility from those who made the statement (if they have any).

Remember, the credibility you establish in your introduction may be more important than the rest of your proposal. Build it! But here, as in all of your proposals, be as brief and specific as you can. Avoid jargon and keep it simple.

## II Problem Statement or Assessment of Need

In the introduction you have told who you are. From the introduction we should now know your areas of interest — the field in which you are working. Now you will zero in on the specific problem or problems that you want to solve through the program you are proposing.

## Pitfalls

There are some common pitfalls into which agencies fall when they try to define problems.

Sometimes an organization will paint a broad picture of all the ills plaguing people in a part of the community. Proposal writers do not narrow down to a specific problem or problems that are solvable, and they leave the funding source feeling that it will take a hundred times the requested budget even to begin to deal with the problems identified. This is overkill. It often comes from the conviction of the applicant that it must draw a picture of a needy community in all its dimensions in order to convince the funding source that there are really problems there. All that this does is to leave the funding source asking "how can this agency possibly hope to deal with all of those problems?" Don't overkill.

Narrow down your definition of the problem you want to deal with to something you can hope to accomplish within a reasonable amount of time and with reasonable additional resources.

## Document the Problem

Document the problem. How do you know that a problem really exists? Don't just assume that "everybody knows this is a problem" . . . That may be true, but it doesn't give a funding source any assurance about your capabilities if you fail to demonstrate your knowledge of the problem. You should use some key statistics here. Don't fill your proposal with tables, charts, and graphs. They will probably "turn off" the reader. If you must use extensive statistics, save them for an appendix, but pull out the key figures for your problem statement. And know what the statistics say.

We saw one proposal where an agency presented demographic (population statistics) pictures of two communities, one in which the program was to be conducted, and another nearby community where there would not be a program. Every statistic (percentage unemployment, ethnic breakdown, number of youth, number of juvenile arrests, etc.) pointed to a vastly greater problem in Community B than Community A yet Community A was the proposed site of the program. Any reviewer would seriously question the program based on those accompanying statistics.

To summarize, you need to do the following:

- make a logical connection between your organization's background and the problems and needs with which you propose to work.
- support the existence of the problem by evidence. Statistics, as mentioned above are but one type of support. You may also get advice from groups in your community concerned about the problem, from prospective clients, and from other organizations working in your community and professionals in the field.

- define clearly the problems with which you intend to work. Make sure that it can be done within a reasonable time, by you, and with a reasonable amount of money.

## III  Program Objectives

One of your concerns throughout your proposal should be to develop a logical flow from one section to another. Whereas you can use your introduction to set the context for your problem statement, you can likewise use the problem statement to prepare the funding source for your objectives.

An objective is a specific, measurable outcome of your program.

Clearly, if you have defined a problem, then your objectives should offer some relief of the problem. If the problem which you identify is a high incidence of drug abuse by youth in your community (substantiated, of course) then an objective of your program should be the reduction of the incidence of drug abuse among youth in your community. If the problem is unemployment, then an objective is the reduction of unemployment.

### Distinguish Between Methods and Objectives

One common problem in many proposals is a failure to distinguish between means and ends — a failure to distinguish methods and objectives.

For example, many proposals read like this: "The purpose of this proposal is to establish a peer-group tutoring program for potential drop-outs in the _____ area of Los Angeles," or "The objective of this program is to provide counseling and guidance services for delinquent youth in _____."

What's wrong with these objectives? They don't speak about outcome! If I support your project for a year, or for two years, and come back at that time and say "I want to see what you have done — what you have accomplished — what can you tell me?" The fact that you have established a service, or conducted some activities, doesn't tell me whether or not you have helped to solve the problem which you defined. I want to know the outcome of your activities. I want to know whether you have, through your tutoring program, reduced the number of drop-outs with whom you have worked, or whether the delinquent youth with whom you worked got into less trouble over the past year. Knowing that you worked at it is not enough!

Some organizations, trying to be as specific as they can, pick a number out of the air as their measurable objective. For example, an agency might say that their objective is to "decrease unemployment among adults in the XYZ community by 10%." The question I ask is "where did they get that figure?" Usually it is made up because it sounds good. It sounds like a real achievement. But it should be made of something more substantial than that. Perhaps no program has ever achieved that high a percentage. Perhaps similar programs have resulted in a range of achievement of from 2% to 6% increase in employment. In that case, 5% would be very good, and 6% would be as good as ever has been done. Ten percent is just plain unrealistic. And it leads me to expect that you don't really know the field very well.

If you are having difficulty in defining your objectives, try projecting your agency a year or two into the future. What differences would you hope to see between then and now? What changes would have occurred? These changed dimensions may be the objectives of your program.

In addition, I want to examine your objectives in a little more detail. Maybe some programs create jobs for people that are very temporary in nature, and they reduce the unemployment problem in the short term, but after a year or two the problem will be back with us, as bad, or worse, than ever. This gets into the question of evaluation, which clearly relates to the setting of measurable objectives, for a good set of criteria for the evaluation of our program, and thus serves another purpose.

## IV  Methods

By now you have told me who you are, the problem(s) you want to work with, your objectives (which promise a solution to or reduction of the problems) and now you are going to tell me how you will bring about these results. You will describe the methods you will use — the activities you will conduct to accomplish your objectives.

### Research

The informed reviewer wants to know why you have selected these methods. Why do you think they will work? This requires you to know a good deal about other programs of similar nature. Who is working on the problem in your community and elsewhere? What methods have been tried in the past, and are being tried now and with what results? In other words, can you substantiate your choice of methods?

One agency recently brought a proposal into [a Grantsmanship Workshop] class that dealt with the provision of counseling services to delinquent youth by professional social workers with MSW degrees. Each of these two professional staff members was to receive a salary in excess of $15,000 per year. The agency was concerned about the limited number of MSW's they could hire within their budget limitations.

A number of questions were raised about this program. One key question was this — why did you decide that professional Social Workers, with MSW degrees and $15,000 salaries were necessary to the success of your program? Do you have any evidence that similar programs have been effective elsewhere? What other models exist that you could work with? Is it possible that para-professionals (non-degreed workers), perhaps even ex-offenders themselves could do the job as well or perhaps better than the trained professionals you want to hire? Do you know of programs using para-professionals in this capacity and have you assessed the results of such programs? How can you complain of lack of sufficient money to employ more than these two highly-trained staff, when you don't know if there is a less expensive, and perhaps more successful model to follow?

The consideration of alternatives is an important aspect of describing your methodology. Showing that you are familiar enough about your field to be aware of different models for solving the problems, and showing your reasons for selecting the model that you have, gives a funding source a feeling of security that you know what you are doing, and adds greatly to your credibility.

One planning technique which you might want to use is this. Take a sheet of paper and divide it into columns. The first column is the "problem" column, the second is headed "objectives," the third "methods" and the fourth is "evaluation." If you list all your objectives in the second column, you can then identify the problem that it relates to, the specific methods in your program that deal with the objective, and the criteria of success in reaching the objective as well as the method of evaluation.

This helps you to see whether you are truly dealing with all of the problems you talked about, whether your objectives relate to the problem(s), whether you have a method of reaching each objective and whether you have set up an evaluation mechanism to deal with your entire program. This leads us into the next proposal component — evaluation.

## V  Evaluation

Evaluation of your program can serve two purposes for your organization. Your program can be evaluated in order to determine how effective it is in reaching the objectives you have established — in solving the problems you are dealing with. This concept of evaluation is geared toward the results of your program.

Evaluation can also be used as a tool to provide information necessary to make appropriate changes and adjustments in your program as it proceeds.

As we have stated, measurable objectives set the stage for an effective evaluation. If you have difficulty in determining what criteria to use in evaluating your program, better take another look at your objectives. They probably aren't very specific.

### Subjective and Objective Evaluations

Also, be sure you understand the difference between subjective and objective evaluations.

Subjective evaluations of programs are rarely evaluations at all. They may tell you about how people "feel" about a program, but seldom deal with the concrete results of a program. For example, we saw an example of an evaluation of an educational program that surveyed opinions about program success held by students, parents, teachers and administrators of the program. This is a pretty "soft" evaluation, and doesn't really give much evidence to support the tangible results of such a program.

In addition, this particular evaluation solicited comments from students when they completed the program, failing to deal with over 50% of the students who started but did not complete the program. Clearly, those students who finished the program are going to react differently, as a group, from those who didn't complete the program. And we might, as an agency, learn a great deal from those who didn't finish. From the nature of this evaluation, one might suppose that the educational institution involved was committed to producing what they thought would "look like" a good evaluation, but it wouldn't pass muster with a critical reviewer.

**Subjectivity** – introducing our own biases into an evaluation – will often come in when we evaluate our own programs. Particularly if we feel that continued funding depends on producing what "looks like" a good evaluation.

One way of obtaining a more objective evaluation, and sometimes a more professionally prepared evaluation, is to look to an outside organization to conduct an evaluation for you. You might go to other non-profit agencies, colleges and universities in your community which will work with you in developing an evaluation for your program. Sometimes it is possible to get an outside organization to develop an evaluation design and proposal for evaluation that can be submitted to a funding source, complete with its own budget, along with your proposal. This not only can guarantee a more objective evaluation, but can also add to the credibility of your total application, since you have borrowed the credibility of the evaluating institution.

It is essential to build your evaluation into your proposal, and to be prepared to implement your evaluation at the same time that you start your program, or before. If you want to determine change along some dimension, then you have got to show where your clients have come from. It is very difficult to start an evaluation at or near the conclusion of a program, for you usually don't know the characteristics of the people you are working with as they existed prior to being in your program.

### An Excellent Program Evaluation

I'd like to give you an example of what I think was a very fine program evaluation. It took a lot of time and resources to conduct, and it may look like a pretty big project in and of itself. That is true. The agency that conducted this evaluation had the resources to do it. But evaluations of this nature may have enough value in and of themselves to be able to be funded quite separate and distinct from the programs to which they are attached.

Some years ago the Los Angeles County Probation Department operated what was called the Group Guidance Program. Group Guidance was a program that employed "streetwise" Probation Officers as gang workers, with the goal of orienting gangs away from criminal behavior and into more productive activities. Some agencies questioned the effectiveness of the program and an evaluation design was created. (This is not a particularly good practice in setting up evaluations, in that evaluations set up to justify the continued existence of a program, and conducted by the agency itself, tend to be biased in favor of the agency.)

What is interesting is the evaluation design itself. It was an attempt to gather information about the presumed reduction in delinquent behavior among gang members involved in the project, and to put this data into an economic context which would justify the cost of the program. This is the basic evaluation design.

Gangs were identified which had reputations of being violent, moderate and quiet. It was proposed that the violent gangs got into far more trouble than the other two, and that this would be reflected in their court records — they would be arrested more often, be in jail and juvenile hall more often, and for longer periods of time, spend more time at correctional facilities, etc. The Probation Department, with access to court records, examined the records of all members of these varied gangs. They identified all contacts that a youth could have with one institution or another and then went to each institution, conferred with their business department, and came away with a cost figure, in dollars and cents, that could be attached to a particular entry on a court record. In other words, it cost X dollars for a youth to spend the night in Juvenile Hall and Y dollars for 24 hours in a Probation Camp. Each gang member's record had a total dollar value assigned to it.

The result of this was the finding that the three kinds of gangs in question did cost the community a varying amount of money, with much higher costs being attributed to the violent gang.

The agency had done a number of things in designating this evaluation. It had established a measurable "index of delinquency" and it had created a "dollar and cents" measure which could demonstrate to the funding source, the Board of Supervisors of the County of Los Angeles, a possible saving which could be realized were the records to show that the decrease in cost for the gangs worked with in the program was greater than the cost of conducting the program itself. Pretty ingenious!

## VI Budget

As with proposals themselves, funding source requirements for budgets differ, with foundations requiring less extensive budgets than federal agencies. . . . [Budget details are omitted.]

## VII Future Funding

This is the last section of your proposal, but by no means the least important. Increasingly, funding sources want to know how you will continue your program when their grant runs out. This is irrelevant for one-time only grant applications, such as requests for vehicles, equipment, etc. But if you are requesting program money, if you are adding to your projects through this proposal, then how will you keep it going next year?

A promise to continue looking for alternative sources of support is not sufficient. You must present a plan that will assure the funding source, to the greatest extent possible, that you will be able to maintain this new program after their grant has been completed. They don't want to adopt you — they don't want you continually on their back for additional funds. Moreover, if you are having problems keeping your current operations supported, you will probably have much more difficulty in maintaining a level of operation which includes additional programs. The funding source may be doing you no favor by supporting a new project, and putting you in the position of having to raise even more money next year than you do now.

What is a good method to guarantee continued support for a project? One good way is to get a local institution or governmental agency to agree to continue to support your program, should it demonstrate the desired results. But get such a commitment in writing. A plan to generate funds through the project itself — such as fees for services that will build up over a year or two, subscriptions to publications, etc., is an excellent plan. The best plan for future funding is the plan that does not require outside grant support.

## STUDY QUESTIONS

1. Why does Kiritz stress program planning for the proposal writer?
2. How important are graphics and layout in his presentation of these instructions?
3. Why is it important that the proposed program has a "measurable outcome"? Can you think of some worthy projects that would not necessarily have a measurable outcome?
4. Kiritz uses one extended example — the Group Guidance Program — to illustrate program evaluation. Might a general explanation have been more effective? Discuss.
5. Kiritz presents a proposal format that has seven parts (eight, if we include the proposal summary). What logic underlies the order of the parts? Could the parts be placed in a different order? Discuss.
6. Kiritz limits himself to instructions for proposals to be submitted to private foundations and the Federal Government. Are his instructions useful for other types of proposals — say, a business proposal?
7. *Writing Assignment:* Write a proposal that uses Kiritz's format as a guide. Modify the format to the specific requirements of your proposal.

# One Highly-Evolved Toolbox

## J. Baldwin

This selection is from *Soft-Tech,* a catalog of tools and sources for tools. *Soft-Tech* is edited by J. Baldwin, the author of the selection below, and Stewart Brand, who won a National Book Award for the *Last Whole Earth Catalog.* Their approach to

J. Baldwin, "One Highly-Evolved Toolbox," in *Soft-Tech,* ed. Baldwin and Brand (New York: Penguin, 1978), pp. 8–17.

technology is sometimes unorthodox, but it is usually sensible, as Baldwin's "One Highly-Evolved Toolbox" makes evident.

As thing-makers, tool freaks and prototypers, Kathleen and I find ourselves custodians of about a ton of versatile hand tools. These have been used by us and friends over the years to help many projects and repairs get done. People keep asking us what tools to get, where to get them, and how to keep them from getting ripped off. Well . . . here goes.

Stand in Sears Tool Dept. and it'll soon be obvious that you don't need one-of-each even if you have the money. Ask a craftsman what to buy, and you'll get as many answers as people you ask, for each has their own favorites and specialized needs. They'll all agree on one thing though: *Buy the Best You Can.* And the more a tool will be used, the better the quality should be. Tools used every day, especially electric tools, should be of commercial or production line grade. You usually can't find these at hardware stores. Industrial supply houses are where to go. Take a friend who can buy wholesale. These tools will be expensive, so we'd better justify the cost.

For many, the best reason to go first class is that good tools are a real pleasure to use and handle. This helps make work less labor. The heavy duty stuff looks brutal. It wasn't made to look good in the box, it was made to do the job and has been perfected over many years. The tough ones have their own kind of beauty that you'll see better as your viewpoint gets aligned with reality. Such tools, of course last longer and are repairable when they finally do wear. They can take a lot more abuse, especially the inevitable overload. They can handle the bigger jobs and poor working conditions that would soon trash cheap versions. And after a few years in your hand, they often get to be old friends.

For tools that get used now and then, middle quality will do. By that I mean Sears better grades and no lower. Really cheapo tools are of no use at all, can be dangerous, and often break the first time you use them. They are also discouraging to use, which might even cause a beginner to give up. Our only regrets have been not buying the best when we could have. Tools that receive great strain, such as gear pullers, should be super top quality only. If you only need one every five years, rent it.

OK so what tools do you need? How do you start the stash? There are a few basic tools that everyone should have available: Hammer, crosscut saw, adjustable wrench, pliers, screwdrivers (get a set), tape measure, hand drill and bits. Beyond these, you'd best gather tools as you need them. Auto work will require a rather complete set of wrenches and a whole boxfull of special tools, some of which are for particular vehicles. Carpentry will require another whole group: planes, chisels, etc. Electrical and plumbing still more. Our rule of thumb is if we need to borrow a common tool more than once, we buy one.

Fleamarkets are a good place to look for expensive items like vises or anvils. Absolutely the best place to get a whole mess of tools at once is to keep alert for a widow selling off her deceased husband's retirement shop. Another place to look is auctions, but you'd better know what you're doing. You should shop around. Recently in the Bay Area, we were quoted prices varying 50% on a tool we wanted! If you want to buy a bunch all at once (which makes sense these days of inflation — tools are a good savings account), some stores will make you a 20% deal. Even Sears can be dealt with, as the sales people work on commission. They and other stores also have unadvertised freight-damaged goods hidden away. These can be good deals, as the damage is often merely cosmetic. You can give a salesman your name (and take his card) and have him call you when a certain tool is on sale or arrives damaged.

Whether in a Big Store or private sale, you should critically inspect each tool for condition. These days, many new tools by reputable (?) manufacturers are faulty. Used ones may be worn beyond repair. Anyway, be pickynit about it; you'll be living with it in your hand. And beware of package deals claimed to be a great saving. The "complete mechanics tool set for $450.00" often includes tools you don't need, and may force you to take inferior items that you would be better off picking up individually.

What do you do about that little voice that whispers, "Buy one, you might need it someday!" Well, it's *possible* you'll be needing them *all* someday but Sears is only the tip of the iceberg. Have you ever seen a *real* hardware catalog? 2000 pages? On the other hand, it often does pay to get a set of tools that greatly increases your capability, such as a welding rig. Another way to go is for a group to buy a set of tools for working on one particular item, such as old Chevy 6 engines, and then everyone in the group that needs a vehicle gets one that uses that engine

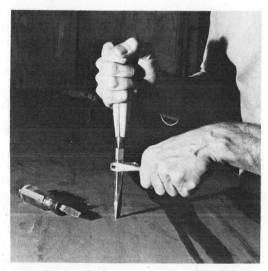

FAT SCREWDRIVERS Big handles, heavy blade, compact size, make Sears #41586 and Irwin our favorite screwdrivers. Square shank allows help with wrench. You can't own too many screwdrivers, as they grow legs easily, and there are so many screw sizes.

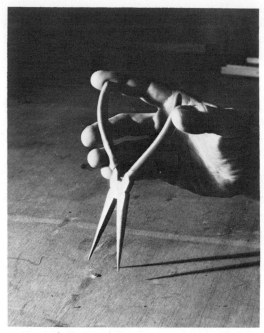

NEEDLENOSE PLIERS WITH SPRING (left) The spring and delicate jaw shape permits very delicate nabbing. You can actually pick up a live ant without damaging it (physically, anyway).

and hence those tools, and the consequent parts pile. (That's being done around here. There must be *dozens* of 56 Chevy pickups and flatbeds within 30 miles.) Some groups get known for specialties: "the Butterfly Mountain people fix tractors." Those communities and families pool their resources and buy a set of expensive heavy duty tools maybe for tractor repair. You have to be pretty mellow to make this work, especially if there is a high turnover of people. But this is a growing trend, and we think a good one. It leads to barter and lessens the need for duplicate sets of specialty equipment.

Our shop is known for its versatility. It's portable; everything fits a 4 × 5 U-Haul trailer. It's been set up in ten different places in 5½ years. The tools were chosen for quality and versatility. With versatility goes a handy ability to work in harmony with other tools, enhancing all. For example, with the drills and vises we have, we can drill a hole at any angle in just about anything. The combinations allow us to easily mass-produce parts like dome struts or Inkleloom frames. This gives a nice potential for making money as well as greatly easing tasks that might be as bad as working in Detroit. Versatility also means needing fewer tools which means less money out, less space for storage, and less tools to keep track of.

For many people, the biggest problem with tools is keeping them together. That was our problem too for awhile, especially at Pacific High School where there were always a number of young people who didn't yet see that tools are in a different category than other possessions. Our answer has been to take the time to try and give people a good feeling about tools being extensions of their own hands, and that tools are the means to getting good shelter and other desirable results. A French poet (whose name I regrettably can't remember) said, "Hammers spend a lot of time sleeping. . . ." We like to see the tools at work. We show people how to use the tools and encourage them to in turn show still others how. Having good tools in the hand, together with that tasty feeling that comes from teaching somebody else, gives the tool borrowers a respect for the whole bit.

We also have all the tools marked with a colored stripe. This not only reduces arguments on job sites where lots of people's tools are at work, but it makes it easy for people of good heart to return strays. We put out the word: "Bring a blue stripe tool to breakfast" and we round 'em up. We also ask that tools be brought back at sundown unless needed that night. There's a place to bring them back to. This is essential. A casual pig-pen shop just can't keep its tools because there "isn't any there, there." As an experiment, we abandoned our collapsed old bureau toolbox and bought a (freight damaged) Sears (the best for the money) rolling mechanics tool chest like you see in big auto shops. We segregated the tools by function and labeled the drawers. The result is that tools are easily looked over and selected and just as easily put to bed. To our great surprise, we found that this chest caused a drastic increase in the number of tools being used and a similar increase in action.We even found that we were using our own tools more! The neat storage made it easy to see who was missing, but people brought them back much more reliably than before anyway. The chest can be locked to control unannounced borrowing which is always a disaster. The overall effect has been that under very poor risk conditions, both sociological and physical, we've only lost about $50.00 worth of tools in 7 years! And this without having to get too heavy or "high school shoppish" about things. In case you wondered, we did try the toolboard-on-the-wall. It didn't work, and nobody we know that's tried it has made it work either, though it is nice to see all those tools hangin'. It has not been necessary to sentence anyone to being tool crib librarian either. We'll admit that it takes some time to develop tool-consciousness in a crew, but it can be done, and peaceably. The tools spend a lot less time sleeping too.

Making a deliberate effort to raise your own tool consciousness can result in some interesting new possibilities in your life. As with most mysterious-appearing phenomena, a bit of learning soon clears things up and you wonder what had been previously keeping you from doing your own repairs and thing-making. Sometimes all it takes is a different point of view. I've remarked that tools are extensions of your hands. No mystery there; a hammer is just a hard fist; a screwdriver, a tough fingernail. But hands usually operate according to instructions from head, so it can also be said that *tools are an extension of your mind.* Looked at this way, the big (expensive) mechanic's cabinet with all the tools of similar function stored together with high visibility becomes even easier to justify. I find it is effective to store the tools by function rather than by name because this is the most useful way to think of the best tool when you are selecting. Hitters, grabbers, slashers, abraders — regardless of what they are called, are there in their places. You take your pick. Often, just looking at them will give you a better idea not only of what tool to use, but how to do the job or how to design the object. That's a big advantage of the neat toolbox. If the tools are "somewhere out on the back porch or maybe in the back seat of the VW" then your mind is deflected from creative thinking into a hunting mode, and the aggravation can easily cause you to lose your ability to get things done.

Easily accessible, functionally sorted tools also give you a ready familiarity with the tools you have. This has two effects. First is that as you get to know your tools, you gain the easy fluid motions that go with using them. You're not afraid of them any more, though respect is increased. This makes you able to work faster with less fatigue just as good form in sports often makes a big difference. It makes you safer too. Safety is also enhanced by having the tools where you can easily inspect them for condition, sharpness and rust. We have found that safety is largely a matter of attitude. The closer tools are to being a working extension of your mind, the safer you'll be. Self-preservation.

The second effect is that you get to "know" all the tools you own without having to consciously think about it. This makes it simple to round up strays, of course, and it's easier to see where there are annoying gaps in your capabilities ("we don't have a lightweight mallet"). More importantly, you begin to think in terms of the tools you have. The eventual result is that you and your entire toolbox and shop become a big, complex tool with many possibilities. You begin to sense what you can do together. Buying tools with overall flexibility of purpose in mind, keeps you from falling into the trap of building a one-function capability with accompanying tendency to conservatively fossilize your creativity. This tendency is strengthened as the value of the tools rises, which is the main reason society doesn't get fast response to its changing needs from large corporations who have sunk enormous capital into shops that make *that* only. Like fat cars. Once in that position, it's difficult to evolve at all, let alone without damage or drastic change of form.

So you begin to build your tool capability into the way you think about making things. As anyone who makes lots of stuff will tell you, the tools soon become sort of an automatic part of the design process. Beginners worry too much about skill and safety, rather like new drivers worry most about jerking the clutch when learning to drive a stick shift. It doesn't take long before more serious aspects take over, and the manipulation problems fade out. But tools can't become part of your design process if you don't know what is available and what the various tools do. In addition to buying tools that I find useful, I spend some time reading catalogs so as to become familiar with tools that I can't afford or don't need at the time. Tool catalogs such as Silvo are rather like my cabinet in appearance so I find it painless to sort of automatically file the information away in the backroom somewhere. Tool dictionaries, especially of older tools, are helpful too.

Some of you are saying about now, "Who wants to get into it that far anyway?" Friends, there are advantages. Obviously, making or repairing things yourself

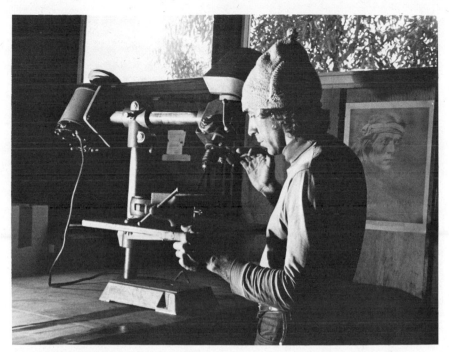

ROCKWELL RADIAL DRILL PRESS Our most-used tool, bar none. This ½-inch drill can extend to drill a hole in the center of a 32″ circle, and can swivel cut to drill big things sitting on the floor. It can drill at any angle, including horizontal. Though not of machine shop accuracy, it will do 99% of the work most people will ask of it. Radial feature costs extra and is worth every penny. You clamp the work down (we use a Versa Vise) and bring the drill to it at the desired angle. We have it mounted on a box that holds accessories and brings typical drill table height to that of other shop benches so we can support long objects being drilled. Dependable too: no repairs in 14 years. It's light enough to carry to a big job.

VERSA VISE These wonderful vises can stand up, lay down, swivel, and come with a clamp base that you can take to the job on the third floor. We have two, and a number of bases (one on the drill press table), allowing us to grip just about anything you could name short of a dead sheep in any position. They can be used as clamps when removed from base, a 1-second operation. Come with pipe jaws, too. Not for heavy metal work or heavy pounding.

can save you money and time. Well, maybe it isn't so obvious. Example: next time your car breaks down, find out how many hours it will likely take to fix it. You don't have the time right? OK, how many hours will you have to work at some job so you can pay that mechanic? For many of you, the hours you have to work to pay the mechanic will be more than the job would take if you did it yourself. Moreover, you don't have to pay yourself, and the job can be done to your standards and at your convenience. If you don't have the skills or the tools, that's what we're talk-

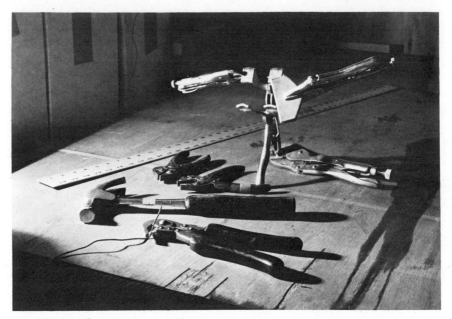

**BIRDHEADS** (Sometimes called "Parrot") If you do a lot of work with heavy wire, these are just the thing. Compound levers let you snip through most wire like it wasn't there. Nose makes working overhead easy.

**4 FOOT RULER** If you work with plywood or 4x8 anything, one of these will save you lots of time and grief. NB. Some new plywood isn't 90° square! Check it always. Best rulers have etched numbers.

**STEEL HANDLED HAMMERS** Steel or fiberglas may not be as aesthetic as wood, but the heads don't fly off when dry weather shrinks the handle. Violent nail pulling won't break them either. Pro carpenters don't like them, claiming that they eventually injure elbows if used every day.

**VISEGRIPS** Buy these by the genuine name Visegrip. They come in an array of sizes and jaw shapes, allowing you to grab what you see with a grip strong enough to crush things. Handy for undoing old rusty machines, and as a portable vice for welding, etc.

**"BERNARDS"** Pliers whose jaws work parallel (there's a nifty wire cutter too). We often use these in pairs for twisting and shaping small parts, and glass breaking. Our most pilfered item too; we've lost a dozen pairs. I can see why.


ing about! Doing it yourself can free you from certain dependencies that you may find smothering. What if the $20.00/hour plumber can't come until next Friday? Repairing pipes is relatively easy. Once you learn how, you not only avoid being at someone's mercy, you have a skill that can help friends or make money for you. How-to-books are tools, in case you haven't guessed.

Another advantage of having some tools that you know how to use, is that as you get an easy, facile familiarity using them, you begin to get a better feel of the ergonomics of other techno-hardware that you use or make. (Ergonomics is the man-machine interface; how the steering wheel feels in the hand and how it tells you what's happening to the wheels; the wrist-breaking poor feel of eggbeaters; the built-into-you feel of a good rifle). Poor ergonomics is one of the main reasons behind the recent public disenchantment with technology. Things are made with the convenience of the machine in mind instead of the human user. The result is hardware that is hard to hold, too cold or too hot, difficult to repair, easy to lose or lose control of, easily broken, etc., etc., etc. The machine is in control of you instead of the other way around. You can do better, yes? Most highly evolved good quality tools are ergonomically good. (Rifles are tools. You peaceable types can put down your neck hairs, it was only an example.) So without having to take a course in the subject, you can gain an informed feel of what is satisfying. As with most problems brought to us by technology, ergonomic problems are often best solved not with more technology but with clear thought and a better-informed intuition.

An informed tool intuition works best if it's augmented by an informed materials and processes intuition. For instance, if you don't know anything about foundry work (casting), it's unlikely that you will come up with ideas that require it. Often, this ignorance (ignore-ance) is easily remedied. A bit of inquiry may well show that what you had considered a black art is actually not one at all. Bronze and aluminum castings, for instance, are made every day in high school art departments by unskilled students using scrap metals from auto wrecking yards. Anyway, things take a form dictated by the possibilities inherent in the material to be used and the tools that can shape it, and the ideas in the head of the worker. It follows that the more you read, and snoop around and experiment and practice, the easier it will all come and the more independent you can be. Freedom rising.

The ultimate is to make your own tools. Tools fitted intimately to you by you. What could be niftier? Blacksmiths are really into that. A good example is found in the books by Alexander Weygers. But tools need not be limited to the shop. How about making your own personal canoe paddle? Or your own left-handed kitchen equipment? You can modify existing tools too. For instance, when we needed to make 7-inch diameter pistons for a small production run of giant raft-inflating hand pumps, we reversed the bit in a hole cutter so it made discs instead. The pistons were then easily and accurately cut from heavy plywood with a great saving in time and material compared to turning them on a lathe that we would have had to borrow. Making the big pump's leather "piston ring" seals proved to be easy after we spent some time talking to craftsmen in a sandal shop. With their advice, we soaked heavy leather discs in Mink Oil and then pressed them into the desired shape with a matched male and female die rammed by our vise. The dies were made on a bandsaw modified with a simple homemade attachment that enabled us to cut bevelled round holes with good accuracy. That attachment was also used to make the next batch of pistons, as it proved faster than the disc-maker on the drill press. Tools making tools making tools.

I can hear some of you saying "small production run! Yuk . . ." Unless you are an artist, and maybe even then, you will sooner or later need a bunch of things all alike. Even with only the most basic tools, you can mass-produce things. The precision and complexity of the produced parts is somewhat dependent upon the adaptability and quality of your tool bank, another reason to intelligently gather good stuff. Large scale mass-production tends to enslave both the workers and the customers. The workers are used as if they were machines. The huge capital outlay for the factory means that there must be a huge and relatively steady demand. This in turn means heavy manipulation of public "desires" and almost always involves politics and coercion.

Small scale production, though, can mean a great reduction in drudgery as well as interesting possibilities in barter. By means of jigs and other simple fixtures that you can figure out yourself, you can be freed from having to measure each part. Hold the piece of wood against the jig and hit it with the drill and all the holes will be in exactly the same place in each part.

We've mass-produced thousands of dome struts and parts for conventional construction. We've also produced simple looms and the aforementioned pumps (which we bartered for a fleet of rafts and got into the whitewater river running business), solar collector parts, signs (during an anti-freeway fight), toys, boxes, electronic parts, concrete forms, fence rails, adobe blocks, tents, shelving, lighting systems, and model parts, to name but a few. The ability to get on a small production run frees you from dependence on larger less efficient manufacturers, their prices, specifications and schedules. It can be rather fun, too, if it doesn't go on for too long. "Shop Yoga," we sometimes call it.

And it can be done without sophisticated expensive equipment if you take the time to think it all out first. The thinking is the most powerful part. You can actually *change* some things out there! Maybe not in a big way, but certainly at a scale that you can understand. You will find that as you work, your understanding of technology will increase a bit, and your fears based on ignorance will decrease. (Your fears based on newfound understanding might well increase, but that's another paper.) In a modest way, you can combat the "machines taking over" by having better control of the technology you live with. It's a good feeling. And it's free.


## STUDY QUESTIONS

1. Baldwin consistently uses such slang words as "tool freaks," "cheapo," and "stash." Does his vocabulary detract from the value of the instructions? Does the vocabulary narrow the potential audience?
2. Although the piece is about hand and power tool selection, the implications suggest a broader purpose as well. Can you identify the broader purpose? Does the attempt to work in a broader purpose detract from the practical information on tools?
3. Do you find principles of tool selection that might be applied to choosing equipment and materials in other fields? Discuss.
4. *Writing Assignment:* Write a guide for tool selection in your job field.

# Business Conduct Guidelines

## IBM

Many organizations provide for their employees handbooks that explain employee responsibilities, company policies, and company procedures. The range of these handbooks is as broad as the world of work itself. The handbook for a franchise food operation might instruct employees on points such as smiling for customers, or making change accurately. At another extreme, companies such as Xerox and IBM might provide their highly paid professional employees with flexible guidelines on how to protect the company image, or how to avoid a conflict of interest. The section below is a portion of an in-house handbook (meaning that it is to be used only within the company) for IBM employees.

### A Letter from the Chairman

Business today is being called upon as never before to explain its actions, provide reasons for its decisions and speak out clearly on where it stands on ethical behavior. I think that's fine. I believe that the vast preponderance of men and women in business live up to a high ethical standard, and have sound answers to the questions they are being asked.

I also believe it essential in this time of questioning and testing that everyone — employees and their families, customers and competitors, friends as well as critics — know just where IBM stands on basic ethical issues. Although IBM long has had detailed guidelines for business conduct which have been read and attested to by employees who must act for the corporation, this book summarizes our fundamental requirements for worldwide business conduct.

*If there is a single, overriding message in this book, it is that* IBM *expects every employee to act, in every instance, according to the highest standards of business conduct.*

Ultimately, in every business decision — as in personal ones — the responsibility is yours. And knowing IBM people as I do, that makes me very comfortable.

Frank Cary, *Chairman of the Board*

IBM from its beginnings has had a reputation for high standards of business conduct. Those standards didn't just happen. They grew out of the beliefs upon which

From International Business Machines Corporation.

IBM was founded and which have been reaffirmed by IBM managers and employees for more than sixty years. The three basic beliefs that guide all IBM actions are:

*Respect for the individual.* Respect for the dignity and the rights of each person in the organization.
*Customer service.* To give the best possible service to the customer.
*Excellence.* The conviction that an organization should pursue all tasks with the objective of accomplishing them in a superior way.

There are also related responsibilities: Providing the user with the best possible products. Obligation to the stockholders. A fair deal for suppliers. Competing fairly. Being a good corporate citizen wherever we do business.

In IBM, the chief executive officer and the senior executives have a primary responsibility to set the standard for business ethics, and to say clearly what they expect of all IBM employees. And all employees are responsible for following the guidelines.

First, there is the law. It must be obeyed. This book will discuss some of the most important laws that affect our business, and what they require of you.

But the law is the minimum. You also must act ethically. This book discusses some of the basic ethical issues you may confront in our business — a broad range of issues with which, generally speaking, all of us should be familiar. It does not, however, go into the level of detail that, for instance, a marketing representative might need to assess a complex competitive situation. IBM employees who work in sales or in purchasing, or in even more specialized areas such as government marketing, regulatory matters or tax procedures, must also be familiar with their own functional or divisional guidelines.

No book of rules, however, can provide all the answers. You are responsible for your actions, and this responsibility will not always be an easy one.

The next time you have an ethical dilemma, you might try this test. Ask yourself: If the full glare of examination by associates, friends, even family were to focus on your decision, would you remain comfortable with it? If you think you would, it probably is the right decision.

## Some Broad Guidelines

### Public Service

IBM encourages its employees to be good citizens, to speak out on public issues and to be active in civic and political life. However, conflicts of interest can arise when an IBM employee is active in public service, and it's important to avoid such conflicts — or even the appearance of one.

There are three ground rules:
*Make it clear that you are employed by* IBM.
*Abstain from participating in any decision or vote involving* IBM. *This would include, for example, purchasing decisions for things such as typewriters and*

*computers. Or a decision by a board of tax assessors on the assessment of* IBM *property.*
*When you do abstain, state clearly that it's because there is a potential conflict of interest.*

If you run for public office or serve as a public official, you cannot be paid by the company for any time spent on political activities or in public office. Requests for time off for public duties, like all requests for time off, must be approved by your manager. Together, you can decide whether the time should be taken without pay, be counted as vacation time or be made up, when that can be arranged.

It's also important that when you speak out on public issues you do so as an individual. Don't give the appearance that you're speaking or acting on behalf of the company.

### Respect for IBM's Assets

Our company's assets are more than physical plant and equipment, more than production machines, computers, typewriters and pencils. They include technologies and concepts, valuable ideas, business and product plans, as well as information about the business. They include drawings, computer programs, surveys and charts.

In past years, there have been significant, well-documented cases of misappropriation of IBM assets, including concepts, technologies and customer lists. These losses represented theft of the ideas, work and creativity of IBM people and of the advantages these would have brought in the marketplace. A number of individuals, including some IBM employees, have been prosecuted in the courts and convicted as a result of such thefts.

Company property can also be given away inadvertently. Sometimes outsiders know about new products, new practices, or business plans before we announce them. They can learn about such things from IBM people through casual conversations at social gatherings. A good many people are interested in our company and our industry — data processing users, stock analysts, and most importantly, competitors. The people who hear things from us may be above suspicion. But they may speak to others — and so on. Until finally IBM's plans for a new product or technology are revealed.

This is not a complex guideline. Respect IBM's assets as you would your own. IBM property, ideas and information belong in your hands or in the plant, laboratory or office — not in the hands of a competitor.

### "Moonlighting"

A conflict of interest also can arise when any employee is involved in activity for personal gain which for *any* reason is in conflict with IBM's business interests. Generally speaking, "moonlighting" is defined as working at some activity for personal gain outside of your IBM job. If you do perform outside work, you have a special responsibility to avoid any conflict with IBM's business interests.

Obviously, you cannot solicit or perform work in competition with IBM product or service offerings. Outside work cannot be performed on IBM time, including "personal" time off. You cannot use IBM equipment, materials, resources or "inside" information for outside work. Nor should you solicit business or clients, or perform outside work, on IBM premises.

## Political Contributions and Questionable Payments

IBM will not make contributions or payments to political parties or candidates. Nor will IBM bribe or make payoffs to government officials, civil servants or anyone. This is a single worldwide policy. If you ever are approached for what you believe is a questionable payment, report the circumstances to your manager as soon as possible.

In many countries political contributions by corporations are illegal. In other countries they are legal, but IBM will not make them in either case. Nor will IBM provide things other than direct cash payments which may be considered contributions. For example, if you want to campaign for a political candidate, the company will give you reasonable amounts of time off from work without pay, commensurate with your duties. IBM encourages people to be involved in politics, but on their own time and at their own expense. If IBM were to pay you while you were campaigning, your salary could be considered a contribution to the candidate or party you were supporting.

## Financial Interests and Insider Information

No two individuals are likely to view an investment or other financial interest in a competitor or supplier in the same light. Its absolute size, its importance in relation to their income and other investments, how much it could possibly influence their business decisions — these all vary.

But everyone understands the consequences of divided loyalty — a situation in which an individual is pulled two ways. So when we consider the possible implications of a financial interest in a competitor or supplier to IBM, the basic question isn't complicated: Could it cause divided loyalty — could it pull us two ways if we had to make a decision about our finances, our job, our career, our employer?

IBM employees should not have any financial interest in a competitor or supplier that could cause divided loyalty, or even the appearance of divided loyalty. Nor should they have any interest that could cause speculation or misunderstanding about *why* they have the interest.

Some of the questions you must ask yourself before you can decide whether you have a financial conflict of interest are:

> *What's your job in* IBM*? For example, could your decisions for* IBM *be affected by your interest in a competitor?*
> *What is the dollar amount of the investment and how does it measure up against your salary and other family income, your other savings and investments, your financial needs?*

*When and where was the investment originally made and under what circumstances? Long before you joined* IBM, *or after you became aware through your job of information about the competitor or supplier?*
*What is the nature and extent of the competition or relationship between* IBM *and the other company or business?*

If your professional or managerial responsibility includes working directly with information about a competitor or supplier, you must not buy or sell any of its stock.

It's always important to establish whether the company in question is a competitor. Many companies have more than one line of business, so just a portion of their operation may be competitive. Product lines change from time to time, IBM's and others, so situations change.

If you have any doubts or questions about the propriety of holding a financial interest in any company, the best course is to discuss it with your manager or IBM counsel.

Finally, we have to recognize that it isn't just money that causes people to have an interest in a company's success. Relationships such as being on a board of directors, or being an employee or advisor of a competitor, could cause divided loyalties.

A specific area of concern in investing is improper use of what's often called "insider information": the use of confidential nonpublic company information for your own financial benefit.

Such improper use may be more than an ethical consideration; it may be a violation of law. The U.S. Securities and Exchange Act of 1934 has sections on insider trading and deceptive practices in stocks and securities, and these sections may apply outside the United States.

Some examples of improper use of insider information are:

*If* IBM *is about to announce a new product or make a purchasing decision and the news could affect the stock of a competitor or supplier, you must not trade in the stock of those companies.*
*If* IBM *is about to make an announcement that could affect the price of its stock, you must not trade in* IBM *stock.*
*If you work with a customer, you normally may own stock in that customer's company. You must not, however, buy or sell that stock based on inside information.*
*If* IBM *is about to build a new facility, you must not invest in land or business near the new site.*
*If you know someone in a business whose customers are mainly persons moving into newly purchased homes, you should not disclose information on new hires and transfers of* IBM *employees to anyone who might offer the service or product of that business to* IBM *employees.*

Of course, you should never attempt to evade any rules against improper investment or misuse of insider information by acting indirectly through anyone, whether a spouse, relative or friend.

### Tips, Gifts and Entertainment

There are two basic guidelines:

> *No* IBM *employee, or any member of his or her immediate family, can accept gratuities or gifts of money from a supplier, customer or anyone in a business relationship. Nor can they accept a gift or consideration that could be perceived as having been offered because of the business relationship. "Perceived" simply means this: If you read about it in your local newspaper, would you wonder whether the gift just might have had something to do with a business relationship?*
>
> *No* IBM *employee can give money or a gift of significant value to a customer, supplier, or anyone if it could reasonably be viewed as being done to gain a business advantage.*

If you are offered money or a gift of some value by a supplier or if one arrives at your home or office, let your manager know immediately. If the gift is perishable, your manager will arrange to donate it to a local charitable organization. Otherwise, it should be returned to the supplier. Whatever the circumstances, you or your manager should write the supplier a letter, explaining IBM's guidelines on the subject of gifts and gratuities.

Of course, it is an accepted practice to talk business over a meal. So it is perfectly all right to occasionally allow a supplier or customer to pick up the check.

Similarly, it frequently is necessary for a supplier, including IBM, to provide education and executive briefings for customers. It's all right to accept or provide some services in connection with this kind of activity — services such as transportation, food or lodging. For instance, transportation in IBM or supplier planes to and from company locations, and lodging and food at company facilities are all right.

It's important to remember that there are a number of local, state and federal laws and regulations governing relations with government customers and suppliers. These may prohibit or modify the customary practices governing IBM relations with commercial accounts.

It's important, too, to have a good understanding of any detailed guidelines that your functional area may have — whether you are in marketing, purchasing or another part of the business. Ultimately, though, the best guideline of all is your common sense. Rely on it.

### Accurate Reporting

Almost every employee reports data of some kind. The engineer filling out a product test report; the salesman reporting on the status of an order; the scientist filling out an expense account; the customer engineer completing a call record — all are reporting information. All should do it accurately and honestly.

Some forms of inaccurate reporting are illegal. Listing a fictitious expense on an expense account or petty cash card, for example, is illegal. You should list on your expense account everything you paid for on a trip that IBM is required to pay for, neither more, nor less.

All reporting of information — whether sales results, hours worked or equal opportunity efforts — should be accurate and timely and should be a fair representation of all the facts. It should not be organized in any way that is intended to mislead or misinform the reader.

## Employee Personal Information

IBM has four basic practices concerning the use of personal information about employees:

> *To collect, use and retain only personal information that is required for business or legal reasons.*
> *To provide employees with a means of ensuring that their personal information in* IBM *personnel records is correct.*
> *To limit the internal availability of personal information about an individual only to those with a clear business need to know.*
> *To release personal information outside* IBM *only with approval of the employee affected, except to verify employment or to satisfy legitimate investigatory or legal requirements.*

But even with these practices, not every case can be covered. What constitutes a legitimate business need for a particular piece of information? Should information about an employee ever be released without his or her knowledge, even if it might be to his or her benefit?

Ultimately, you must balance the right of the organization to use information for valid business purposes with the individual's right to privacy. Your own conscience and judgment and the advice of your management and of IBM Personnel all should be considered in this delicate area.

## Information About Customers, Prospects, and Suppliers

Information about any organization or individual must not be misued. IBM collects only necessary business information about customer, prospect, and supplier organizations and their employees. Information which an organization identifies as confidential may not be received or stored in an IBM file without properly approved written agreements.

Only persons with a business need to know should have access to information. Whenever feasible, information should be analyzed only in the aggregate, to avoid identifying individual persons or organizations.

## STUDY QUESTIONS

1. What is the purpose in placing a letter from the chairman of the board as a preface to the handbook?

2. The handbook instructs you to test an ethical dilemma with this question: "If the full glare of examination by associates, friends, even family were to focus on your decision, would you remain comfortable with it?" Is this an effective test? Why or why not?

3. Do all the guidelines boil down to the simple warning "be honest"? Explain.

4. If the essential message is "be honest," why not just include that warning directly, and eliminate all the trimmings?

5. Does this in-house document give you a view of IBM that differs from the presentation of the company in advertising or public information brochures? If so, what is the difference, and how do you account for it?

# Success Through Communication

## Michael Korda

Each year "how to" books on topics that range from Zen Buddhism to macramé are best sellers. The popularity of these books would indicate that Americans either desire self-improvement or that they at least like to read about it. The following excerpt is from Michael Korda's *Success!* (a best seller), which is a guide on achieving success in the highly competitive corporate world.

Dr. Herbert H. Clark, a psychologist from Johns Hopkins University, recently made the somewhat startling discovery that it takes the average person about 48 percent longer to understand a sentence using a negative than it does to understand a positive, or affirmative, sentence. This is scientific confirmation of something that every successful person knows: *the secret of good communication is positive affirmation.* It is not what you won't or can't do that interests people, but what you will or can.

If you have ever left a meeting with the feeling that you failed to get your point across, or made no impact, if you think other people have a wrong impression of the kind of person you really are, if you know exactly what you want to say but it comes out in a way that doesn't satisfy you and leaves others uninterested — you have a communications problem. If you want to be a success, you're going to have to overcome it. Nor is this impossibly difficult.

Winston Churchill was born with a cleft palate. In his youth he stammered, stuttered and was taken for an idiot, even by his own family. He painfully set out to correct his defect, and went on to become perhaps the greatest orator (perhaps the last orator) of the twentieth century. King George VI was not only born with a

From Michael Korda, *Success!* (New York: Random House, 1977), pp. 127–138.

speech defect but was also incurably shy as a young man, and stammered so badly that it was painful to wait for him to finish a sentence. When his elder brother, King Edward VIII, abdicated, George VI forced himself to make speeches. By the end of his reign he was an accomplished speaker.

## Body Signals

In meetings, where the ability to communicate is of paramount importance and is the standard by which one is judged, many people find that they simply cannot get a clear shot to say what they want to say. Just as they start to talk, someone else begins talking. Or at the exact instant when they are about to make their point, someone interrupts and moves the discussion to some other topic. They wait for the right moment to make their move, then realize that it's too late.

There are various ways to set the stage for yourself. Elderly men, for example, are prone to throat-clearing as a preliminary to speech. It is a warning device, indicating that silence is requested, and most people understand the signal perfectly well. Many people rely on facial cues, but these are often ignored, and can easily be misinterpreted. Winking, grimacing, and sticking your tongue out may well attract attention, but not necessarily the kind of attention you want. The face is simply too small an object to dominate most groups, and its range of expression is ambiguous and subtle. It is better to use your whole body as a cue. Consider these simple yet effective body signals:

- Instead of sitting at a table with your elbows on it and your head down, sit well back away from it, perhaps even lean back in a relaxed, but alert, posture if the chair allows you to do so. Then when you want to speak, sit up straight, move forward with your whole body and put your hands down on the table with a positive motion. This is a recognizable and effective way of signaling your intention to talk.
- If you wear glasses, it may help to take them off and hold them in front of you, in the direction of the most important person at the meeting. This is a widely understood signal.
- Chain smokers can sometimes make an elaborate display of putting out a cigarette prior to speaking. Pipe smokers can knock their ashes out into an ashtray, creating a noise which is bound to attact attention.
- If you don't smoke and don't wear glasses, my suggestion would be to buy a pair of glasses and have clear lenses put into them. They make an ideal instrument for signaling your intention to speak, and you can always attract attention by putting them on or taking them off in order to read something.
- Half-moon reading spectacles have the additional advantage of making it possible for you to glare over the top of them at people who interrupt you. But not everyone can get away with this. Unless you are capable of making a very severe face, forget it.

- Auditory signals are sometimes effective — throat clearing, blowing your nose or cracking your knuckles.
- Many men rub their hands through their hair and down the back of their neck preparatory to speaking, a gesture that gives the impression of concentration.
- Women with long hair can ostentatiously push it back, exposing their full face and their ears as a sign that they want attention.
- Women with short hair can remove their earrings, a signal that they mean to get down to real business.
- Those who wear a ring can remove it and place it on the table in front of them. Women who wear several rings can reduce a meeting to complete silence by taking them off one by one.
  The following suggestions are for men only:
- At any meeting in which "serious" business is being discussed, it may be useful to write notes in pen on a yellow legal pad while other people are speaking. Do this as ostentatiously as possible, with a great many thick black underlinings, which will not only attract attention to you and show you're taking the meeting seriously, but will also make everyone else nervous. Then when you're ready to speak, put the cap back on your pen with as loud a click as you can achieve, and raise it like a pointer. This seldom fails to ensure an instant silence. On the whole, a bright metal pen is best. Most of the Parker fountain pens make a very satisfactory clicking noise when you put the cap back on them. A ball-point pen that can be sharply clicked will serve the same purpose for very much less money, though it has rather less class.
- Jingling the change in your pocket, or your keys, may work if nothing else is available.

The main thing is to get people's attention and give them a solid, comprehensible clue that you have something to say and intend to speak.

## Bridging

Now that you have won yourself a clear shot, how do you begin? A suggestion: it helps to link what you are going to say to the last thing that was said. An abrupt beginning sometimes confuses your listeners, and also suggests that you haven't been listening to what everyone else was saying, but have merely been waiting for an opportunity to speak out. Begin, in effect, by repeating what the person just before you said, summarize the discussion as it stands before your intervention, then go on to relate your arguments to what has gone before. In other words, create a "bridge" between the last speaker and yourself. An example of such a "bridge" might be this:

1. *Bridge:* "I think what Lynn just said — that we have to increase export sales — is perfectly true."

2. *Summary:* "Up to now we've been discussing what markets to go into, and I think we've reached a consensus to take the rifle approach instead of the shotgun approach, which makes sense."
3. *Your own message:* "Now, from my department's point of view, I think there are several things we could do immediately to get things under way . . ."

This is invariably the correct way to introduce yourself into a discussion, as opposed to waiting for a pause in the conversation and shouting, "Hey, I've got an idea!" Once you have begun, go slowly. Use your pen, glasses or cigarette to hold the attention of the others, especially when you're pausing for dramatic effect. This signal cue serves as an indication that you haven't finished speaking, and helps prevent interruption.

## How to Structure Your Points

While you're actually talking, don't look at one person all the time. Switch your attention around to different people. When your remarks make it possible to refer to someone else by name, do so ("I think what I'm about to say will be of special interest to Ed . . ."), looking as relaxed and pleasant as you can. Unless you are the chairman of the board, do not try to be funny, and never try to be funny at anybody's expense even if you *are* the chairman of the board. Your aim is to hold people's attention, and draw them into agreeing with you.

If you have trouble making your thoughts clear, you should concentrate on reducing everything to very short sentences, broken by as many significant dramatic pauses as you can manage. This, by the way, is Henry Kissinger's approach to public speaking, and while it does not qualify him to stand among the great orators of all time, it never fails to hold the audience's attention and get his points across. Avoid complex sentences, and break everything down into numbers — you may even want to use your fingers — i.e.:

"They wanted a ten-year deal (*pause*).

We were against this for the following reasons (*pause*):

One . . . [*pause; statement*]

Two . . . [*pause; statement*]

Three . . . [*pause, statement*]

Instead we counteroffered a five-year deal (*pause*).

What are the advantages of this (*pause and signal cue*)?

One . . . [*pause; statement*]

Two . . . [*pause; statement*]

Three . . . [*pause; statement*]

"I think they are now willing to accept this (*pause*), and I think at this point we should go ahead (*pause and signal cue*).

"It's a good deal for us, as I believe I've demonstrated, and as I think some of you, like Ed, have already said, and (*pause*) I wouldn't look this particular gift horse (*chuckle at cliché*) in the mouth (*pause and signal cue*)."

It is often effective to emphasize the numbers by using your fingers. (Do not break down whatever you have to say into more than five sections, or you'll overwhelm your listeners and get into some curious finger displays.) For instance:

"Number one (*pause*): How much will it cost us?

"Number two (*pause*): Should we use our existing personnel and facilities or build up a new department?

"Number three (*pause*): I've been observing what other companies have been doing, and I think we can learn from them."

Note in the above example that it is very useful to ask a question, then answer it yourself. The more questions you raise and answer, the fewer questions other people will ask, since after a while they will assume you have already asked and answered the important ones.

## The Positive Approach

Always try to put things affirmatively — avoid negatives. Thus, instead of saying, "If we don't get the sales department to explore new domestic areas first, we'll never be ready for exporting," say, "*First,* let's explore new domestic sales areas, then we'll have a solid base to go after export sales."

Whenever possible, refer to other people present by name, and give them credit for what they've said, even when you disagree, as in : "I think Lynn put the need to increase export sales very well, BUT (*pause*) I think we have to examine the effect this is going to have on our (*pause; with emphasis*) *domestic* sales first."

Negative statements not only interrupt the flow of your remarks, they antagonize and depress your listeners. Remember: nobody is interested in what you can't do, don't know or won't agree to. Emphasize the positive, and let the negative points make themselves known unobtrusively, if at all.

## Winding Up Strategies

When you have made your points, do not just stop and fall silent or allow your remarks to trail off. If people are to remember what you have said, you must anchor it in their minds. You must also ensure that what you have said becomes the theme of the following discussion. At long meetings, it may be useful to wait until just before the coffee break or the lunch break to make your remarks. You can then suggest breaking after you've finished speaking, which will effectively prevent any immediate opposing remarks, and leave whatever you have said on people's minds for the next hour or so. If you cannot do this, it is always possible to look around the room and say, "I'd be very grateful if anybody would like to comment on this, or ask any questions." This establishes your control over the meeting, and will usually moderate what might otherwise be fiercely expressed opposition to what you have said. Nobody likes to take up this kind of invitation, since it implies accepting a challenge on someone else's terms. Truly skilled success players will, of course, have arranged in advance for a friend to ask the right kind of question. This is the fox's approach to winding up a presentation. A more lionlike approach is to watch out for someone who doesn't seem to be listening, and put a question directly to him, e.g., "What do *you* think, Ed?" when it's almost certain that Ed has been daydreaming for the last half-hour. This will not only embarrass Ed but will also prompt him to agree with you hastily, rather than to risk making himself look foolish by arguing against something he hasn't even heard.

## Clichés

When you get to the end of what you're saying, quickly summarize your argument, putting it into a solid, clear and memorable sentence. Don't be afraid of clichés. On the contrary; they help to anchor what you've said in people's minds precisely because they are familiar. If you think that going into export sales too fast is a mistake, in the example above, say, "I think going after exports while we've still got domestic sales problems is putting the cart before the horse." Your thought attached to a familiar cliché is instantly understood and certain to be remembered. It is a simple and reliable formula:

THOUGHT + APPROPRIATE CLICHÉ = INSTANT COMMUNICATION

When it comes to clichés, use down-to-earth ones. Avoid literary allusions. Don't be afraid to say things like:
"The ball is in our court."
"Fifty percent of something is better than a hundred percent of nothing."
"It's the last mile that counts."
"You can't make bricks without straw."
"The name of the game is profit."
I heard all of the above used by the president of a major manufacturing company within the space of a few minutes, and most effectively. Each one followed a precise summary of his point of view, and served to pin down in his listeners' minds exactly what he meant.

## "I have no problem with that"

In a one-to-one conversation, you can apply most of these same strategies with very good effect. Sit as close as you can to the other person and keep your voice reasonably low. Don't mumble, of course, or whisper, but remember that nobody likes to be shouted at. The lower your voice is, the harder the other person will have to work to hear and pay attention to what you're saying.

Here, it is permissible, and even a good idea, to interrupt from time to time, particularly if the other person is talkative. You can do this with extreme politeness, simply by indicating your agreement with what he's saying. However, you must be very careful to signal when you no longer agree. It is vital not to let anyone go away under the impression that you have accepted his point of view if that is not the case.

This frequently happens to poor communicators. They are reluctant to talk, and therefore listen silently to things they have no intention of accepting, thinking they can make this clear later on with a letter or at the conclusion of the conversation. But this leads them into positions where they appear to be either stupid or treacherous.

Communicate your opinion continuously, whether you do it verbally or nonverbally. That way the other person knows where he stands. If nothing else, nod your head when you're affirming agreement, and shake it very slightly when you don't agree. Frown when you're doubtful. Be careful about mixing signals. A frown usually signifies that you don't agree with something the other person has said, but a frown combined with an up-and-down motion of the head means "It's a serious matter but I'm on your side." Keep your hands still when the other person is talking. It's all right to use them for emphasis when you're talking, but they merely distract the other person when he has the floor. It's useful to make an occasional comment, if only because it reminds the other person that the conversation is taking place between two equals.

My friend Morton L. Janklow, a distinguished New York attorney, has a superb desk-side manner, and uses a phrase of which I've become very fond. When you say something he doesn't find objectionable, he will comment, "I have no problem with that." This is encouraging and indicates both that he is listening and that the point you have made is one that can be negotiated. Note that his phrase very carefully stops short of complete agreement. The implication is merely that some common ground can be found in this area at some later date.

When you say something that he does *not* agree with, Morty takes off his glasses and says, "Let's leave that one until later." He does not say, "You're wrong!" or "Not on your life!" or "You must be out of your mind!" Instead, he very politely puts you on notice that on this particular point he is not in agreement, and reserves the right to come back to it in detail later on.

This is a helpful rule of communication. Get the things that you're agreed on out of the way first, then go on to the ones that are going to cause trouble. Don't go through a long list of points, agreeing to some but fighting out the others on the spot in the order in which they occur. When there is disagreement, make a note of it, indicate that you'll come back to it later on, and go on to dispose of the points on which there *is* agreement. You will find that the residue of things on which you haven't reached agreement is easier to deal with once you've gone over the affirmative points of agreement, and that a great deal of them will be found to be trivial after they've been isolated in a separate category.

## STUDY QUESTIONS

1. On page 367, Korda says that "If you . . . don't wear glasses, my suggestion would be to buy a pair of glasses and have clear lenses put into them. They make an ideal instrument for signaling your intention to speak. . . ." Can you take this advice seriously? Do such statements undercut the effectiveness of the piece? Discuss.

2. Although Korda doesn't discuss written communication specifically, his point

about "bridging" (page 368) is a useful technique in writing transitions. Are any of his other points applicable to written communication?

3. Although "how to" information tries to avoid the superfluous, Korda includes several anecdotes with his instructions. Do these distract the reader from the task at hand? Why or why not?

4. How are these instructions essentially different from, say, instructions on fixing a flat tire?

5. On page 371, Korda says to use clichés to make a point. Of course, English courses usually advise students to avoid clichés. Is there an explanation to reconcile this contradiction?

6. *Writing Assignment:* Using your shrewd insights on higher education, write a guide on how college students can survive the freshman year. (You might include such topics as "how to appear to be interested in what the professor says" and "how to convince the professor that he should raise your grade.")

# Index